Power Electronics for Technology

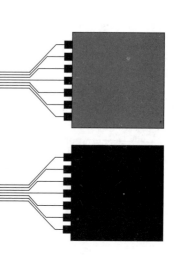

Power Electronics for Technology

Ashfaq Ahmed

Purdue University–Calumet

Prentice Hall
Upper Saddle River, New Jersey   Columbus, Ohio

Library of Congress Cataloging-in-Publication Data

Ahmed, Ashfaq.
Power electronics for technology/Ashfaq Ahmed.
    p.   cm.
  Includes index.
  ISBN 0-13-231069-4
  1. Power electronics.   I. Title.
TK7881.15.A46   1999
621.31′7—dc21   98-15732
                                                  CIP

Cover Photo: © PhotoEdit
Editor: Charles E. Stewart, Jr.
Production Editor: Alexandrina Benedicto Wolf
Cover Design Coordinator: Karrie M. Converse
Production and Editing Coordination: The Clarinda Company
Cover Designer: TOPEgrafix, Inc.
Production Manager: Deidra M. Schwartz
Marketing Manager: Ben Leonard

This book was set in ITC Garamond Light by The Clarinda
Company and was printed and bound by R. R. Donnelly
& Sons.
The cover was printed by Phoenix Color Corp.

© 1999 by Prentice-Hall, Inc.
Simon & Schuster/A Viacom Company
Upper Saddle River, NJ 07458

Printed in the United States of America

10 9 8 7 6 5 4 3 2 1

ISBN: 0-13-231069-4

Prentice-Hall International (UK) Limited, *London*
Prentice-Hall of Australia Pty. Limited, *Sydney*
Prentice-Hall Canada Inc., *Toronto*
Prentice-Hall Hispanoamericana, S.A., *Mexico*
Prentice-Hall of India Private Limited, *New Delhi*
Prentice-Hall of Japan, Inc., *Tokyo*
Simon & Schuster Asia Pte. Ltd., *Singapore*
Editora Prentice-Hall do Brasil, Ltda., *Rio de Janeiro*

O! My God increase my knowledge

To my wonderful parents Basheer and Farhath,

my loving wife Nadira, and my beautiful children

Saba, Asad, and Fahad

Contents

4 Thyristor Devices   73

5 Single-Phase Uncontrolled Rectifiers   122

6 Single-Phase Controlled Rectifiers   150

7 Three-Phase Uncontrolled Rectifiers   188

8 Three-Phase Controlled Rectifiers   216

12 **Static Switches**   410

Preface

This book introduces the subject of power electronics, that is, the switching control and conversion of electrical power using semiconductor devices. The increasing demand for efficient conversion and control of electrical power has made this area of electrical engineering extremely important and has resulted in the development of new power devices, circuits, and control schemes that continue to extend power electronics technology to new areas of application.

The main objective of this book is to present an easily understood explanation of the principles of power electronics to satisfy the needs of two- and four-year programs in electrical engineering technology (EET). The book is primarily intended to be used as a text in an introductory course about semiconductor switches and their applications to power control. Its scope extends from the fundamental principles and concepts to the general applications of power electronics.

Chapter 1 gives an introduction and an overview of power electronics. Chapters 2, 3, and 4 are devoted to introducing the major power semiconductor devices and their behavior and voltage-current characteristics. Chapters 5 to 12 introduce the reader to the analysis and operation of various power conversion circuits—AC/DC, DC/DC, DC/AC, and AC/AC—that have applications at high power levels. The topics covered focus on the application of power electronics devices in rectifiers, inverters, AC voltage controllers, cycloconverters, and DC choppers. Emphasis is placed on circuit topology and function. Voltage and current waveform analyses of the outputs begin with simple resistive loads and continue to more practical inductive loads. The equations that govern the behavior of these circuits are formulated to provide the fundamentals for analyzing power electronics circuits.

The text assumes that the students have a good understanding of basic mathematics as well as a basic background in circuit theory and solid-state electronics fundamentals. The material presented here has been developed from class notes for a power electronics course taught by the author in the Electrical Engineering Technology Department at Purdue University-Calumet, Hammond, Indiana. The text is considered appropriate for a one-semester junior-level course.

Features

- The book includes many worked examples to illustrate concepts introduced in the text. These examples also form the basis for several problems at the end of each chapter for student practice.
- Because the book is primarily intended for students in a technology program, emphasis is placed on applications.
- This textbook should fill the gap between books written for engineering students and books written at the technician level.
- Material is presented methodically by giving the background theory before moving on to specific applications.
- Only a basic knowledge of mathematics is required. The mathematical rigor has intentionally been limited to ensure that the student learning the subject for the first time does not find the information made ambiguous by complex theory.
- A generous quantity of illustrations provides most important information in diagram form.

Acknowledgments

The author wishes to thank Prentice Hall editor, Charles Stewart, and his assistant, Kim Yehle, for their assistance, patience, and flexibility during the progress of this text. I would like to thank the following reviewers for their invaluable feedback: Venkata Anandu, Southwest Texas State University; Shamala Chickamenahalli, Wayne State University; Mohammad Dabbas, ITT Technical Institute–Florida; and Alexander E. Emanuel, Worcester Polytechnic Institute. The author's deepest appreciation goes to his wife, Nadira, for her support, understanding, and encouragement during the many hours of work involved in the successful completion of this project.

Ashfaq Ahmed, P.E.
Munster, Indiana

Power Electronics

Learning Objectives

After completing this chapter, the student should be able to

- define the term power electronics
- list the advantages of using a switch to control electric power
- list the various types of power semiconductor devices
- determine power loss in real switches
- list the various types of power electronic circuits
- list typical applications of power electronics

1.1   Introduction

Applications of solid-state electronics in the electrical power field are steadily increasing, and a course in power electronics is now a common feature of many electrical engineering technology curricula. The term *power electronics* has been used since the 1960s, after the introduction of the silicon controlled rectifier (SCR) by General Electric. Power electronics has shown rapid growth in recent years with the development of power semiconductor devices that can switch large currents efficiently at high voltages. Since these devices offer high reliability and small size, power electronics has expanded its range and scope to such applications as lighting and heating control, regulated power supplies, variable-speed DC and AC motor drives, static VAR compensation, and high-voltage DC transmission systems.

1.2   What Is Power Electronics?

The broad field of electrical engineering can be divided into three major areas—electric power, electronics, and control. Power electronics deals with the application of power semiconductor devices, such as thyristors and transistors, for the conversion and control of electrical energy at high power levels. This conversion is usually from AC to DC or vice versa, while the parameters controlled are voltage, current, or frequency. For example, simple rectification from AC to DC is power conversion, but if voltage level adjustment is applied to rectification, both conversion and control of electrical power are involved. Therefore, power electronics can be considered to be an interdisciplinary technology involving three basic fields, power, electronics, and control, as shown in Figure 1.1.

Figure 1.1
Power electronics: a combination of power, electronics and control

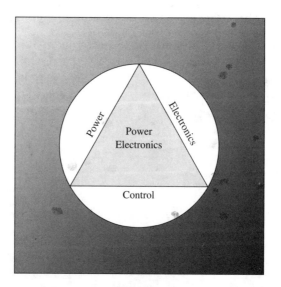

This book will cover the use of power semiconductor devices in applications such as rectification, inversion, frequency conversion, DC and AC drives, and power supplies. In power electronics, devices such as diodes, transistors, thyristors, and triacs are used mainly as switches to perform the on-off action that is basic to power electronics circuits.

1.3   Why Power Electronics?

Transfer of electrical power from a source to a load can be controlled either by varying the supply voltage (by using a variable transformer) or by inserting a regulator (such as a rheostat, variable reactor or switch). Semiconductor devices used as switches have the advantage of being relatively small, inexpensive, and efficient, and they can be used to control power automatically. An additional advantage of using a switch as a control element (compared to using adjustable resistance provided by rheostat or potentiometer) is shown in the following section.

1.3.1   A Rheostat as a Control Device

Figure 1.2 shows a rheostat controlling a load. When R_1 is set at zero resistance, full power is delivered to the load. When R_1 is set for maximum resistance, the power delivered is close to zero. For maximum power transfer to the load, R_1 must equal R_L. Under this condition, the rheostat consumes as much power as the load—the efficiency of conversion is only 50%. Moreover, the rheostat must be physically larger than the load to dissipate additional heat.

Figure 1.2
A rheostat controlling a load

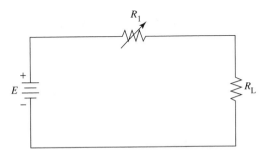

In applications where the power to be controlled is large, the efficiency of conversion is important. Poor efficiency means large losses, an economic consideration, and it also generates heat that must be removed from the system to prevent overheating.

Example 1.1 A DC source of 100 V is supplying a 10-Ω resistive load. Find the power delivered to the load (P_L), the power loss in the rheostat (P_R), the total power supplied by the source (P_T), and the efficiency η, if the rheostat is set at:
a) 0 Ω
b) 10 Ω
c) 100 Ω

Solution a) voltage across the load $V_L = 100$ V
      power supplied to the load $P_L = 100^2/10 = 1$ KW
      power dissipated in the rheostat $P_R = 0$ W
      power supplied by the source $P_T = P_L + P_R = 1$ KW

      efficiency $\eta = \dfrac{P_L}{P_T} * 100 = 100\%$

b) voltage across the load $V_L = 10 * 100/20 = 50$ V
      power supplied to the load $P_L = 50^2/10 = 250$ W
      power dissipated in the rheostat $P_R = 250$ W
      power supplied by the source $P_T = P_L + P_R = 500$ W

      efficiency $\eta = \dfrac{P_L}{P_T} * 100 = 50\%$

c) voltage across the load $V_L = 10 * 100/110 = 9$ V
      power supplied to the load $P_L = 9^2/10 = 8.1$ W
      power dissipated in the rheostat $P_R = 91 * 19/100 = 82.8$ W
      power supplied by the source $P_T = P_L + P_R = 90.9$ W

      efficiency $\eta = \dfrac{P_L}{P_T} * 100 = 8.9\%$

It is clear from this example that the efficiency of power transfer from the source to the load is very poor—note that it is only 50% in case (b).

1.3.2 A Switch as a Control Device

In Figure 1.3, a switch is used to control the load. When the switch is on, maximum power is delivered to the load. The power loss in the switch is zero since it has no voltage across it. When the switch is off, no power is delivered to the load. Again, the switch has no power loss since there is no current through it. The efficiency is 100% because the switch does not waste power in either of its two positions.

The problem with this method is that unlike a rheostat, a switch cannot be set at intermediate positions to vary the power. However, we can create the same effect by periodically turning the switch on and off. Transistors and SCRs used as switches can be automatically turned on and off hundreds of times a second. If we need more power, the electronic switch is set on for longer periods and off for shorter periods. When less power is needed, it is set off longer.

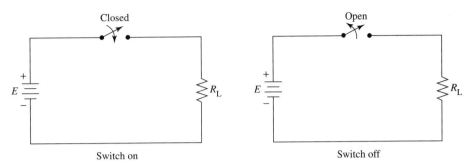

Figure 1.3
A switch controlling a load

Example 1.2 A DC source of 100 V is supplying a 10-Ω resistive load through a switch. Find the power supplied to the load (P_L), the power loss in the switch (P_S), and the total power supplied by the source (P_T), if the switch is:
a) closed
b) open
c) closed 50% of the time
d) closed 20% of the time

Solution a) With the switch closed,
voltage across the load	$V_L = 100$ V
power delivered to the load	$P_L = 100^2/10 = 1$ KW
power loss in the switch	$P_S = 0$ W
power supplied by the source	$P_T = 1$ KW

b) With the switch open,
voltage across the load	$V_L = 0$ V
power delivered to the load	$P_L = 0$ W
power loss in the switch	$P_S = 0$ W
power supplied by the source	$P_T = 0$ W

c) With the switch closed 50% of the time (see Figure 1.4),
average voltage across the load	$V_L = 50$ V
average power delivered to the load	$P_L = 50^2/10 = 250$ W
power loss in the switch	$P_S = 0$ W
power supplied by the source	$P_T = 250$ W

d) With the switch closed 20% of the time,
average voltage across the load	$V_L = 20$ V
average power delivered to the load	$P_L = 20^2/10 = 40$ W
power loss in the switch	$P_S = 0$ W
power supplied by the source	$P_T = 40$ W

Figure 1–4
See Example 1.2

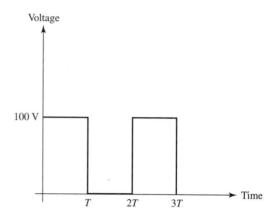

As this example shows, all the power supplied by the source is delivered to the load. The efficiency of power transfer is 100%. Of course, in this example the switch is assumed to be ideal, but when we use a transistor as a switch, the result is very close to ideal circuit operation.

1.4   Power Semiconductor Switches

Power semiconductor switches are the most important elements in a power electronics circuit. The major types of semiconductor devices used as switches in power electronics circuits are:

diodes

bipolar junction transistors (BJT)

metal-oxide semiconductor field-effect transistors (MOSFET)

insulated-gate bipolar transistors (IGBT)

silicon controlled rectifiers (SCR)

triacs

gate-turnoff thyristors (GTO)

MOS-controlled thyristors (MCT)

In power electronics, these devices are operated in the switching mode. These switches can be made to operate at high frequencies to convert and control electrical power with high efficiency and high resolution. The power loss in the switch itself is very small since either the voltage is nearly zero when the switch is on or the current is nearly zero when the switch is off.

We will treat these switches as ideal (the limitations of an actual switch are covered in the next section). An ideal switch satisfies the following conditions:

1. It turns on or turns off in zero time.

2. When the switch is on, the voltage drop across it is zero.

3. When the switch is off, the current through it is zero.

4. It dissipates zero power.

In addition, the following conditions are desirable:

5. When on, it can carry a large current.

6. When off, it can withstand high voltage.

7. It uses little power to control its operation.

8. It is highly reliable.

9. It is small in size and weight.

10. It is low in cost.

11. It needs no maintenance.

1.5   Power Losses in Real Switches

An ideal switch is shown in Figure 1.5. The power loss generated in the switch is the product of the current through the switch and the voltage across the switch. When the switch is off, there is no current through it (although there is a voltage V_S across it), and therefore there is no power dissipation. When the switch is on, it has a current (V_S/R_L) through it, but there is no voltage drop across it, so again there is no power loss. We also assume that for an ideal switch the rise and fall time of the current is zero. That is, the ideal switch changes from the off state to the on state (and vice versa) *instantaneously*. The power loss during switching is therefore zero.

Figure 1.5
Power losses in an ideal switch

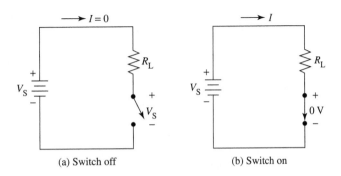

(a) Switch off   (b) Switch on

Unlike an ideal switch, an actual switch, such as a bipolar junction transistor, has two major sources of power loss: conduction loss and switching loss.

1.5.1 Conduction Loss

When the transistor in Figure 1.6(a), is off, it carries a *leakage current* (I_{LEAK}). The power loss associated with leakage current is $P_{OFF} = V_S * I_{LEAK}$. However, since the leakage current is quite small and does not vary significantly with voltage, it is usually neglected and thus the transistor power loss is essentially zero. When the transistor is on, as in Figure 1.6(b), it has a small voltage drop across it. This voltage is called the *saturation voltage* ($V_{CE(SAT)}$). The transistor's power dissipation or conduction loss due to the saturation voltage is:

$$P_{ON} = V_{CE(SAT)} * I_C \qquad \qquad \textbf{1.1}$$

where

$$I_C = \frac{V_S - V_{CE(SAT)}}{R_L} \approx \frac{V_S}{R_L} \qquad \qquad \textbf{1.2}$$

Equation 1.1 gives the power loss due to conduction if the switch remains on indefinitely. However, to control the power for a given application, the switch is turned on and off in a periodic manner. Therefore, to find the average power loss we must consider the duty cycle:

$$P_{ON(avg)} = V_{CE(SAT)} * I_C * \frac{t_{ON}}{T} = V_{CE(SAT)} * I_C * d \qquad \qquad \textbf{1.3}$$

Similarly,

$$P_{OFF(avg)} = V_S * I_{LEAK} * \frac{t_{OFF}}{T} \qquad \qquad \textbf{1.4}$$

Here, the duty cycle *d* is defined as the percentage of the cycle in which the switch is on:

$$d = \frac{t_{ON}}{t_{ON} + t_{OFF}} = \frac{t_{ON}}{T} \qquad \qquad \textbf{1.5}$$

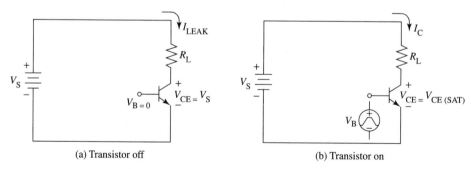

(a) Transistor off   (b) Transistor on

Figure 1.6
Power losses in a transistor switch

1.5.2 Switching Loss

In addition to the conduction loss, a real switch has switching losses because it cannot change from the on state to the off state (or vice versa) instantaneously. A real switch takes a finite time $t_{SW(ON)}$ to turn on and a finite time $t_{SW(OFF)}$ to turn off. These times not only introduce power dissipation but also limit the highest switching frequency possible. The transition times $t_{SW(ON)}$ and $t_{SW(OFF)}$ for real switches are usually not equal, with $t_{SW(ON)}$ generally being larger. However, in this discussion we will assume that $t_{SW(ON)}$ is equal to $t_{SW(OFF)}$. Figure 1.7 shows switching waveforms for (a) the voltage across the switch and (b) the current through it. When the switch is off, the voltage across it is equal to the source voltage. During turn-on, which takes a finite time, the voltage across the switch decreases to zero. During the same time, the current through the switch increases from zero to I_C. The transistor has a current through it and a voltage across it during the switching time; therefore, it has a power loss.

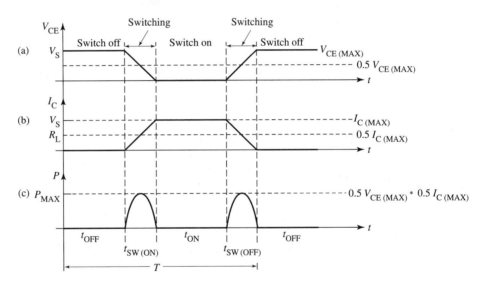

Figure 1.7
Waveforms during switching operation: (a) voltage across the switch; (b) current through the switch; (c) power dissipated in the switch

To find the power dissipated in a transistor during the switching interval, we multiply the instantaneous value of I_C and the corresponding value of V_{CE}. The instantaneous power curve is shown in Figure 1.7(c). The energy dissipated in the switch is equal to the area under the power waveform. Note that the maximum power is dissipated when both the current and the voltage are passing through their midpoint values. Therefore, the maximum power loss when switching from the off state to the on state is:

$$P_{SW\ ON(max)} = 0.5\ V_{CE(max)} * 0.5\ I_{C(max)} \qquad \textbf{1.6}$$

It is interesting to note that the power curve looks essentially like a rectified sine wave. The average value of this waveform is

$$P_{SW\ ON(avg)} = 0.637 * P_{SW\ ON(max)}$$
$$= 0.637 * 0.5\ V_{CE(max)} * 0.5\ I_{C(max)}$$
$$= 0.167\ V_{CE(max)} * I_{C(max)}$$

or

$$P_{SW\ ON(avg)} = \frac{1}{6}\ V_{CE(max)} * I_{C(max)} \qquad\qquad \textbf{1.7}$$

The energy loss (power * time) during turn-on will be $P_{SW\ ON(avg)} * t_{SW(ON)}$

$$W_{SW\ ON} = \frac{1}{6}\ V_{CE(max)} * I_{C(max)} * t_{SW(ON)} \qquad \text{(joules)} \qquad \textbf{1.8}$$

A similar analysis gives the energy loss during turnoff as

$$W_{SW\ OFF} = \frac{1}{6}\ V_{CE(max)} * I_{C(max)} * t_{SW(OFF)} \qquad \text{(joules)} \qquad \textbf{1.9}$$

The total energy loss in one cycle due to switching is given by

$$W_{SW} = W_{SW\ ON} + W_{SW\ OFF} + \frac{1}{6}\ V_{CE(max)} * I_{C(max)} * [t_{SW(ON)} + t_{SW(OFF)}]$$

The average power dissipation in the switch will be

$$P_{SW} = \frac{W_{SW}}{T} = W_{SW} * f$$

$$P_{SW} = \frac{1}{6}\ V_{CE(max)} * I_{C(max)} * [t_{SW(ON)} + t_{SW(OFF)}] * f$$

where T is the switching period and f is the pulse repetition rate (frequency of switching). Note that

$$T = t_{ON} + t_{SW(ON)} + t_{OFF} + t_{SW(OFF)}$$

If we let

$$t_{SW(ON)} = t_{SW(OFF)} = t_{SW}$$

then

$$P_{SW} = \frac{1}{6}\ V_{CE(max)} * I_{C(max)} * (2\ t_{SW}) * f \qquad\qquad \textbf{1.10}$$

The total power loss in switch is

$$P_T = P_{ON(avg)} + P_{OFF(avg)} + P_{SW} \qquad\qquad \textbf{1.11}$$
$$\approx P_{ON(avg)} + P_{SW}$$

$$= d * V_{CE(sat)} * I_C + \frac{1}{3} * V_{CE(max)} * I_{C(max)} * t_{SW} * f \qquad\qquad \textbf{1.12}$$

Example 1.3   In Figure 1.5, V_S is 50 V, R_L is 5 Ω, and the switch is ideal with no switching loss. If the on-state voltage drop is 1.5 V and the leakage current is 1.5 mA, calculate the power loss in the switch when it is:
a) on
b) off

Solution   a) conduction current $= \dfrac{50 - 1.5}{5} = 9.7$ A

power loss during on state $P_{ON} = 1.5 * 9.7 = 14.55$ W

b) power loss during off state $P_{OFF} = 50 * 0.0015 = 75$ mW

For normal load conditions, the power dissipation during the off state can be neglected in comparison to the power loss during the on state.

Example 1.4   Calculate the maximum and average power loss for the switch in Example 1.3 if the switching frequency is 500 Hz with a duty cycle of 50%.

Solution   switching period $T = 1/500 = 2$ ms
duty cycle $d = 50\%$
on time $t_{ON} = 1$ ms
off time $t_{OFF} = 1$ ms

average power loss during on state $= P_{ON} * \dfrac{t_{ON}}{T} = 14.55 * 0.5 = 7.27$ W

average power loss during off state $= P_{OFF} * \dfrac{t_{OFF}}{T} = 0.075 * 0.5 = 0.037$ W

average power loss for one cycle $= P_{ON(avg)} + P_{OFF(avg)} = 7.27 + 0.037 = 7.3$ W

The maximum power dissipation (from Example 1.3) $= 14.55$ W.

Example 1.5   In Figure 1.6, V_S is 120 V, R_L is 6 Ω, and the transistor is ideal with no conduction loss. If $t_{SW(ON)} = t_{SW(OFF)} = 1.5$ μs, calculate the average switching power loss at a switching frequency of 1 kHz.

Solution $I_{C(max)} = 120/6 = 20$ A

$P_{SW\ ON(avg)} = \dfrac{1}{6}\ V_{CE(max)} * I_{C(max)} = (120 * 20)/6 = 400$ W

The energy loss is

$W_{SW\ ON} = P_{SW\ ON(avg)} * t_{SW(ON)} = 400 * 1.5(10^{-6}) = 0.6(10^{-3})$ J

The energy loss during turnoff is

$W_{SW\ OFF} = 0.6(10^{-3})$ J

The total energy loss in one cycle due to switching is given by

$W_{SW} = W_{SW\ ON} + W_{SW\ OFF} = 2 * 0.6(10^{-3}) = 1.2(10^{-3})$ J

The average power dissipation during switching will be

$$P_{SW} = W_{SW} * f = 1.2(10^{-3}) * 1000 = 1.2 \text{ W}$$

Example 1.6   A transistor switch with the following characteristics controls power to a 25-kW load, as shown in Figure 1.6:

$$I_{RATED} = 50 \text{ A}$$
$$V_{RATED} = 500 \text{ V}$$
$$I_{LEAKAGE} = 1 \text{ mA}$$
$$V_{CE(SAT)} = 1.5 \text{ V}$$

turn-on time $t_{SW(ON)} = 1.5 \text{ μS}$

turnoff time $t_{SW(OFF)} = 3.0 \text{ μS}$

The source voltage V_S is 500 V, and R_L is 10 Ω. If the switching frequency is 100 Hz with a duty cycle of 50%, find:

a) on-state power loss
b) off-state power loss
c) maximum power loss during turn-on
d) energy loss during turn-on
e) energy loss during turnoff
f) on-state energy loss
g) off-state energy loss
h) total energy loss
i) average power loss

Solution $I_{C(max)} = 500/10 = 50 \text{ A}$
$T = 1/100 = 10 \text{ mS}$

With $d = 50\%$

$$t_{ON} = t_{OFF} = 5 \text{ mS}$$

a) on-state power loss = $1.5 * 50 = 75 \text{ W}$

b) off-state power loss = $500 * 1(10^{-3}) = 0.5 \text{ W}$

c) maximum power loss during turn-on = $\left(\dfrac{500}{2}\right)\left(\dfrac{50}{2}\right) = 6250 \text{ W}$

d) energy loss during turn-on = $\dfrac{1}{6} 500 * 50 * 1.5(10^{-6}) = 6.25 \text{ mJ}$

e) energy loss during turnoff = $\dfrac{1}{6} 500 * 50 * 3.0(10^{-6}) = 12.5 \text{ mJ}$

f) on-state energy loss = $75 * 5(10^{-3}) = 375 \text{ mJ}$

g) off-state energy loss = $0.5 * 5(10^{-3}) = 2.5 \text{ mJ}$

h) total energy loss per cycle = (6.25 + 12.5 + 375 + 2.5) mJ = 396.25 mJ

i) average power loss = $\dfrac{396.25(10^{-3})}{10(10^{-3})}$ = 39.6 W

Example 1.7   In Example 1.6, repeat parts (d) through (i) if the switching frequency is increased to 100 kHz with a 50% duty cycle.

Solution $T = 1/(100 * 10^3) = 10\ \mu S$

Since the period is quite small, we do not neglect $t_{SW(ON)}$ and $t_{SW(OFF)}$:

$T - t_{SW(ON)} - t_{SW(OFF)} = (10 - 1.5 - 3)\ \mu S = 5.5\ \mu S$

Therefore, with $d = 50\%$,

$t_{ON} = t_{OFF} = 2.75\ \mu S$

d) energy loss during turn-on = 6.25 mJ (from Example 1.6(d))

e) energy loss during turnoff = 12.5 mJ (from Example 1.6(e))

f) on-state energy loss = $75 * 2.75(10^{-6}) = 0.206$ mJ

g) off-state energy loss = $0.5 * 2.75(10^{-6}) = 1.375$ mJ

h) total energy loss per cycle = (0.206 + 1.375 + 6.25 + 12.5) mJ = 20.33 mJ

i) average power loss = $\dfrac{20.33(10^{-3})}{10(10^{-6})}$ = 2033 W

It is clear from these examples that at low switching frequency, the on-state power loss dominates the total losses. As we increase the switching frequency, the switching power loss becomes dominant. The average power dissipation also becomes very high (2033 W) at higher frequency. Obviously, the 50 A transistor cannot dissipate the heat generated and will overheat. The maximum frequency at which the switch can operate therefore depends not only on the power dissipation in the switch but also on the switching speed. The current rating of the switch must also be increased at higher frequencies.

■──

1.6   Types of Power Electronics Circuits

Power electronics circuits (or converters, as they are commonly called) can be divided into the following categories:

1. **Uncontrolled rectifiers (AC to DC):** an uncontrolled rectifier converts a single-phase or three-phase AC voltage to a fixed DC voltage. Diodes are used as the rectifying element to provide power conversion.

2. Controlled rectifiers (AC to DC): a controlled rectifier converts a single-phase or three-phase fixed AC voltage to a variable DC voltage. SCRs are used as the rectifying element, providing both power conversion and control.

3. DC choppers (DC to DC): a DC chopper converts a fixed DC voltage to a variable DC voltage.

4. AC voltage controllers (AC to AC): an AC voltage controller converts a fixed AC voltage to a variable AC voltage at the same frequency. There are two basic methods used in AC voltage controllers—on-off control and phase control.

5. Inverters (DC to AC): an inverter converts a fixed DC voltage to a fixed or variable single-phase or three-phase AC voltage and frequency.

6. Cycloconverters (AC to AC): A cycloconverter converts a fixed voltage and frequency AC to a variable voltage and frequency AC. This conversion can also be obtained indirectly by first rectifying AC to DC, then inverting back to AC at the desired frequency.

7. Static switches (AC or DC): A power device (SCR and TRIAC) can be operated as an AC or DC switch, thereby replacing traditional mechanical and electromagnetic switches.

Table 1.1 Some applications of power electronics

Power converter	Applications
Uncontrolled rectifier	DC source for electronic circuits
Controlled rectifier	DC motor speed control from an AC source Speed control of portable power tools High-voltage DC transmission
DC chopper	DC motor speed control from a DC source Switching power supply
AC voltage controller	Light dimmer switch Control of heaters Speed control of domestic appliances Reactive power control Smooth starting of induction motors
Inverter	Uninterruptible power supply (UPS) Speed control of three-phase AC motors Induction heating
Cycloconverter	Speed control of AC motors Constant frequency source for aircraft
Static switch	Replacement for mechanical and electromagnetic switches

1.7 Applications of Power Electronics

Power electronics finds application in any field that requires electric power conversion and control. Power electronics systems are therefore found in a wide range of consumer and industrial equipment—from small, less-than-one-horse-power motors used in domestic appliances to several-hundred-horsepower-capacity industrial motor drives; from low-power regulated DC power supplies to more than one-thousand-megawatt high-voltage DC transmission systems; from low-power light dimmers to hundreds-of-megawatt-capacity static VAR compensators in power systems. Table 1.1 provides some idea of the magnitude and importance of power electronics.

1.8 Problems

1.1 What is power electronics?

1.2 What is the most efficient method of controlling electric power?

1.3 What is the disadvantage of using a rheostat to control electric power to a load?

1.4 Explain why a switch is superior to a rheostat for controlling electric power to a load.

1.5 A 20-Ω rheostat is connected to a 30-Ω load resistor. If the source voltage is 120 V, find the power dissipated by the rheostat.

1.6 A switch controls a 20-Ω heater connected to a 208 V AC source. If the switch is on, find the power consumed by the heater and switch.

1.7 Repeat Problem 1.6 if the switch is off.

1.8 An ideal switch controls a 20-Ω load connected to a 120 V AC source. If the switch is on 20% of the time, find the power consumed by the load.

1.9 Figure 1.8 shows an ideal switch with no switching loss. If the on-state voltage drop is 2.0 V and the leakage current is 1 mA, calculate the power loss in the switch when it is:
a) on
b) off

Figure 1.8
See Problem 1.9

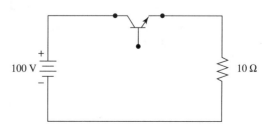

1.10 List the characteristics of an ideal switch.

1.11 List some common applications of power electronics in the following areas:
a)   residential
b)   industrial
c)   commercial
d)   electric utility systems
e)   telecommunications
f)   aerospace
g)   transportation

1.12 If the source voltage in Figure 1.8 is 150 V and the load resistance is 1 Ω, calculate conduction loss, switching loss, and total losses for the duty cycles and frequencies given in Table 1.2. Use $V_{CE(SAT)} = 1.1$ V, $t_{SW(ON)} = 1.0$ μS, and $t_{SW(OFF)} = 1.5$ μS.

Table 1.2

Frequency	Duty cycle	Conduction loss	Switching loss	Total loss
1 kHz	20%			
1 kHz	50%			
1 kHz	75%			
2 kHz	20%			
2 kHz	50%			
2 kHz	75%			

1.13 If the source voltage in Figure 1.6 is 120 V and the load resistance is 10 Ω, calculate switching power loss when $t_{SW} = 1$ μs and the transistor is turned on and off at a frequency of 5 kHz.

1.14 A rheostat controls power supplied by a 100 V DC source to a 10-Ω resistive load. The rheostat resistance is varied in steps from zero to 100 Ω. Write a computer program that calculates the power supplied from the source, the power dissipated in the rheostat, the power consumed by the load, and the efficiency. Tabulate and plot your results.

1.9 Equations

$$P_{ON} = V_{CE(SAT)} * I_C \qquad\qquad \textbf{1.1}$$

$$I_C = \frac{V_S - V_{CE(SAT)}}{R_L} \approx \frac{V_S}{R_L} \qquad\qquad \textbf{1.2}$$

$$P_{ON(avg)} = V_{CE(SAT)} * I_C * \frac{t_{ON}}{T} = V_{CE(SAT)} * I_C * d \qquad\qquad \textbf{1.3}$$

$$P_{OFF(avg)} = V_S * I_{LEAK} * \frac{t_{OFF}}{T} \qquad\qquad \textbf{1.4}$$

$$d = \frac{t_{ON}}{t_{ON} + t_{OFF}} = \frac{t_{ON}}{T} \qquad\qquad \textbf{1.5}$$

$$P_{SW\ ON(max)} = 0.5\ V_{CE(max)} * 0.5\ I_{C(max)} \qquad\qquad \textbf{1.6}$$

$$P_{SW\ ON(avg)} = \frac{1}{6}\ V_{CE(max)} * I_{C(max)} \qquad\qquad \textbf{1.7}$$

$$W_{SW\ ON} = \frac{1}{6}\ V_{CE(max)} * I_{C(max)} * t_{SW(ON)} \qquad\qquad \textbf{1.8}$$

$$W_{SW\ OFF} = \frac{1}{6}\ V_{CE(max)} * I_{C(max)} * t_{SW(OFF)} \qquad\qquad \textbf{1.9}$$

$$P_{SW} = \frac{1}{6}\ V_{CE(max)} * I_{C(max)} * (2\ t_{SW}) * f \qquad\qquad \textbf{1.10}$$

$$P_T = P_{On(avg)} + P_{OFF(avg)} + P_{SW} \qquad\qquad \textbf{1.11}$$

$$P_T = d * V_{CE(SAT)} * I_C + \frac{1}{3} * V_{CE(max)} * I_{C(max)} * t_{SW} * f \qquad\qquad \textbf{1.12}$$

Power Diodes

Learning Objectives

After completing this chapter, the student should be able to

- describe the characteristics and operation of the diode
- analyze diode circuits
- determine power loss in diodes
- list the principal ratings of diodes
- describe how to test diodes
- describe how to protect diodes
- explain how diodes can be connected in series and in parallel to increase their rating

2.1   Introduction

Power diodes play an important role in power electronics circuits. They are used mainly in uncontrolled rectifiers to convert AC to fixed DC voltages and as *freewheeling* diodes to provide a path for the current flow in inductive loads. Power diodes are similar in function to ordinary PN junction diodes; however, power diodes have larger power-, voltage-, and current-handling capabilities.

2.2   The PN Junction Diode

Diodes can be made of either of two semiconductor materials, silicon and germanium. Power diodes are usually constructed using silicon. Silicon diodes can operate at higher current and at higher junction temperatures, and they have greater reverse resistance.

The structure of a semiconductor diode and its symbol are shown in Figure 2.1. The diode has two terminals—an anode A terminal (P junction) and a cathode K terminal (N junction). When the anode voltage is more positive than the cathode, the diode is said to be forward-biased, and it conducts current readily with a relatively low voltage drop. When the cathode voltage is more positive than the anode, the diode is said to be reverse-biased, and it blocks the current flow. The arrow on the diode symbol shows the direction of conventional current flow when the diode conducts.

Figure 2.1
The structure and symbol of a diode

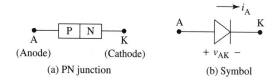

(a) PN junction

(b) Symbol

2.3   The Voltage-Current Characteristic of a Diode

Figure 2.2 shows the V-I characteristic of a diode. When forward-biased, the diode begins to conduct current as the voltage across its anode (with respect to its cathode) is increased. When the voltage approaches the so-called knee voltage, about 1 V for silicon diodes, a slight increase in voltage causes the current to increase rapidly. This increase in current can be limited only by resistance connected in series with the diode.

Figure 2.2
The V-I characteristic of a
diode

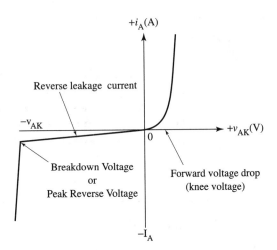

When the diode is reverse-biased, a small amount of current called the *reverse leakage current* flows as the voltage from anode to cathode is increased; this simply indicates that a diode has a very high resistance in the reverse direction. This large resistance characteristic is maintained with increasing reverse voltage until the reverse breakdown voltage is reached. At breakdown, a diode allows a large current flow for a small increase in voltage. Again, a current-limiting resistor must be used in series to prevent destruction of the diode.

2.4   The Ideal Diode

In power electronics we deal with high voltages and currents. Therefore, the detailed characteristic of a diode (for example, Figure 2.2) is not important; we can treat a diode as an ideal element. Figure 2.3 shows the ideal characteristic of a diode. Note that when the diode is forward-biased, it has no voltage across it. The current through the diode then depends on the source voltage and other circuit elements. When the diode is reverse-biased, it has no current through it. The voltage across the diode then depends upon the source voltage and other circuit elements.

Figure 2.3
An ideal diode
characteristic

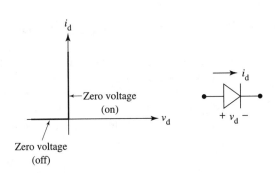

This characteristic of an ideal diode makes it similar to a switch that conducts current in only one direction. The switch can turn on and turn off by itself, depending on the polarity of the voltage. Figure 2.4(a) shows a forward-biased diode, and Figure 2.4(b) shows its switch-equivalent circuit. When a diode's anode is more positive than its cathode, it can be considered to act like a closed switch. Figure 2.4(c) shows a reverse-biased diode, and Figure 2.4(d) shows its switch equivalent. When the diode's anode is more negative than its cathode, it can be considered to act like an open switch.

Figure 2.4
Switch-equivalent circuits of an ideal diode

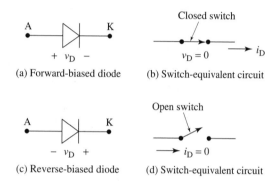

(a) Forward-biased diode   (b) Switch-equivalent circuit

(c) Reverse-biased diode   (d) Switch-equivalent circuit

2.5   The Schottky Diode

The Schottky diode is a low-voltage, high-speed device that works on a different principle from that of the PN junction diode. It is constructed without the usual PN junction. Instead, a thin barrier metal (such as chromium, platinum, or tungsten) is interfaced with the N-type semiconductor. This construction results in a low on-state voltage (about 0.6 V) across the diode when it conducts. Furthermore, it can turn off much faster than a PN junction diode, so switching frequency can be high. However, the reverse leakage current is much higher, and the reverse breakdown voltage is lower compared with that of a PN junction diode. Schottky diodes are therefore used as rectifiers in low-voltage applications where the efficiency of conversion is important. These diodes are also widely used in switching power supplies that operate at frequencies of 20 kHz or higher.

Example 2.1   A Schottky diode rated at 40 V and 25 A has an on-state voltage of 0.5 V and a reverse leakage current of 50 nA. Find the on-state and off-state power loss at rated conditions.

Solution   on-state power loss = 25 * 0.5 = 12.5 W

off-state power loss = 40 * 50 (10^{-9}) = 2 μW

Example 2.2   Repeat Example 2.1 for a PN junction diode rated at 40 V and 25 A, with an on-state voltage of 1.1 V and a reverse leakage current of 0.5 nA.

Solution   on-state power loss = 25 * 1.1 = 27.5 W

off-state power loss = 40 * 0.5 (10^{-9}) = 0.02 μW

The on-state power loss of the Schottky diode is less than half that of the PN junction diode, which is quite significant when efficiency is a concern. There is, of course, a higher power dissipation in the off state due to a higher leakage current; however, the total power loss is still less than that of the PN junction diode.

2.6   Diode Circuit Analysis

2.6.1   Diodes in DC Circuits

To analyze diode circuits, the state of the diode (on or off) must first be found. The diode can then be replaced by the switch-equivalent circuit shown in Figure 2.4. However, in some circuits it may be difficult to figure out which switch equivalent to use (for example, in circuits with more than one source or with more than one diode in series). In these circuits it is helpful to replace the diode(s) mentally with a resistive element and note the resulting current direction due to the applied voltage. If the resulting current is in the same direction as the arrow on the diode symbol, the diode is on.

Example 2.3   For the circuit shown in Figure 2.5, find the diode current (I_D), the diode voltage (V_D), and the voltage across resistor (V_R).

Figure 2.5
See Example 2.3

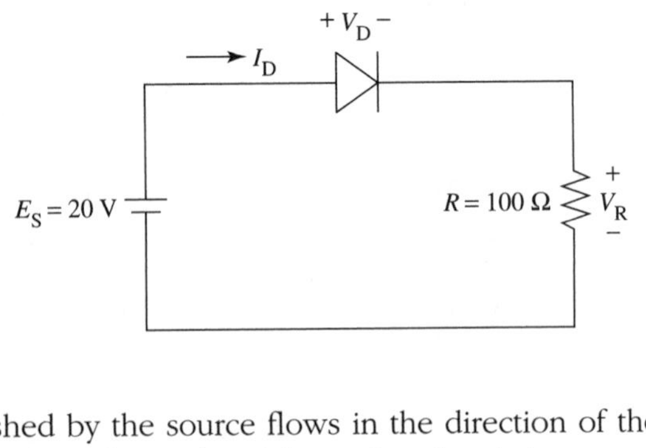

$+ V_D -$

I_D

$E_S = 20$ V

$R = 100\ \Omega$ $+ V_R -$

Solution   Since the current established by the source flows in the direction of the diode's arrow, the diode is on and can be replaced by a closed switch.

voltage across the diode $V_D = 0$ V

voltage across the resistor $V_R = E_S - V_D = 20 - 0 = 20$ V

current through the diode $I_D = V_R/R = 20/100 = 0.2$ A

Example 2.4   Reverse the diode in Figure 2.5 and repeat Example 2.3.

Solution   The direction of current is now opposite to the arrow. The diode is off and can be replaced by an open switch.

current through the diode $I_D = 0$ A

voltage across the resistor $V_R = I_D * R = 0$ V

voltage across the diode $V_D = E_S - V_R = 20 - 0 = 20$ V

Example 2.5   For the circuit shown in Figure 2.6, find the current (I) and voltages V_0, V_1, and V_2.

Figure 2.6
See Example 2.5

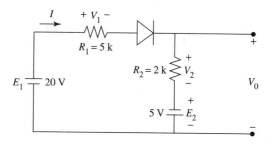

Solution   The two sources are aiding each other in the closed loop; the diode is on and can be replaced by a closed switch.

Applying Kirchhoff's voltage law (KVL),

$E_1 - V_1 - V_2 + E_2 = 0$

$E_1 - I(R_1) - I(R_2) + E_2 = 0$

Solving for I,

$$I = \frac{E_1 + E_2}{R_1 + R_2} = \frac{25}{7k} = 3.5 \text{ mA}$$

$V_1 = I * R_1 = 17.5$ V

$V_2 = I * R_2 = 7.0$ V

$V_0 = V_2 - E_2 = 7 - 5 = 2$ V

2.6.2   Diodes in AC Circuits

AC circuits have a voltage that varies with time. Therefore, there may be times when the AC voltage forward-biases a diode and times when it reverse-biases the same diode. Circuit analysis can be done separately for positive and negative half-cycles. It must be noted when the voltage polarity across the diode forward-biases it and when it reverse-biases it. The diode can then be replaced by its switch-equivalent circuit.

Figure 2.7
Diode in an AC circuit

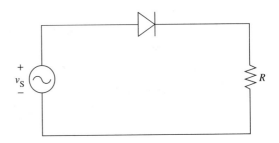

Example 2.6   Find the switch-equivalent circuit of a diode with an AC source voltage V_S, as shown in Figure 2.7.

Solution   During the positive half-cycle, the anode is more positive than its cathode, and therefore the diode is forward-biased. We can replace the diode with a closed switch.

During the negative half-cycle, the anode is more negative than its cathode, and therefore, the diode is reverse-biased. We can replace the diode with an open switch.

Example 2.7   For the circuit shown in Figure 2.8, draw the waveforms of the voltage across the resistor (V_R) and the voltage across the diode (V_D).

Figure 2.8
See Example 2.7

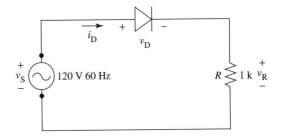

Solution   During the positive half-cycle, the diode is forward-biased and can therefore be replaced by a closed switch. The voltage across the diode is zero, and the voltage across the resistor is the same as the source voltage. During the negative half-cycle, the diode is reverse-biased and can therefore be replaced by an open switch. The voltage across the resistor is zero, and the voltage across the diode is the same as the source voltage.

The waveforms of V_R and V_D are shown in Figure 2.9.

Figure 2.9
Waveforms of V_R and V_D

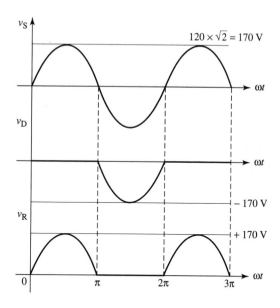

2.7 Diode Losses

The total power loss that occurs in a diode is the sum of the on-state, off-state, and switching losses:

$$P_T = P_{ON} + P_{OFF} + P_{SW} \qquad \textbf{2.1}$$

where

$$P_{ON} = V_F * I_F * \frac{t_{ON}}{T} \qquad \textbf{2.2}$$

$$P_{OFF} = V_R * I_R * \frac{t_{OFF}}{T} \qquad \textbf{2.3}$$

$$P_{SW} = P_{SW(ON)} + P_{SW(OFF)} \qquad \textbf{2.4}$$

$$P_{SW(ON)} = \frac{1}{6} V_{F(MAX)} * I_{F(MAX)} * t_F * f \qquad \textbf{2.5}$$

$$P_{SW(OFF)} = \frac{1}{6} V_{F(MAX)} * I_{F(MAX)} * t_R * f \qquad \textbf{2.6}$$

In these equations,

V_F = forward voltage

I_F = forward current

V_R = reverse voltage

I_R = reverse leakage current

t_{ON} = time of diode conduction

t_{OFF} = time during which diode is reverse-biased

t_F = switching time in forward direction

t_R = switching time in reverse direction

Example 2.8   In the circuit shown in Figure 2.8, V_S = 400 V, f = 10 kHz, d = 50%, and I_D = 30 A. If the diode has the following characteristics, find the total power loss in the diode.

V_F = 1.1 V

I_R = 0.3 mA

t_F = 1 μs

t_R = 0.1 μs

Solution $T = 1/f = 0.1$ ms

$T = t_{ON} + t_F + t_{OFF} + t_R$

$$t_{ON} = t_{OFF} = \left(\frac{T - t_F - t_R}{2}\right) = \left(\frac{100 - 1 - 0.1}{2}\right) \text{μs} = 49.45 \text{ μs}$$

$$P_{ON} = 1.1 * 30 * \frac{49.45 \ (10^{-6})}{0.1 \ (10^{-3})} = 16.32 \text{ W}$$

$$P_{OFF} = 400 * 0.3 \ (10^{-3}) * \frac{49.45 \ (10^{-6})}{0.1 \ (10^{-3})} = 0.06 \text{ W}$$

$$P_{SW} = 2\left[\frac{1}{6} * 400 * 30 * \frac{1.1 \ (10^{-6})}{0.1 \ (10^{-3})}\right] = 44 \text{ W}$$

$P_T = 16.32 + 0.06 + 44 = 60.38$ W

Example 2.9   Repeat Example 2.8 with f = 100 kHz.

Solution $T = 1/f = 10$ μs

$$t_{ON} = t_{OFF} = \frac{10 - 1.1}{2} \text{μs} = 4.95 \text{ μs}$$

$$P_{ON} = 1.1 * 30 * \frac{4.95 \ (10^{-6})}{10 \ (10^{-6})} = 16.32 \text{ W}$$

$$P_{OFF} = 400 * 0.3(10^{-3}) * \frac{4.95 \ (10^{-6})}{10 \ (10^{-6})} = 0.06 \text{ W}$$

$$P_{SW} = 2\left[\frac{1}{6} * 400 * 30 * \frac{1.1 \ (10^{-6})}{10 \ (10^{-6})}\right] = 440 \text{ W}$$

$P_T = 16.32 + 0.06 + 440 = 456.40$ W

The total power loss is almost eight times that in Example 2.8.

2.8 Principal Ratings for Diodes

2.8.1 Peak Inverse Voltage (PIV)

The *peak inverse voltage* rating of a diode is the maximum reverse voltage that can be connected across a diode without breakdown. If the PIV rating is exceeded, the diode begins to conduct in the reverse direction and can be immediately destroyed. PIV ratings extend from tens of volts to several thousand volts, depending on the construction. The PIV rating is also called the *peak reverse voltage* (PRV) or *breakdown voltage* V(BR).

2.8.2 Maximum Average Forward Current ($I_{f(avg)max}$)

The maximum average forward current is the maximum current a diode can safely handle when forward-biased. Power diodes are presently available in ratings from a few amperes to several hundred amperes. If a diode is to be used economically, it must be operated near its maximum forward current rating.

2.8.3 Reverse Recovery Time (t_{rr})

The reverse recovery time of a diode is of great significance in high-speed switching applications. A real diode does not instantaneously switch from a conduction to a nonconduction state. Instead, a reverse current flows for a short time, and the diode does not turn off until the reverse current decays to zero, as shown in Figure 2.10. The diode initially conducts a current I_F; when the diode is reverse-biased, this current decreases and reverse current I_R flows. The time interval during which reverse current flows is called the reverse recovery time. During this time, charge carriers that were stored in the junction when forward conduction terminated are removed.

Figure 2.10
Reverse recovery characteristics

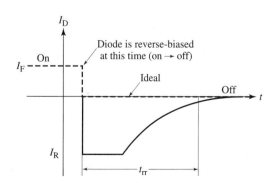

Diodes are classified as "fast recovery" or "slow recovery" types based on their reverse recovery times. Recovery times range from a few microseconds in a PN junction diode to several hundred nanoseconds in a fast-recovery diode like a Schottky diode. The PN junction diode is normally sufficient for rectification of a 60 Hz AC signal. Fast-recovery diodes with low t_{rr} are used in high-frequency applications such as inverters, choppers, and uninterruptible power supplies (UPS).

2.8.4   Maximum Junction Temperature ($T_{J(max)}$)

This parameter defines the maximum junction temperature that a diode can withstand without failure. The rated temperatures of silicon diodes typically range from $-40°C$ to $+200°C$. Operation at lower temperatures generally results in better performance. Diodes are usually mounted on heat sinks to improve their temperature rating.

2.8.5   Maximum Surge Current (I_{FSM})

The I_{FSM} (forward surge maximum) rating is the maximum current that the diode can handle as an occasional transient or from a circuit fault.

2.9   Diode Protection

A power diode must be protected against overvoltage, overcurrent and transients.

2.9.1   Overvoltage

When a diode is forward-biased, the voltage across it is low and poses no problems. A reverse-biased diode acts like an open circuit. If the voltage across the diode exceeds its breakover voltage, it breaks down, resulting in a large current flow. With this high current and large voltage across the diode, it is quite likely that the power dissipation at the junction will exceed its maximum value, destroying the diode. It is a common practice to select a diode with a peak reverse voltage rating that is 1.2 times higher than the expected voltage during normal operating conditions.

2.9.2   Overcurrent

Manufacturer's data sheets provide current ratings based on the maximum junction temperatures produced by conduction losses in diodes. In a given circuit, it is recommended that the diode current be kept below this rated value. Overcurrent protection is then accomplished by using a fuse to ensure that the diode current does not exceed a level that will increase the operating temperature beyond the maximum value.

2.9.3   Transients

Transients can lead to higher-than-normal voltages across a diode. Protection against transients usually takes the form of an RC series circuit connected across the diode. This arrangement, shown in Figure 2.11, snubs or reduces the rate of change of voltage and is commonly called a *snubber circuit.*

Figure 2.11
A snubber circuit

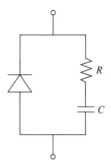

2.10   Testing a Diode

An ohmmeter can be used to test power diodes easily and safely. The diode–switch-equivalent circuits shown in Figure 2.4 can be used to determine how diodes can be tested. Connect the ohmmeter so that it forward-biases the diode; this should give a low resistance reading. The actual reading will depend on the current flow through the diode from the internal battery of the ohmmeter. Reversing the leads should give much higher resistance or even an infinite reading. A high resistance reading in both directions suggests an open diode, while a very low resistance reading in both directions suggests a shorted diode.

2.11   Series and Parallel Operation of Diodes

The maximum power than can be controlled by a single diode is determined by its rated reverse voltage and by its rated forward current. In high-power applications, a single diode may have insufficient power-handling capability. To increase power capability, diodes are connected in series to increase the voltage rating or in parallel to increase the current rating. A series/parallel arrangement of diodes can be used for high-voltage and high-current applications.

2.11.1 Series Connection of Diodes

In very high-voltage applications, the reverse-voltage rating of a single diode may not be sufficient. A series connection of two or more diodes (see Figure 2.12), is then used to increase the voltage rating. However, the reverse voltage may not be equally divided between the two diodes: the diode with the lower leakage current can have excessive reverse voltage across it. Even if we use same number-type diodes, their V-I characteristics may not be identical, as illustrated in Figure 2.13. The current rating of the diodes in series is the same as the current rating of one of the diodes. In the reverse direction, both series diodes have the same reverse leakage current, but, as shown, they have different values for reverse voltage. In such a case diode D_1 may exceed its reverse-voltage rating.

Figure 2.12
Series connection of diodes

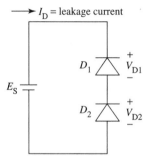

Figure 2.13
V-I characteristics of two diodes

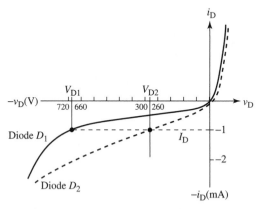

Forced voltage sharing can be obtained by connecting voltage-sharing resistors of appropriate value across each series diode. Figure 2.14 shows the effect of placing resistors across the diode. To be effective, the resistors must conduct a current much greater than the leakage current of the diodes. These sharing resistors will consume power during reverse-bias operation, so it is important to use as large a resistance as possible.

In addition, there can be excessive reverse voltage across a diode due to different reverse recovery times. A capacitor connected in parallel with each diode (see Figure 2.15) will protect the diode from voltage transients.

The value of the voltage-sharing resistor can be obtained as follows: The source current is

$$I_S = \frac{V_{D1}}{R} + I_{D1} = \frac{V_{D2}}{R} + I_{D2}$$

Solving for R,

$$R = \frac{V_{D1} - V_{D2}}{I_{D2} - I_{D1}}$$

2.7

The power dissipated in R is

$$P_R = I^2_{R1} * R + I^2_{R2} * R$$

Figure 2.14
Series connection of diodes with resistors added

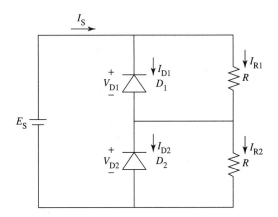

Figure 2.15
Series connection of diodes with resistors and capacitor added

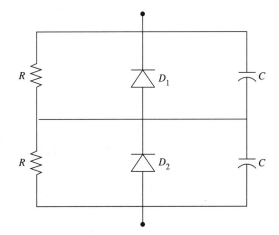

Example 2.10 Two diodes with voltage ratings of 800 V and reverse leakage currents of 1 mA are connected in series across an AC source whose peak value is $V_{S(max)} = 980$ V. The reverse characteristics are as shown in Figure 2.13. Determine
a) the reverse voltage across each diode
b) the value of the voltage-sharing resistor, so that the voltage across any diode is no more than 55% of $V_{s(max)}$
c) the total source current and power loss in the resistors

Solution a) With no force sharing, the current through the diodes is the leakage current. Therefore, at 1 mA, from Figure 2.13,

$$V_{D1} = 700 \text{ V}$$
$$V_{D2} = 280 \text{ V}$$

b) With forced voltage sharing, such that

$$V_{D1} = 55\% * 980 = 539 \text{ V}$$
$$V_{D2} = 900 - 495 = 441 \text{ V}$$

We obtain from the graph

$$I_{D1} = 0.7 \text{ mA}$$
$$I_{D2} = 1.4 \text{ mA}$$

Using Equation 2.7,

$$R = \frac{V_{D1} - V_{D2}}{I_{D2} - I_{D1}}$$

$$= 140 \text{ K}$$

c) The current through R is

$$I_{R1} = 539/140K = 3.85 \text{ mA}$$
$$I_{R2} = 441/140K = 3.15 \text{ mA}$$
$$\text{source current} = 0.00385 + 0.0007 = 4.55 \text{ mA}$$

or

$$\text{source current} = 0.00315 + 0.0014 = 4.55 \text{ mA}$$

The power dissipated in R is

$$P_R = I^2{}_{R1} * R + I^2{}_{R2} * R = 2.1 + 0.44 = 2.54 \text{ W}$$

2.11.2   Parallel Connection of Diodes

If the load current is greater than the current rating of a single diode, then two or more diodes can be connected in parallel (see Figure 2.16) to achieve a higher forward-current rating. Diodes connected in parallel do not share the current equally due to different forward-bias characteristics. The diode with the lowest forward voltage drop will try to carry a larger current and can overheat. Figure 2.17 shows the V-I on-state characteristics of two diodes. If these two diodes are connected in parallel at a give voltage, a different current flows in each diode. The total current flow is the sum of I_{D1} and I_{D2}. The total current rating of the pair is not the sum of the maximum current rating for each but is a value that can be just larger than the rating of one diode alone.

Figure 2.16
Parallel connection of diodes

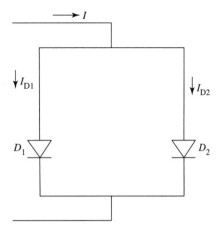

Figure 2.17
V-I characteristics of two diodes

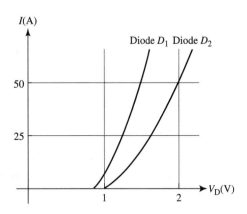

Parallel diodes can be forced to share current by connecting a very small resistor in series with each diode. In Figure 2.18, the current-sharing resistor R establishes values of I_{D1} and I_{D2} that are nearly equal. Although current sharing is very effective, the power loss in the resistors is very high. Furthermore, it causes an increase in voltage across the combination. Unless using a parallel arrangement is absolutely necessary, it is better to use one device with an adequate current rating.

Figure 2.18
Parallel connection of
diodes with added resistors

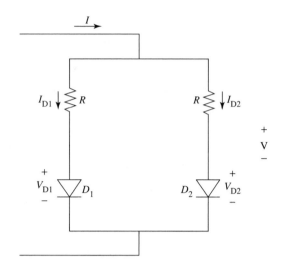

The value of the current-sharing resistor can be obtained as follows:

$$V = V_{D1} + I_{D1} * R = V_{D2} + I_{D2} * R$$

Solving for R,

$$R = \frac{V_{D2} - V_{D1}}{I_{D1} - I_{D2}}$$

The power dissipated in R is

$$P_R = I^2_{D1} * R + I^2_{D2} * R \qquad\qquad\textbf{2.10}$$

The voltage across the diode combination is

$$V = V_{D1} + I_{D1}\,R = V_{D2} + I_{D2}\,R \qquad\qquad\textbf{2.11}$$

Example 2.11 Two diodes having the characteristics shown in Figure 2.17 are connected in parallel. The total current through the diodes is 50 A. To enforce current sharing, two resistors are connected in series. Determine:

a) the resistance of the current-sharing resistor, so that the current through any diode is no more than 55% of I

b) the total power loss in the resistors

c) the voltage across the diode combination (V)

Solution a) With forced current sharing, such that

$$I_{D1} = 55\% * 50 = 27.5 \text{ A}$$
$$I_{D2} = 50 - 27.5 = 22.5 \text{ A}$$

We obtain from Figure 2.17,

$$V_{D1} = 1.3 \text{ V}$$
$$V_{D2} = 1.6 \text{ V}$$
$$V = V_{D1} + I_{D1} * R = V_{D2} + I_{D2} * R$$
$$= 1.3 + 27.5 \, R = 1.6 + 22.5 \, R$$

Solving for R,

$$R = 0.06 \, \Omega$$

b) The power dissipated in R is

$$P_R = I^2_{D1} * R + I^2_{D2} * R = 27.5^2 * 0.06 + 22.5^2 * 0.06 = 75.8 \text{ W}$$

c) The voltage across the diode combination is

$$V = V_{D1} + I_{D1} \, R = V_{D2} + I_{D2} \, R$$
$$= 1.3 + 27.5 * 0.06 = 1.6 + 22.5 * 0.06$$
$$= 2.95 \text{ V}$$

2.12 Problems

2.1 What type of semiconductor material is used in power diodes?

2.2 What are the main advantages of silicon diodes?

2.3 What condition forward-biases a diode?

2.4 What condition reverse-biases a diode?

2.5 What is the voltage across an ideal diode that is forward-biased?

2.6 Draw the switch-equivalent circuit of a forward-biased diode.

2.7 Draw the switch-equivalent circuit of a reverse-biased diode.

2.8 Define the PIV rating of a diode.

2.9 How can a diode be tested using an ohmmeter?

2.10 For the circuit shown in Figure 2.19, find I_D. What is the maximum reverse voltage across the diode?

Figure 2.19
See Problem 2.10

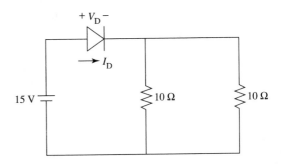

2.11 For the circuit shown in Figure 2.20, find I_1, I_2, I_D, and V_D.

Figure 2.20
See Problem 2.11

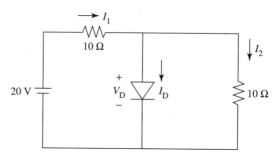

2.12 For the circuit shown in Figure 2.21, find I, I_{D1}, and I_{D2}

Figure 2.21
See Problem 2.12

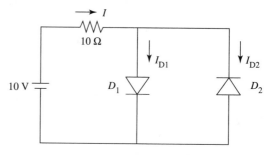

2.13 For the circuit shown in Figure 2.22, drawn the waveform for I.

Figure 2.22
See Problem 2.13

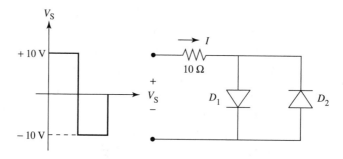

2.14 For the circuit shown in Figure 2.23, find (a) the maximum forward current that the diode should handle (b) the maximum reverse voltage that the diode should withstand.

Figure 2.23
See Problem 2.14

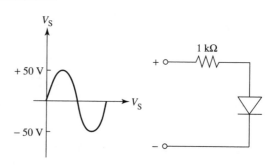

2.15 From various manufacturers' data sheets, obtain the following ratings for some currently manufactured power diodes: (a) peak reverse voltage (b) maximum forward current (c) maximum operating junction temperature (d) maximum on-state voltage drop at rated current (e) maximum reverse leakage current (f) maximum switching frequency (g) turn-on time (h) turnoff time

2.16 Write a computer program to plot the V-I characteristic of a diode.

2.13   Equations

$$P_T = P_{ON} + P_{OFF} + P_{SW} \qquad\qquad \textbf{2.1}$$

$$P_{ON} = V_F * I_F * \frac{t_{ON}}{T} \qquad\qquad \textbf{2.2}$$

$$P_{OFF} = V_R * I_R * \frac{t_{OFF}}{T} \qquad\qquad \textbf{2.3}$$

$$P_{SW} = P_{SW(ON)} + P_{SWOFF} \qquad\qquad \textbf{2.4}$$

$$P_{SW(ON)} = \frac{1}{6} V_{F(MAX)} * I_{F(MAX)} * t_F * f \qquad\qquad \textbf{2.5}$$

$$P_{SW(OFF)} = \frac{1}{6} V_{F(MAX)} * I_{F(MAX)} * t_R * f \qquad\qquad \textbf{2.6}$$

$$R = \frac{V_{D1} - V_{D2}}{I_{D2} - I_{D1}} \qquad\qquad \textbf{2.7}$$

$$P_R = I^2_{R1} * R + I^2_{R2} * R \qquad\qquad \textbf{2.8}$$

$$R = \frac{V_{D2} - V_{D1}}{I_{D1} - I_{D2}} \qquad\qquad \textbf{2.9}$$

$$P_R = I^2_{D1} * R + I^2_{D2} * R \qquad\qquad \textbf{2.10}$$

$$V = V_{D1} + I_{D1} R = V_{D2} + I_{D2} R \qquad\qquad \textbf{2.11}$$

Power Transistors

Chapter Outline

Learning Objectives

After completing this chapter, the student should be able to

- describe the characteristics and operation of the bipolar junction transistor (BJT)
- list the power losses in a BJT
- describe how to test BJTs
- describe how to protect BJTs
- list the principal ratings of BJTs
- explain how BJTs can be connected in series and in parallel to increase their ratings
- describe the characteristics and operation of the metal-oxide

- semiconductor field-effect transistor (MOSFET)
- list the power losses in a MOSFET
- describe how to protect MOSFET
- explain how MOSFETs can be connected in series and in parallel to increase their ratings
- describe the characteristics and operation of the insulated-gate bipolar transistor (IGBT)
- list the power losses in an IGBT

3.1 Introduction

Transistors with high voltage and current ratings are known as power transistors. A transistor is a three-layer PNP or NPN semiconductor device with two junctions. Transistors have two basic types of applications: amplification and switching. In power electronics, where the main objective is the efficient control of power, transistors are invariably operated as switches. They are mainly used in chopper and inverter applications.

Diodes are uncontrollable switches having only two terminals. They only respond to switch the voltage across them. Transistors, on the other hand, have three terminals. Two terminals act like switch contacts, while the third is used to turn the switch on and off. Thus, the control circuit can be independent of the circuit being controlled.

Two types of power transistors are extensively used in power electronics circuits: the bipolar junction transistor (BJT) and the metal-oxide semiconductor field-effect transistor (MOSFET). Until the development of the power MOSFET, the BJT was the device of choice in power electronics applications.

The switching speed of a BJT is many times slower than that of a MOSFET of similar size and rating. A BJT is a current-controlled device, and a large base current is required to keep the device in the on state. In addition, to obtain fast turnoff, a higher reverse base current is required. These limitations make the base drive circuit design more complex and therefore more expensive than that of the MOSFET. Power MOSFETs, on the other hand, are voltage-controlled devices. They are preferable to BJTs in high-frequency applications where switching power loss is important. However, the on-state voltage drop of the power MOSFET is higher than that of a BJT of similar size and rating. Therefore, in high-voltage applications where on-state losses are to be minimized, a BJT is preferred at the expense of poor high-frequency performance.

The invention of insulated-gate bipolar transistor (IGBT) was partly driven by the limitations of BJTs and MOSFETs. IGBTs provide high-voltage capability, low on-state losses, simple drive circuitry, and relatively fast switching speeds. IGBTs are therefore becoming an ideal choice for high-voltage applications where conduction losses must be kept low.

3.2   Power Bipolar Junction Transistors (BJTs)

Power transistors are available in both NPN and PNP types. However, we will concentrate on the NPN device, since it has a higher current and voltage rating than the PNP device. The structure and symbol of an NPN transistor are shown in Figure 3.1. This type of transistor is called a *bipolar junction transistor* (BJT). BJT is usually referred to as a transistor.

A transistor has three terminals: the *base* (B), the *collector* (C) and the *emitter* (E). The collector and the emitter in a transistor are not reversible. In fact the transistor's characteristics and ratings change significantly when these two terminals are reversed. If the arrowhead on the emitter points toward the base, the transistor is a PNP transistor. If the arrow points away from the base, it is an NPN transistor.

When a transistor is used as a switch to control power from the source to the load, terminals C and E are connected in series with the main power circuit, while terminals B and E are connected to a driving circuit that controls the on and off action. A small current through the base-emitter junctions turns on the collector-to-emitter path. This path may carry many times more current than the base-emitter junction.

Figure 3.1
An NPN transistor and its symbol

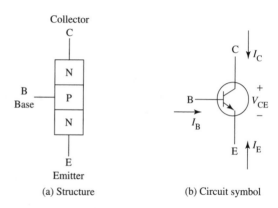

(a) Structure   (b) Circuit symbol

3.2.1 BJT V-I Characteristics

Since most applications of power transistors have a common emitter connection, the characteristic will be explained for that configuration. Figure 3.2 shows the V-I characteristic of a transistor. There are three regions of operation, the cutoff, saturation, and active regions. If the base current I_B is zero, the collector current I_C is negligibly small and the transistor is in the cutoff region, which is the off state of the transistor. In this region both the collector-base and base-emitter junctions are reverse-biased and the transistor behaves as an open switch. On the other hand, if the base current I_B is sufficient to drive the transistor into saturation (the collector current is very high and V_{CE} is approximately zero), then the transistor behaves as a closed switch. In the saturation region, both junctions are forward-biased. In the active region of operation, the base-emitter junction is forward-biased while the collector-base junction is reverse-biased. The active region is used for amplification and is avoided in switching applications.

Note that the V-I characteristic does not show any reverse region. A BJT cannot block more than about 20 V in the reverse direction. Therefore BJTs are not used to control AC power, unless a reverse shunting diode is connected between the emitter and the collector to protect the transistor from reverse voltages.

Figure 3.2
BJT *V-I* characteristic

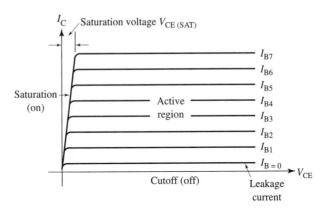

Since transistors are used mainly as switches, the idealized transistor characteristic is of prime importance. Figure 3.3 shows the V-I characteristic of a BJT

Figure 3.3
Idealized characteristic of a transistor

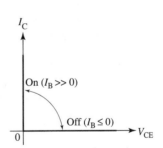

operating as a switch. When the transistor is off, there is no collector current no matter what the value of V_{CE} is. When the transistor is on, the voltage V_{CE} is zero no matter what the value of the collector current is. A transistor has excellent characteristics as an ideal switch.

3.2.2 Biasing a Transistor

When a transistor is used as a controlled switch, the base current is provided by the control circuit connected between the base and the emitter. The collector and the emitter form the power terminals of the switch. Figure 3.4 shows how an NPN transistor is biased. The input base current I_B determines whether the transistor switch will be off (with no current to the load R_C) or on (allowing current to flow).

Figure 3.4
Biasing a transistor

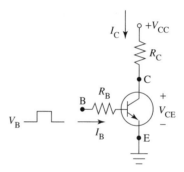

Figure 3.5 shows the DC load line, which represents all possible operating points. Point P_1 is the ideal operating point for the switch when it is on. Here the collector current I_C is equal to V_{CC}/R_C, and the voltage across the collector-emitter is zero.

Point P_4 is the ideal operating point for the switch when it is off. Here the collector current I_C is zero and the voltage across the collector-emitter is equal to the supply voltage V_{CC}.

The line drawn between points P_1 and P_4 is the load line. The intersection of the load line with the base current is the operating point of the transistor. The operating point is determined by the circuit that is external to the transistor, i.e., V_{CC} and R_C.

Point P_2, where the load line intersects the $I_B = 0$ curve, is the actual operating point at cutoff. At this point, the collector current is the leakage current. The voltage across the collector-emitter terminal can be found by applying Kirchoff's voltage law (KVL) around the output loop:

$$V_{CC} - I_C R_C - V_{CE} = 0$$
$$V_{CE} = V_{CC} - I_C R_C$$

3.1

Point P_3, where the load line intersects the $I_B = I_{B(sat)}$ curve, is the actual operating point when the BJT is on. Point P_3 is called the saturation point. An on transistor has a small voltage drop across its collect-emitter terminals; this voltage is called the saturation voltage $V_{CE(sat)}$. The collector current is at a maximum here and is given by

$$I_{C(sat)} = \frac{V_{CC} - V_{CE(sat)}}{R_C} \approx \frac{V_{CC}}{R_C} \qquad \textbf{3.2}$$

The minimum base current required to ensure satisfactory operation is given by

$$I_B = \frac{I_{C(sat)}}{\beta} = \frac{V_{CC}}{\beta R_C} \qquad \textbf{3.3}$$

where β is the DC current gain given by I_C/I_B.

Any value of I_B higher than the value calculated using Equation 3.3 will ensure a saturated on state. In fact, to accommodate any changes in I_C above the required value, it is desirable to use a little higher value of base current than is obtained by the above formula. A high base current also reduces the turn-on time and therefore reduces power dissipation.

Figure 3.5
DC load line

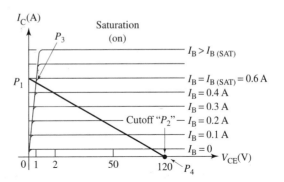

All operating points between cutoff and saturation are in the active regions of a transistor. In this region, both I_C and V_{CE} are relatively high, resulting in high power dissipation in the transistor.

3.2.3   Power Loss in BJTs

There are four sources of power loss in a transistor: conduction or on-state loss, leakage or off-state loss, turn-on switching loss, and turnoff switching loss.

Power losses in the transistor when it is on or off are low, but there are appreciable power losses during switching and they cannot be neglected. Moreover, the switching losses depend on the frequency—they are directly proportional. In fact, in applications using BJTs, the switching losses tend to limit the highest switching frequency achievable.

3.2.3.1 Power Loss in the BJT in the On State

Because base current is required to keep a power transistor on, the power loss in the base drive circuit should be considered. The power loss in the base is

$$P_B = V_{BE(SAT)} I_B$$

The power loss in the collector is

$$P_C = V_{CE(SAT)} I_C$$

The total power loss in the transistor in the on state is given by

$$P_{ON} = V_{CE(SAT)} I_C + V_{BE(SAT)} I_B$$

If the base power loss is small compared with the collector power losses,

$$P_{ON} \approx V_{CE(SAT)} I_C \qquad \text{3.4}$$

3.2.3.2 Power Loss in the BJT in the Off State

When the BJT is in the off state, the power loss P_{OFF} is given by

$$P_{OFF} = V_{CE} * I_C \approx V_{CC} * I_{leakage} \qquad \text{3.5}$$

3.2.3.3 Energy Loss in the BJT During Turn-on

Energy loss in the transistor during turn-on is given by

$$W_{(SW\text{-}ON)} = \frac{V_{CC} \, I_{C(max)}}{6} \, t_r \qquad \text{3.6}$$

where t_r is the collector current rise time (typically 1–2 μs).

3.2.3.4 Energy Loss in the BJT During Turnoff

Energy loss in the transistor during turnoff is given by

$$W_{(SW\text{-}OFF)} = \frac{V_{CC} \, I_{C(max)}}{6} \, t_f \qquad \text{3.7}$$

where t_f is the collect-emitter voltage fall time (typically 1–3 μs).

If the transistor is switched on and off in a periodic manner, the average power that results depends on the frequency of these switching operations. The total average power dissipation in the transistor is given by

$$P_{T(avg.)} = \{P_{ON} * t_{ON} + P_{OFF} * t_{OFF} + W_{(SW\text{-}ON)} + W_{(SW\text{-}OFF)}\} f \qquad \text{3.8}$$

Example 3.1 In Figure 3.4, V_{CC} is 120 V, R_C is 20 Ω, and the V-I characteristic of the transistor is as shown in Figure 3.5. Find the load current and the power loss for the following base currents:

a) $I_B = 0.6$ A
b) $I_B = 0.4$ A
c) $I_B = 0.2$ A
d) $I_B = 0.0$ A

Solution a) For $I_B = 0.6$ A,

$$V_{CE} \approx 1 \text{ V}$$
$$I_C = (120 - 1)/20 = 5.95 \text{ A}$$
$$P_{ON} = 1 * 5.95 = 5.95 \text{ W}$$

b) For $I_B = 0.4$ A,

$$V_{CE} = 2 \text{ V}$$
$$I_C = (120 - 2)/20 = 5.9 \text{ A}$$
$$P_{ON} = 2 * 5.9 = 11.8 \text{ W}$$

c) For $I_B = 0.2$ A,

$$V_{CE} = 50 \text{ V}$$
$$I_C = (120 - 50)/20 = 3.5 \text{ A}$$
$$P = 50 * 3.5 = 175 \text{ W}$$

d) For $I_B = 0.0$ A,

$$I_C = 0 \text{ A}$$
$$V_{CE} = 120 \text{ V}$$
$$P_{OFF} = 120 * 0 = 0 \text{ W}$$

Note that in parts (a) and (b) of this example the base current is sufficient to ensure operation in the saturated region. In this region, the voltage across the transistor is very small and the load current is very close to its maximum value for the given load resistance. However, with a decrease in base current to 0.2 A, the transistor is no longer in the saturated region. There is now a large increase in the voltage across the transistor, with a corresponding decrease in load current. More importantly, the power loss in the transistor during conduction is very high under this condition.

Example 3.2 In Figure 3.4, $V_{CC} = 208$ V, $R_C = 20$ Ω, $V_{CE(SAT)} = 0.9$ V, $V_{BE(SAT)} = 1.1$ V, and $\beta = 10$. Find:

a) I_C
b) I_B
c) the power loss in the collector (P_C)
d) the power loss in the base (P_B)

Solution a) $$I_C = \frac{V_{CC} - V_{CE(SAT)}}{R_C} = \frac{208 - 0.9}{20} = 10.36 \text{ A}$$

b) $I_B = I_C/\beta = 10.36/10 = 1.36$ A

c) $P_C = V_{CE(SAT)}I_C$
$= 0.9 * 10.36$
$= 9.32$ W

d) $P_B = V_{BE(SAT)} \, I_B$
$= 1.1 * 1.36$
$= 1.5$ W

Example 3.3   In Figure 3.4, V_{CC} = 200 V, R_C = 20 Ω, t_r = 1.0 μs, and t_f = 1.5 μS. If the switching frequency is 5 kHz, find:
a) the turn-on energy loss
b) the turnoff energy loss
c) the switching power loss

Solution $I_{C(max)} \approx \dfrac{V_{CC}}{R_C} = \dfrac{200}{20} = 10$ A

a) $W_{(SW\text{-}ON)} = \dfrac{(200)\ (10)\ 1(10^{-6})}{6} = 333.3$ μJ

b) $W_{(SW\text{-}OFF)} = \dfrac{(200)\ (10)\ 1.5(10^{-6})}{6} = 500$ μJ

c) $W_{(SW)} = (333.3 + 500.0)$ μJ $= 833.8$ μJ

$P_{SW(avg.)} = 833.3(10^{-6}) * 5(10^{-3}) = 4.2$ W

3.2.4   Testing a Transistor

The state of a transistor can be tested using an ohmmeter. Figure 3.6 shows the schematics of an NPN transistor along with its diode analogy. This analogy can be used to see how to test a transistor with an ohmmeter.

Figure 3.6
Diode analogy of an NPN transistor

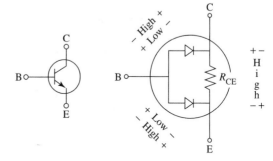

When forward-biased, the base-emitter and base-collector junctions should give a relatively low resistance, while both junctions should register a much higher resistance if reverse-biased. The resistance between the collector and the emitter (R_{CE}) can also be tested. In both directions, this resistance is much higher than the forward resistance of either junction. In silicon transistors, R_{CE} may read infinity on some ohmmeters. Faulty power transistors often appear shorted from collector to emitter, even when both junctions test all right.

The type of transistor can also be found by simply noting the polarity of the ohmmeter leads as applied to the base-emitter junction. If the positive lead is connected to the base and the negative lead to the emitter, a low resistance reading suggests an NPN transistor, while a high reading indicates a PNP transistor.

3.2.5 BJT Protection

A transistor must be protected against excessive currents and voltages to prevent damage. When transistors are used at high frequencies, the switching losses are high because of the transition through the active region. Therefore, thermal conditions are especially important.

3.2.5.1 Overcurrent Protection

When the transistor is in the on-state, the collector current I_C increases as the voltage V_{CE} increases. Therefore, the power dissipation ($V_{CE} * I_C$) also increases and the junction temperature rises. This results in a decrease in the effective resistance, since a BJT has a negative temperature coefficient. The decrease in resistance causes a higher collector current, more power dissipation, and a further rise in temperature, which continues the process. This positive feedback can lead to thermal runaway that will eventually destroy the transistor.

Since thermal runaway takes only a few microseconds, fuses cannot be used to protect a BJT. A better way to protect the transistor is to turn the device off when V_{CE} and I_C rise above a reference level.

Protection against severe faults can be ensured by using a shorting switch in parallel with the transistor. The shunt switch is turned on by a signal from the control circuit upon detection of a severe fault, thus providing an alternate path for the fault current.

3.2.5.2 Overvoltage Protection

If the collector-base-reverse-bias voltage exceeds a certain limit (>1000 V) when the transistor is off, the minority carrier flow can cause avalanche breakdown, resulting in very high current. The resulting high power dissipation can easily damage the transistor. An antiparallel diode connected across the transistor can be used for protection.

3.2.5.3 Reverse Blocking Voltage Protection

A transistor does not have reverse blocking capability. Therefore, it must be shorted by an antiparallel diode if it is used in an AC circuit.

3.2.5.4 Snubber Circuits

A snubber circuit is used to limit the voltage on the device during switching transients. A typical snubber circuit for a BJT is shown in Figure 3.7. It consists of a diode (D), a resistor (R), and a capacitor (C). When the transistor is on, the voltage across it and the snubber circuit is close to zero. During turnoff, the diode turns on and the capacitor starts charging. When the transistor is off, the capacitor is charged to full blocking voltage. Therefore, the capacitor diode combination slows the rate of rise of voltage across the transistor. The capacitor discharges the next time the transistor is on. The resistor limits the peak value of the discharge current through the transistor.

Figure 3.7
Snubber circuit

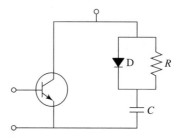

3.2.6   Power Transistor Ratings

3.2.6.1   Collector-Emitter Saturation Voltage ($V_{CE(SAT)}$)

When the transistor is fully on, it has a small voltage drop across its collector-to-emitter terminals. This voltage drop is called the *saturation voltage* ($V_{CE(SAT)}$). This on-state voltage is usually very small—in the 1–2 V range. The conduction power loss in the transistor depends on this voltage.

When using a power transistor as a switch, the base current must be high enough to force the transistor into the collector saturation region. This is desirable to reduce $V_{CE(SAT)}$ and thus reduce power dissipation during the on time.

3.2.6.2   DC Current Gain (h_{fe})

The ratio of the DC value of the collector current I_C to the corresponding DC value of the base current I_B is called the DC current gain (h_{FE} or β).

$$h_{FE} = I_c/I_B \qquad\qquad\qquad\qquad\qquad \textbf{3.9}$$

Unlike signal transistors, for which h_{FE} of more than 200 is common, the current gain of a power transistor is usually between 5 and 50 at the rated current, so these devices are usually connected in a Darlington configuration (see Section 3.2.9).

3.2.6.3   Switching Speeds

Power transistors switch on and off much faster than thyristors. They may switch on in less than 1 μs and turn off in less than 2 μs. Therefore, a power transistor can be used in applications where the frequency is as high as 100 kHz.

3.2.6.4   Forward Blocking Voltage

A transistor has a maximum collector-to-emitter voltage that it can withstand; above this voltage, breakdown at the collector junction will occur. This voltage is specified as $V_{CE(SU)}$ or V_{CEO}, that is, maximum V_{CE} with the base open. Power transistors with ratings as high as 1400 V are available.

3.2.6.5 Collector Current Rating

The collector current rating is the maximum permissible continuous collector current ($I_{C(MAX)}$).

3.2.6.6 Maximum Permissible Junction Temperature

The maximum permissible junction temperature (T_{jMAX}) is typically 125°C.

3.2.6.7 Power Dissipation

The maximum power rating of a transistor is specified as $P_{D(max)}$.

3.2.7 Safe Operating Area (SOA)

To ensure the safe operation of the transistor, manufacturers specify boundaries on the V_{CE} curve versus the I_C curve to define the safe operating area (SOA). A typical SOA is shown in Figure 3.8.

A transistor can withstand a maximum collector-to-emitter voltage $V_{CE(SU)}$. This voltage is the maximum voltage limit shown by the vertical line 1 in Figure 3.8. Since normal operation is above the cutoff region, line 2 defines that boundary. Similarly, the saturation region defines the line 3 boundary.

The maximum allowable collector current I_{Cmax} forms the upper boundary (4) of the SOA. The maximum power dissipation $P_{max} = V_{CE} * I_C$ forms the boundary indicated by line 5 in Figure 3.8. The final boundary line 6 of the SOA depends on *secondary breakdown* (discussed in the next section), which occurs when both voltage and current are high during turnoff.

Figure 3.8
Safe operating area (SOA) of a BJT

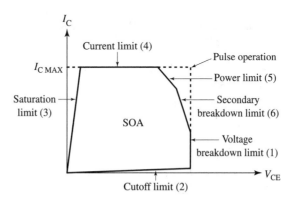

The SOA is important mainly for power transistors that operate in the active region. When used as a switch, a BJT operates in the cutoff and saturation region, and the transistor is in the active region only for a short period during switching. For pulse operation, the SOA is extended to a rectangle within limits.

3.2.8 Secondary Breakdown

BJTs fail under certain high-voltage and high-current conditions. If high voltage and high current occur simultaneously during turnoff, power dissipation will cause the device to fail. Figure 3.9 shows the turnoff characteristic of a power transistor. If we remove the base current (I_B) to turn off the transistor, the voltage across the transistor (V_{CE}) increases. When it reaches the DC supply voltage (V_{CC}), the collector current (I_C) falls. The power dissipation (P) during turnoff is shown in Figure 3.9 by the dashed line. Note that the peaks of V_{CE} and I_C occur simultaneously, and this may lead to transistor failure. Snubbers can be used with power transistors to avoid the simultaneous occurrence of peak voltage and peak current.

Figure 3.9
Secondary breakdown

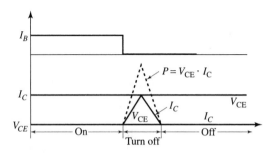

3.2.9 Darlington Connection

The current gain (β) of a power transistor can be as low as 5. To obtain higher current gains, a Darlington connection of two BJTs (see Figure 3.10) can be used. Current gain in the hundreds can be obtained in a high-power Darlington transistor. To turn the Darlington switch on, it is only necessary to provide a very small input at the base of Q_1, to enable it to provide a higher base current

Figure 3.10
The Darlington connection

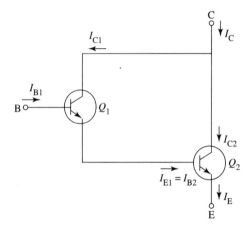

to Q_2. The base current needed to drive the pair is much less than what is necessary to drive Q_2 alone. A smaller base current allows a reduction in the current rating of the base drive circuit. The pair of transistors can be fabricated on one chip, or two discrete transistors can be physically connected to form a Darlington pair.

The Darlington configuration has two disadvantages. The first is that this configuration results in a higher on-state voltage drop ($V_{CE(SAT)}$). A Darlington pair can have an on-state voltage drop between 2 and 5 V, depending on its current and voltage rating. On-state losses are therefore higher. The second disadvantage of the Darlington connection is its slower switching speed.

Example 3.4   Find the overall gain of the Darlington connection.

Solution $$\beta_1 = I_{C1}/I_{B1}$$

$$\beta_2 = I_{C2}/I_{B2}$$

Overall,

$$\beta = I_C/I_{B1}$$

Now

$$I_C = I_{C1} + I_{C2}$$

Therefore,

$$\beta = \frac{I_{C1} + I_{C2}}{I_{B1}} = \frac{I_{C1}}{I_{B1}} + \frac{I_{C2}}{I_{B1}}$$

$$= \beta_1 + \frac{I_{C2}}{I_{B1}} \frac{I_{B2}}{I_{B2}}$$

$$= \beta_1 + \frac{I_{C2}}{I_{B2}} \frac{I_{B2}}{I_{B1}}$$

Now

$$I_{B2} = I_{E1} = I_{C1} + I_{B1}$$

so

$$\beta = \beta_1 + \beta_2 \frac{I_{C1} + I_{B1}}{I_{B1}}$$

$$= \beta_1 + \beta_2 (\beta_1 + 1) = \beta_1 + \beta_2 + \beta_1 \beta_2$$
$$\cong \beta_1 \beta_2$$

Example 3.5   In Figure 3.10, Q_1 is rated at 20 A with $\beta_1 = 20$, and Q_2 is rated at 100 A with $\beta_2 = 10$. Find the base current required to turn the Darlington pair on. What is the base current requirement if Q_2 is used alone?

Solution   Overall,

$$\beta = 20 * 10 = 200$$
$$I_C = I_{C1} + I_{C2} = 20 + 100 = 120$$
$$I_{B1} = I_C/\beta = 120/200 = 0.6 \text{ A}$$

If Q_2 is used alone,

$$I_B = 100/10 = 10 \text{ A}$$

Clearly, the base drive circuit for the Darlington pair will be quite small.

3.2.10   Series and Parallel Connection of Transistors

For high-voltage and high-current applications, a single transistor may not be sufficient. Then transistors can be connected in series or in parallel to increase the blocking voltage or conduction current, respectively. Proper consideration should be given to the sharing of voltage or current between devices to make sure that the individual components operate within their limits. Care must be taken when BJTs are connected in parallel to share the current, since they suffer from thermal runaway. Their forward voltage drop decreases with increasing temperature, causing a diversion of current to a single device.

3.3   Power Metal-Oxide Semiconductor Field-Effect Transistors (MOSFETs)

A power MOSFET is similar to a small signal MOSFET except for its higher voltage and current ratings. It is a fast-switching transistor, characterized by a high input impedance, suited to low-power (up to a few kilowatts), high-frequency (up to 100 kHZ) applications. It has important applications in switching power supplies, in which high switching frequency means that components are smaller and cheaper, and in speed control of small motors using pulse width modulation.

MOSFETs are available in both the N-channel and P-channel type. However, N-channel devices are available with higher current and voltage ratings. Figure 3.11 shows the symbol of a N-channel MOSFET. It has three terminals: the gate G, the source S and the drain D. The source is always at the potential nearest the gate. The drain is connected to the load. To turn the device on, the drain is made positive with respect to the source and a small positive voltage (V_{GS}) is applied to the gate. Having no voltage at the gate turns the switch off; that is, the gate voltage controls the on and off conditions.

In both the on and the off states, the input resistance is extremely high and the gate current is essentially zero due to the insulated gate. This allows gate control circuits that are simple and efficient compared with those necessary to drive a BJT. The MOSFET is capable of faster transitions between the on and off states than a BJT and thus has replaced it in high switching frequency applications.

Figure 3.11
Circuit symbol of an
N-channel power MOSFET

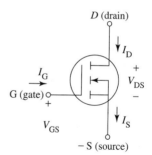

At higher switching frequencies, MOSFET switching losses are negligible compared with those of a BJT. However, conduction, i.e., on-state voltage drop, is high (approximately 4 V at rated current), therefore conduction loss is high.

3.3.1   MOSFET Voltage-Current Characteristics

The V-I characteristic of a power MOSFET is shown in Figure 3.12. It shows the relationship between drain-source voltage (V_{DS}) and drain current (I_D) for different values of V_{GS}. As the gate voltage increases from zero, the drain current does not increase significantly. The MOSFET turns on when V_{GS} exceeds a value called the *threshold voltage* (V_{TH}), which is commonly 2–4 V for high-voltage MOSFETs. The device is considered to operate in the *enhancement mode,* since the application of a positive voltage greater than V_{TH} produces a conducting N-channel. This channel basically acts like a resistance and provides a path for current flow from drain to source. The gate source voltage controls the drain current. The higher the value of V_{GS}, the higher the drain current. However, for a given value of V_{GS}, there is a limit to the maximum current that can flow. If we keep on increasing V_{DS}, the drain current (I_D) will increase rapidly until we reach a saturation value (I_{DSS}). After that, there will be no further significant increase in current for that particular value of V_{GS}. If the power MOSFET is used as a switch, it should be operated in the unsaturated region to ensure a small

Figure 3.12
V-I characteristic of a
power MOSFET

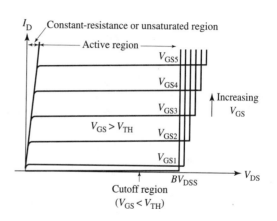

voltage drop across the device when it is in the on state. Once the saturation value is reached, a further increase in V_{DS} will only cause an increased voltage drop across the device and increased power dissipation in it, without an increase in current.

For a given gate voltage, there are three separate operating regions on the V-I characteristic: the cutoff region, the active region, and the constant-resistance (or ohmic, or unsaturated) region.

The *cutoff region* or off state of the MOSFET exists as long as $V_{GS} < V_{TH}$. This condition holds for any value of drain-to-source voltage (V_{DS}) until the breakdown voltage BV_{DSS} is reached. At this voltage, current can increase rapidly and can damage the device. A MOSFET must therefore be operated such so that the drain-to-source voltage V_{DS} is maintained below the value of BV_{DSS}.

For values of V_{GS} greater than V_{TH}, the MOSFET can operate in the active region or in the ohmic region.

In the *active region*, the MOSFET is operated as an amplifier. Here $V_{GS} > V_{TH}$. For a given V_{GS}, the drain current remains nearly constant irrespective of the drain-source voltage. The drain current (I_D) is controlled by V_{GS}, so the voltage V_{DS} and the drain current (I_D) can both be high simultaneously. The associated power loss $V_{DS} * I_D$ can be large, so this region is not used in power electronics applications.

The region of interest in power electronics is the *ohmic region*, where the drain current increases in direct proportion to the drain-source voltage and the MOSFET is in the on state. Here $V_{DS} > 0$. The ratio of voltage (V_{DS}) to current (I_D) in this region is called the on-state drain-to-source resistance ($R_{DS(ON)}$), and it is almost constant. This region is similar to the saturation region of the BJT.

$$V_{DS(ON)} = R_{DS(ON)} \, I_D \qquad\qquad\qquad \textbf{3.10}$$

A typical value of $R_{DS(ON)}$ is 0.5 Ω. To make sure that the MOSFET remains in the ohmic region for all desired values of I_D, it is best to use a higher value of V_{GS} than is necessary in the active region. This value is around 10 V. However, V_{GS} should not exceed 20 V—if it does, the MOSFET will be destroyed. A simple way to limit V_{GS} is to connect a 20-V Zener diode across the gate-source terminal.

3.3.2   The MOSFET Transfer Characteristic

When the MOSFET is used as a switch, its basic function is to control the drain current by the gate voltage. Figure 3.13 shows the transfer characteristic, which is a plot of drain current (I_D) versus gate voltage (V_{GS}) at a fixed drain voltage. The actual curve can be approximated by the linearized characteristic indicated by the dashed line in Figure 3.13. The drain current is zero until we reach the threshold voltage, then the current increases linearly with voltage. The slope is the transconductance g_m.

$$I_D = 0 \qquad\qquad\qquad \text{for} \qquad V_{GS} < V_{TH}$$
$$I_D = g_m \, (V_{GS} - V_{TH}) \qquad \text{for} \qquad V_{GS} > V_{TH}$$

Figure 3.13
Transfer characteristic

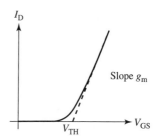

3.3.3   The Ideal MOSFET Characteristic

A MOSFET power switch has the ideal characteristic shown in Figure 3.14. With no signal applied to the gate, the device is off. The drain current (I_D) is zero and the voltage V_{DS} is equal to the value of the supply voltage. A voltage at the gate (V_{GS}), turns the device on, and the drain current is limited by the load resistance. The voltage (V_{DS}) across the MOSFET is zero.

The two MOSFET states correspond to the two states of an on-off switch. Although a MOSFET is not an ideal switch, it is sufficiently close to the ideal requirements to be a very useful and practical device.

Figure 3.14
Ideal characteristic

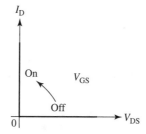

3.3.4   A MOSFET as a Switch

When a power MOSFET is used as a switch and is in the on condition, it is forced to operate in the ohmic region. This ensures that voltage drop across the device is low, so that the drain current is determined by the load and the power loss in the device is small.

The condition for MOSFET operation in the ohmic region is given by:

$$V_{DS} \leq V_{GS} - V_{TH} \quad \text{and} \quad V_{DS} > 0 \qquad\qquad \textbf{3.11}$$

Therefore for switching applications, the on-state resistance ($R_{DS(ON)}$) is a very important parameter, since it determines the conduction power loss for a given value of the load (drain) current. The lower the value of $R_{DS(ON)}$, the lower the on-state voltage drop, the lower the power dissipation, and the higher the current-carrying capability of the device.

The forward voltage drop is

$$V_F = I_D * R_{DS(ON)}$$

Internal power dissipation is

$$P = I_D^2 * R_{DS(ON)}$$

When the MOSFET is off, the drain current is zero and the voltage V_{DS} equals the value of the supply voltage. Under this condition the resistance between the drain and source R_{DS} is very high.

The condition for MOSFET operation in the cutoff region is given by:

$$V_{DS} \geq 0 \quad \text{and} \quad V_{GS} < V_{TH} \qquad \qquad \textbf{3.12}$$

3.3.5 MOSFET Losses

There are four sources of power losses in the switching MOSFET: the conduction or on-state loss, the off-state loss, the turn-on switching loss, and the turnoff switching loss.

3.3.5.1 Conduction or On-State Loss

A MOSFET has relatively high on-state losses given by:

$$P_{ON} = I_D^2 \, R_{DS(ON)} \, \frac{t_{ON}}{T} \qquad \qquad \textbf{3.13}$$

where T is the total period.

3.3.5.2 Off-State Loss

The off-state loss is given by

$$P_{OFF} = V_{DS(max)} \, I_{DSS} \, \frac{t_{OFF}}{T} \qquad \qquad \textbf{3.14}$$

3.3.5.3 The Turn-on Switching Loss

The energy loss in the MOSFET when it switches from the off state to the on state is given by

$$W_{ON} = \frac{V_{DS(max)} \, I_D \, t_r}{6} \qquad \qquad \textbf{3.15}$$

where t_r is the rise time of the drain current (I_D).

3.3.5.4 The Turnoff Switching Loss

The energy loss in the MOSFET when it switches from the on state to the off state is given by:

$$W_{OFF} = \frac{V_{DS(max)} \, I_D \, t_f}{6} \qquad \qquad \textbf{3.16}$$

where t_f is the fall time of the drain current (I_D).

3.3.5.5   Switching Power Loss

The switching power loss is

$$P_{SW} = (W_{ON} + W_{OFF}) * f \qquad\qquad\qquad\qquad \textbf{3.17}$$

where f is the switching frequency.

3.3.5.6   Total Power Loss in the MOSFET

$$P_T = P_{ON} + P_{OFF} + P_{SW} \qquad\qquad\qquad\qquad \textbf{3.18}$$

It is important to point out that the total power loss in a MOSFET is higher than in a BJT at low switching frequency, due to the higher conduction loss of the MOSFET. However, as the switching frequency is increased, BJT switching losses increase more than those of the MOSFET. Therefore, for high-frequency applications, it is desirable to use a MOSFET.

Example 3.6   In Figure 3.15, the DC supply voltage V_S is 120 V and the load resistance R_L is 10 Ω. The MOSFET parameters are t_r = 1.5 μS and $R_{DS(ON)}$ = 0.1 Ω. If the duty cycle d = 0.6 and the frequency of switching is 25 kHz, find

a) the power loss in the on state
b) the power loss during the turn-on interval

Figure 3.15
See Example 3.6

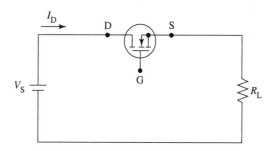

Solution

$$\begin{aligned} I_D &= V_S/\{R_L + R_{DS(ON)}\} \\ &= 120/10 + 0.1 \\ &= 12 \text{ A} \end{aligned}$$

a) the switching period $T = 1/f = 1/25\text{K} = 40 \text{ μS}$
   on time $t_{ON} = d * T = 0.6 * 40 \text{ μS} = 24 \text{ μS}$
   energy loss during on-time $\begin{aligned} W_{ON} &= I_D{}^2 R_{DS(ON)} * t_{ON} \\ &= 12^2 * 0.1 * 24(10^{-6}) = 345.6 \text{ μJ} \end{aligned}$

   power loss during on-time $\begin{aligned} P_{ON} &= W_{ON} * f = 345.6(10^{-6}) * 2(10^3) \\ &= 8.6 \text{ W} \end{aligned}$

b) energy loss during turn-on $W_{ON} = \dfrac{120 * 12}{6} 1.5 * 10^{-6} = 360 \text{ μJ}$

   power loss during turn-on $P_{ON} = W_{ON} * f = 360(10^{-6}) * 25(10^3) = 0.9 \text{ W}$

Example 3.7   A MOSFET has the following parameters: $I_{DSS} = 2$ mA, $R_{DS(ON)} = 0.3$ Ω, duty cycle $d = 50\%$, $I_D = 6$ A, $V_{DS} = 100$ V, $t_r = 100$ ns, and $t_f = 200$ ns. If the frequency of switching is 40 kHz, find the total power loss.

Solution

$$T = 1/f = 1/40(10^3) = 25 \ \mu S$$

$$t_{ON} = t_{OFF} = 12.5 \ \mu S$$

$$P_{ON} = \frac{6^2 * 0.3 * 12.5 \ (10^{-6})}{25 \ (10^{-6})} = 5.4 \ W$$

$$P_{OFF} = \frac{100 * 2(10^{-3}) * 12.5 \ (10^{-6})}{25 \ (10^{-6})} = 0.1 \ W$$

$$P_{SW(ON)} = \frac{100 * 6 * 100 \ (10^{-9})}{6} * 40(10^3) = 0.4 \ W$$

$$P_{SW(OFF)} = \frac{100 * 6 * 200 \ (10^{-9})}{6} * 40(10^3) = 0.8 \ W$$

$$P_T = 5.4 + 0.1 + 0.4 + 0.8 = 6.7 \ W$$

Example 3.8   Repeat Example 3.7 for a switching frequency of 100 kHz.

Solution

$$T = 1/f = 1/100(10^3) = 10 \ \mu s$$

$t_{ON} = t_{OFF} = 5 \ \mu s$ (since the duty cycle is still 50%)

P_{ON} and P_{OFF} are independent of frequency. Therefore, they remain the same.

$$P_{SW(ON)} = \frac{100 * 6 * 100 \ (10^{-9})}{6} * 100(10^3) = 1.0 \ W$$

$$P_{SW(OFF)} = \frac{100 * 6 * 200 \ (10^{-9})}{6} * 100(10^3) = 2.0 \ W$$

$P_T = 5.4 + 0.1 + 1.0 + 2.0 = 8.5$ W (27% more than in Example 3.7)

3.3.6   The Internal Body Diode of a Power MOSFET

If we reverse-bias the source (make the source positive with respect to the drain), the MOSFET cannot block this voltage; that is, the MOSFET has no reverse-voltage blocking capability. This is due to the intrinsic antiparallel diode within its structure. It provides a direct internal path for the current to flow in the reverse direction (from source to drain) across the junction, which becomes forward-biased. This integral diode, called the *body diode* and shown in Figure 3.16, is very useful for most switching applications, since it provides a free-wheeling current path.

Figure 3.16
Body diode

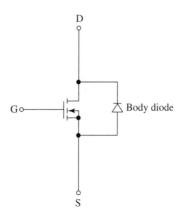

3.3.7 MOSFET Protection

A MOSFET, like all other semiconductor devices, must be protected against overvoltages, overcurrents, and transients. MOSFETs can be protected by removing the gate voltage, thus turning the device off. In fact, MOSFETs are available with built-in current and temperature sensors and gate-drive circuits that remove the gate voltage in case of overcurrents or transients.

3.3.7.1 Overvoltages

Overvoltages do not affect a MOSFET when it is on, since the device acts like a short circuit. However, in the off state, overvoltages that appear across the drain-source and across the gate-source affect the MOSFET directly.

In the off state, the MOSFET will operate in the active region if V_{DS} exceeds BV_{DSS}. In this region, the voltage (V_{DS}) and the current (I_D) can be high simultaneously, and the associated power loss $V_{DS} * I_D$ can damage the MOSFET. This situation is easily avoided by making sure that the supply voltage is less than the MOSFETs breakover voltage. Protection against external overvoltages is accomplished by connecting a nonlinear resistor called a *varistor* across the MOSFET. In case of an overvoltage close to BV_{DSS} across the MOSFET, the resistance of the varistor decreases and allows a path for the current flow.

3.3.7.2 Overcurrents

An overcurrent in a power MOSFET will cause the junction temperature to exceed its normal value of 150°C. This overheating will eventually cause destruction of the MOSFET. One simple means of protection is to ensure that the current flow through the MOSFET does not exceed 75% of its rated value. This method allows a safety factor of about 25% if the source voltage increases or if the load impedance is decreased inadvertently. Some manufacturers use built-in current sensors. An overcurrent that results in a rise in temperature can be sensed, and if the temperature exceeds a predetermined level, the MOSFET can be turned off by removing the gate signal.

3.3.8 Safe Operating Area (SOA)

The safe operating area (SOA) shows the operational limits of the MOSFET. Manufacturers' data sheets provide the maximum allowable drain current ($I_{D(max)}$) that the device can handle. The power dissipation must be limited so that the junction temperature T_j does not exceed its maximum allowable value of 150°C. The maximum drain-to-source voltage limit also defines the boundary of acceptable operation. A typical SOA is shown in Figure 3.17. The limits for current and power are higher for pulsed operation than for DC. These limits also depend on the pulse duration, as shown in the figure. The shorter the conduction time, the larger the permitted power dissipation. For a sufficiently short pulse duration, it is even possible to operate at point P, where maximum values of V_{DS} and I_D occur simultaneously. There is no secondary breakdown phenomenon in power MOSFETs, so they can withstand simultaneous application of high current and voltage without undergoing destructive failure.

Figure 3.17
Safe operating area of a MOSFET

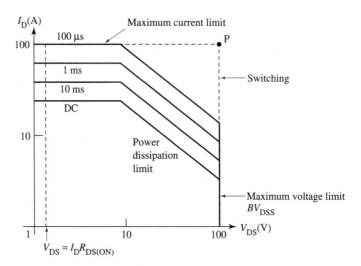

3.3.9 MOSFETs in Series and in Parallel

MOSFETs with high power ratings are still not available. Therefore, they are connected in series or in parallel to increase their voltage or current ratings.

3.3.9.1 MOSFETs in Series

MOSFETs are available with voltage ratings as high as 1200 V. However, if a single MOSFET's voltage rating is lower than the source voltage, we can connect two or more MOSFETs in series to handle the higher voltage. A resistor connected in parallel can be used to share voltage equally.

3.3.9.2 MOSFETs in Parallel

MOSFETs can be connected in parallel to share the load current if the current rating of a single MOSFET is less than that required by the load. MOSFETs connected in parallel share the current equally due to the positive temperature coefficient of $R_{DS(ON)}$, so there is no need for series current-sharing resistors.

If two MOSFETs are connected in parallel, the device with the lower resistance $R_{DS(ON)}$ will try to draw more current than the other. Since $P = I_D^2 R_{DS(ON)}$, the device conducting the higher current will dissipate more power, raising the junction temperature, and in turn increasing the on-state resistance ($R_{DS(ON)}$), which will limit the current. This self-limiting characteristic forces an even distribution of current in the two MOSFETs.

Example 3.9   In Figure 3.18, find the power dissipated by each MOSFET.

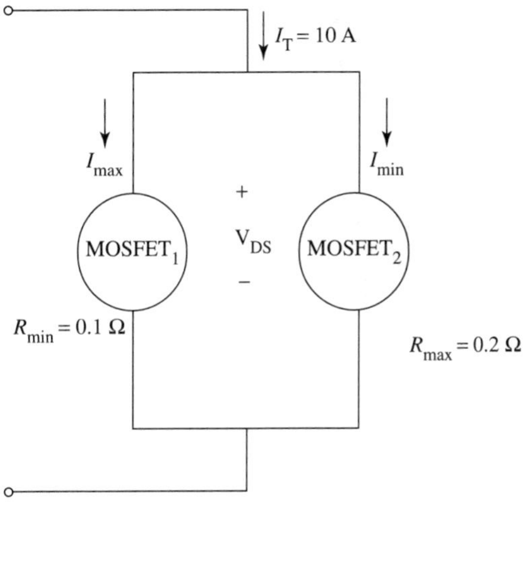

Figure 3.18
See Example 3.9

Solution   Using the current divider rule,

$$I_{max} = \frac{R_{max}}{R_{min} + R_{max}} I_T$$

Substituting the given values,

$$I_{max} = \frac{0.2}{0.3} * 10 = 6.67 \text{ A}$$

$$I_{min} = 10 - 6.67 = 3.33 \text{ A}$$

$$P_1 = I_{max}^2 * R_{min} = 6.67^2 * 0.1 = 4.45 \text{ W}$$

$$P_2 = I_{min}^2 * R_{max} = 3.33^2 * 0.2 = 2.22 \text{ W}$$

3.4 Insulated-Gate Bipolar Transistors (IGBTs)

The Insulated-Gate Bipolar Transistor (IGBT) combines the low on-state voltage drop and high off-state voltage characteristics of the BJT with the excellent switching characteristics, simple gate-drive circuit, and high input impedance of the MOSFET. IGBTs are available in current and voltage ratings well beyond what is normally available for power MOSFETs. For example, the POWEREX IGBT CM 1000HA-28H has a voltage rating of 1400 V and current rating of 1000A. IGBTs are replacing MOSFETs in high-voltage applications where conduction losses must be kept low. Although the switching speeds of IGBTs are higher (up to 50 kHz) than those of BJTs, they are lower than those of MOS-FETs. Therefore, the maximum switching frequencies possible with IGBTs are between those of BJTs and MOSFETs. Unlike the MOSFET, the IGBT has no internal reverse diode, so its reverse-voltage blocking capability is very poor. The maximum reverse voltage it can withstand is less than 10 V.

Figure 3.19 shows the circuit symbol of an N-channel IGBT and its equivalent connection of MOSFET and BJT. An IGBT has three terminals: the gate, the collector, and the emitter.

Figure 3.19
(a) The IGBT (b) Equivalent
MOSFET-BJT connection

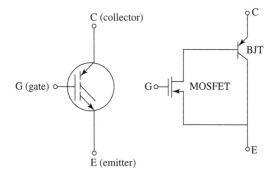

3.4.1 Operating Principles

The operation of the IGBT is very similar to that of the power MOSFET. To turn on the IGBT, the collector terminal (C) must be positively biased with respect to the emitter terminal (E). A positive voltage V_G applied to the gate will turn on the device when the gate voltage exceeds the threshold voltage ($V_{GE(TH)}$). The IGBT is turned off by simply removing the voltage signal from the gate terminal.

3.4.2 The IGBT Voltage-Current Characteristic

The IGBT V-I characteristic shown in Figure 3.20 is a plot of the collector current (I_C) versus the collector-to-emitter voltage (V_{CE}). When there is no voltage applied to the gate, the IGBT is in the off state. In this state, the current (I_C) is zero and the voltage across the switch is equal to the source voltage. If a voltage $V_G > V_{GE(TH)}$ is applied to the gate, the device turns on and allows current I_C to flow. This current is limited by the source voltage and the load resistance. In the on state, the voltage across the switch drops to zero.

Figure 3.20
IGBT V-I characteristic

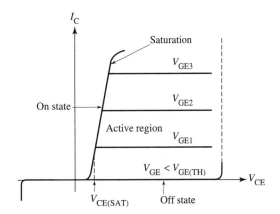

3.4.3 The Ideal IGBT Characteristic

An ideal IGBT, when on, has no voltage across it while the current is determined by $I_C = V_S/R_L$. When off, it can block any positive or negative voltage. Figure 3.21 shows the ideal V-I characteristic.

Figure 3.21
Ideal IGBT V-I characteristic

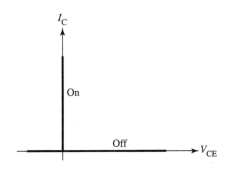

Example 3.10 In Figure 3.22, the source voltage is 220 V and the load resistance is 5 Ω. The IGBT is operated at a frequency of 1 kHz. Find the on time for the pulse if the power required to the load is 5 kW.

Figure 3.22
See Example 3.10

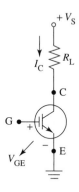

Solution

$$T = 1/f = 1 \text{ ms}$$

$$V_{L(avg.)} = \frac{V_S \, t_{ON}}{T}$$

$$P_L = \frac{V_{L(avg.)}^2}{R_L}$$

Therefore,

$$P_L = \frac{V_S^2 \, t_{ON}}{R_L T}$$

$$t_{ON} = \frac{P_L R_L \, T}{V_S^2} = \frac{5000 * 5 * 1(10^{-3})}{220^2} = 0.52 \text{ ms}$$

3.4.4 IGBT Losses

The energy loss in the IGBT during the turn-on process is given by:

$$W_{ON} = \frac{V_{CE(max)} \, I_{C(max)} \, t_{ON}}{6} \qquad \textbf{3.19}$$

The average power dissipated due to the turn-on process is

$$P_{ON} = W_{ON} * f_{SW} \qquad \textbf{3.20}$$

where f_{SW} is the IGBT switching frequency.

The energy loss during the turnoff time (t_{OFF}) is the same as that for turn-on and is given by:

$$W_{OFF} = \frac{V_{CE(max)} \, I_{C(max)} \, t_{OFF}}{6} \qquad \textbf{3.21}$$

Example 3.11   In Figure 3.22, $V_S = 220$ V, $R_L = 10$ Ω, $f_{SW} = 1$ kHz, and $d = 0.6$. If the IGBT has the following data:

$$t_{ON} = 2.5 \text{ } \mu S, \text{ } t_{OFF} = 1 \text{ } \mu S, \text{ and } V_{CE(SAT)} = 2.0 \text{ V}$$

find:
a) average load current
b) conduction power loss
c) switching power loss during turn-on
d) switching power loss during turnoff

Solution   a) $I_{C(max)} = \dfrac{V_S - V_{CE(SAT)}}{R_L} = \dfrac{220 - 2}{10} = 21.8$ A

$I_{C \text{ (avg.)}} = d * I_{C(max)} = 0.6 * 21.8 = 13.08$ A

b) conduction power loss $= V_{CE(SAT)} * I_{C(avg.)}$
$$= 2 * 13.08$$
$$= 26.16 \text{ W}$$

c) switching power loss during turn-on $= \dfrac{V_{CE(max)} * I_{C(max)} * t_{ON}}{6} * f_{SW}$

$$= \dfrac{220 * 21.8 * 2.5(10^{-6})}{6} * 10^3$$

$$= 2 \text{ W}$$

d) switching power loss during turn-off $= \dfrac{V_{CE(max)} * I_{C(max)} * t_{OFF}}{6} * f_{SW}$

$$= \dfrac{220 * 21.8 * 1(10^{-6})}{6} * 10^3 = 0.8 \text{ W}$$

3.5   Unijunction Transistors (UJT)

The unijunction transistor (UJT) is a three-terminal device with an emitter (E) and two bases: base one (B_1) and base two (B_2). The emitter is a P-type material, while the main body of the UJT is an N-type material. Therefore, a PN junction exists between the emitter and the body of the UJT. The UJT is used for generating trigger pulses for larger devices such as SCRs and triacs. Figure 3.23 shows the structure, the schematic symbol and the V-I characteristics of the UJT. Its control terminals (E and B_1) are also its power terminals. Terminal B_2 is used for biasing.

With no emitter bias at terminal E, $V_{EB1} = 0$ and the UJT has a certain internal resistance between B_2 and B_1. This resistance is called the *interbase resistance* (R_B), and at 25°C it is of the order of kilohms. It is made up of two resistances, R_{B1} and R_{B2} (see Figure 3.24). The ratio of R_{B1} to R_B is called the *standoff ratio* η, because its value determines the reverse bias that the equivalent PN diode junction experiences.

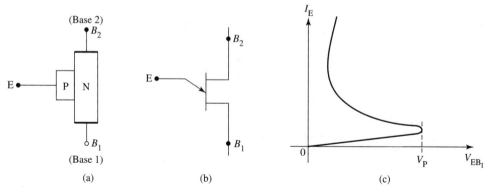

Figure 3.23
The UJT (a) structure; (b) schematic symbol; (c) characteristic

$$\eta = \frac{R_{B1}}{R_B} = \frac{R_{B1}}{R_{B1} + R_{B2}} \qquad\qquad \textbf{3.22}$$

The value of η lies between 0.5 and 0.8. A typical value for most UJTs is 0.6.

3.5.1 Biasing the UJT

Figure 3.24 shows how the UJT is normally biased. If terminals B_2 and B_1 are positively biased with a voltage source V_{BB}, the resistances R_{B2} and R_{B1} act as a voltage divider, such that the voltage at the η point is

$$V_{RB1} = \eta \ V_{BB} \qquad\qquad \textbf{3.23}$$

Therefore, to forward-bias the diode D and turn the UJT on, the emitter voltage V_E must be greater than a value called the *peak voltage* (V_p), which is given by

Figure 3.24
Biasing a UJT

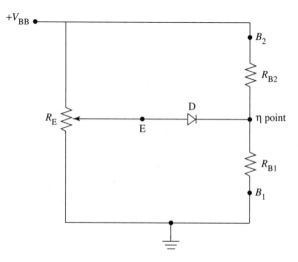

$$V_p = V_b + \eta \ V_{BB}$$ **3.24**

where V_b is the PN-junction barrier potential (0.7 V for silicon). After the UJT turns on, the UJT acts like a forward-biased diode and resistance R_{B1} reduces to a very low value, almost zero. R_{B2} is not affected by this and is fixed at its original off-state value. Figure 3.25(a) shows the equivalent circuit of the UJT when it is on.

If V_E is less than V_P, the emitter is reverse-biased, the UJT turns off, and only a small reverse leakage current will flow. Figure 3.25(b) shows the equivalent circuit of the UJT when it is off.

Figure 3.25
Equivalent circuits for a UJT:
(a) on; (b) off

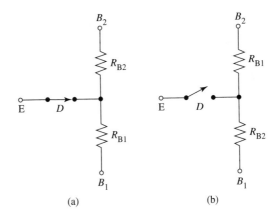

(a) (b)

3.5.2 Testing a UJT

The equivalent circuit of a UJT (Figure 3.25), a diode, and two resistors can be used to show how to test the UJT with an ohmmeter. There are five resistances to check when testing a UJT:

1. B_1 to B_2 (R_B)
2. E to B_1 with D forward-biased (R_{B1F})
2. E to B_1 with D reverse-biased (R_{B1R})
4. E to B_2 with D forward-biased (R_{B2F})
5. E to B_2 with D reverse-biased (R_{B2R})

R_{B1R} and R_{B2R} should be extremely large compared with the other three resistances. R_B is the same in either direction; it is the third-largest resistance (about 5–10 kΩ for most UJT). R_{B1F} and R_{B2F} have the smallest values. These resistances vary widely since the resistance of a forward-biased diode depends on the current through it. Assuming equal current flow, R_{B2F} is about 20% smaller than R_{B1F}.

3.5.3 Using a UJT Circuit to Trigger a Thyristor

A UJT is an ideal device to use in relaxation oscillators for firing an SCR. Figure 3.26 shows the UJT oscillator circuit. When the switch S is closed, capacitor C begins to charge through resistor R at a rate that depends on the time constant $T = RC$. When the capacitor voltage ($V_C = V_E$) reaches the peak voltage V_p, the UJT will fire. This places a low resistance across C, and it very quickly discharges through R_1. This turns the UJT off, and C begins to charge again to repeat the cycle.

Figure 3.26
A UJT circuit for triggering a thyristor

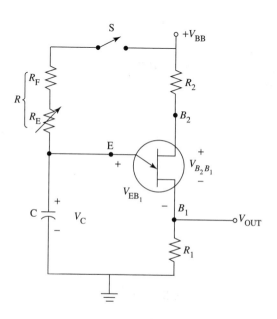

The burst of current can be used to trigger a thyristor or turn on a transistor. The frequency of oscillation is given approximately by:

$$f = 1/T \qquad\qquad \textbf{3.25}$$

where T, the time it takes to turn on the UJT, is given by:

$$T = RC \ln \frac{1}{1-\eta} \approx RC \qquad\qquad \textbf{3.26}$$

Example 3.12   A 15-V source is connected across B_2-B_1. If $\eta = 0.6$, find the emitter voltage needed to turn the silicon UJT on.

Solution   The voltage across R_{B1} is

$$V_{RB1} = 0.6 * 15 = 9 \text{ V}$$

Therefore the voltage at the η point is 9 V with respect to ground. This places +9 V on the cathode of the silicon. It takes about 0.7 V to forward-bias and turn this diode on. Therefore, the emitter voltage must be greater than $V_P = 0.7 + 9 = 9.7$ V to fire the UJT. Note that if the emitter voltage is less than 9.7 V, it will reverse-bias the device and only a small reverse leakage current will flow.

Example 3.13 In Figure 3.26, $V_{BB} = 15$ V, $R = 45$ kΩ, $C = 0.1$ μF, $R_2 = 270$ Ω, $R_1 = 90$ Ω, η = 0.6, and the device is silicon.
a) Draw the output waveform.
b) Find the frequency.

Solution a) The voltage-divider circuit of R_2, R_1, and the UJTs R_{B2} and R_{B1} set up 0.6 * 15 = 9 V at the UJT's η point. When the capacitor charges to a voltage greater than 9.7 V, the UJT turns on. This places the resistance R_1 across C, and it quickly discharges, turning off the UJT. C begins to charge again, and the cycle repeats. The voltage across the capacitor V_C is shown in Figure 3.27. Its sawtooth waveform is caused by relatively slow charging and quick discharging. During the quick discharge, a large current through R_1 causes a voltage spike at the output V_{OUT}, as shown in Figure 3.27.

Figure 3.27
Output waveform

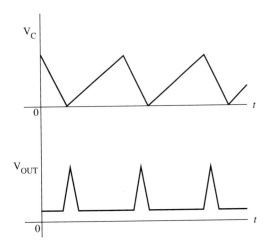

b) $\qquad f = \dfrac{1}{RC} = \dfrac{1}{45(10^3) * 0.1(10^{-6})} = 222$ Hz

■───

3.6 Problems

3.1 List the terminals of a bipolar junction transistor.

3.2 Describe how the base-emitter junction and the collector-base junction must be biased to turn on the BJT.

3.3 Describe how the base-emitter junction and the collector-base junction must be biased to turn off the BJT.

3.4 Which terminal is at the most negative voltage in an on PNP transistor?

3.5 Which two BJT terminals act like the switch contacts?

3.6 Which two MOSFET terminals act like the contacts of a switch?

3.7 What two UJT terminals are the control terminals?

3.8 What is terminal B_2 used for on a UJT?

3.9 In Figure 3.4, $V_{CC} = 200$ V, $R_C = 20$ Ω, $\beta = 20$, $V_{CE(SAT)} = 1.0$ V, and $V_{BE(SAT)} = 1.2$ V. Find:
(a) the minimum value of I_B required to ensure a saturated on state
(b) the on-state power loss in the transistor

3.10 In Figure 3.4, $V_{CC} = 200$ V and $R_C = 20$ Ω. The BJT is turned on and off at a frequency of 5 kHz, $t_{SW(ON)}$ is 1 µS and $t_{SW(OFF)}$ is 1.5 µS. Find the total switching power loss.

3.11 A BJT switch controls DC power to a 5-Ω resistive load. The DC source voltage $V_S = 120$ V, $V_{CE(SAT)} = 1.2$ V, $V_{BE(SAT)} = 1.5$ V, the base resistance = 1.5 Ω, and the base bias voltage $V_{BB} = 5$ V. If the frequency is 5 kHz with $t_{SW(ON)} = 1$ µs and $t_{SW(OFF)} = 1.5$ µS, find
(a) β
(b) power loss in the BJT

3.12 A BJT switch controls DC power to a 5-Ω resistive load. If the DC source voltage $V_S = 120$ V, $V_{CE(SAT)} = 1.2$ V, and turn-on time is 1 µs, find:
(a) the BJT on-state losses
(b) the energy loss in the BJT during the turn-on process

3.13 In Figure 3.4, $V_{CC} = 200$ V, $R_C = 20$ Ω, $R_B = 5$ Ω, $\beta = 30$, and $V_{BE} = 0.6$ V when the switch is on. Find the minimum input voltage required to turn the switch on.

3.14 In Figure 3.4, $V_{CC} = 300$ V and $R_C = 20$ Ω. The BJT is turned on and off at a frequency of 2 kHz, $t_{SW(ON)}$ is 10 µs, $t_{SW(OFF)}$ is 1.2 µs, and $V_{CE(SAT)} = 1.6$ V. Find the total switching power loss.

3.15 A MOSFET switch controls power to a 5-Ω resistive load. The DC source voltage $V_S = 120$ V, $R_{DS(ON)} = 0.1$ Ω, switching frequency = 25 kHz, $t_{ON} = 150$ ns, and the duty cycle = 0.6. Find:
(a) the energy loss during the on time
(b) the power loss in the switch when it is on

3.16 An IGBT switch controls power to a 15-Ω resistive load. The DC source voltage V_S = 440 V, $V_{CE(SAT)}$ = 1.5 V, the switching frequency = 2 kHz, t_{ON} = 20 ns, and the duty cycle = 0.6. Find

(a) the IGBT minimum current rating
(b) the on-state power loss
(c) the turn-on power loss

3.7 Equations

$$V_{CE} = V_{CC} - I_C R_C \tag{3.1}$$

$$I_{C(SAT)} = \frac{V_{CC} - V_{CE(SAT)}}{R_C} \approx \frac{V_{CC}}{R_C} \tag{3.2}$$

$$I_B = \frac{I_{C(SAT)}}{\beta} = \frac{V_{CC}}{\beta\, R_C} \tag{3.3}$$

$$P_{ON} \approx V_{CE(SAT)}\, I_C \tag{3.4}$$

$$P_{OFF} = V_{CE} {}^* I_C \approx V_{CC} {}^* I_{leakage} \tag{3.5}$$

$$W_{(SW\text{-}ON)} = \frac{V_{CC}\, I_{C(max)}}{6}\, t_r \tag{3.6}$$

$$W_{SW(OFF)} = \frac{V_{CC}\, I_{C(max)}}{6}\, t_f \tag{3.7}$$

$$P_{T(avg.)} = [P_{ON} {}^* t_{ON} + P_{OFF} {}^* t_{OFF} + W_{(SW\text{-}ON)} + W_{(SW\text{-}OFF)}]\, f \tag{3.8}$$

$$h_{FE} = I_c / I_B \tag{3.9}$$

$$\beta_{DP} = \beta_1 + \beta_2 + \beta_1\, \beta_2$$

$$V_{DS(ON)} = R_{DS(ON)}\, I_D \tag{3.10}$$

$$V_{DS} \leq V_{GS} - V_{TH} \text{ and } V_{DS} > 0 \tag{3.11}$$

$$V_{DS} \geq 0 \text{ and } V_{GS} < V_{TH} \tag{3.12}$$

$$P_{ON} = I_D{}^2\, R_{DS(ON)}\, \frac{t_{ON}}{T} \tag{3.13}$$

$$P_{OFF} = V_{DS(max)}\, I_{DSS}\, \frac{t_{OFF}}{T} \tag{3.14}$$

$$W_{ON} = \frac{V_{DS(max)}\, I_D\, t_r}{6} \tag{3.15}$$

$$W_{OFF} = \frac{V_{DS(max)}\, I_D\, t_f}{6} \tag{3.16}$$

$$P_{SW} = (W_{ON} + W_{OFF}) {}^* f \tag{3.17}$$

$$P_T = P_{ON} + P_{OFF} + P_{SW} \tag{3.18}$$

$$W_{ON} = \frac{V_{CE(max)} \; I_{C(max)} \; t_{ON}}{6}$$

3.19

$$P_{ON} = W_{ON} * f_{SW}$$

3.20

$$W_{OFF} = \frac{V_{CE(max)} \; I_{C(max)} \; t_{OFF}}{6}$$

3.21

$$\eta = \frac{R_{B1}}{R_B} = \frac{R_{B1}}{R_{B1} + R_{B2}}$$

3.22

$$V_{RB1} = \eta \; V_{BB}$$

3.23

$$V_p = V_b + \eta \; V_{BB}$$

3.24

$$f = 1/T$$

3.25

$$T = RC \ln \frac{1}{1 - \eta} \approx RC$$

3.26

Thyristor Devices

4

Chapter Outline

Learning Objectives

After completing this chapter, the student should be able to

- define the term thyristor
- describe the operation of the silicon controlled rectifier (SCR)
- interpret the V-I characteristic curve of an SCR
- define some important electrical parameters associated with SCRs
- explain how to test an SCR
- list the principal ratings of SCRs
- describe how SCRs can be connected in series and in parallel to increase their ratings

- list the power losses in SCRs
- explain how to protect SCRs
- describe the basic forms of SCR gate-triggering circuits
- describe some common SCR commutation circuits
- describe the operation of the GTO
- describe the operation of the triac
- describe the operation of breakover trigger devices such as diacs and SCSs
- describe the operation of the MOS-controlled thyristor

4.1   Introduction

Thyristors are four-layer PNPN power semiconductor devices used as electronic switches. Their main advantage is that they can convert and control large amounts of power in AC or DC systems while using very low power for control. This chapter will introduce the thyristor family, which includes the silicon controlled rectifier (SCR), the gate-turnoff thyristor (GTO), the triac, the diac, the silicon controlled switch (SCS), and the MOS-controlled thyristor (MCT). The SCR is the most important member of the family and is emphasized in this chapter. SCRs are widely used in such applications as regulated power supplies, static switches, choppers, inverters, cycloconverters, heaters, lighting, and motor control.

4.2   The Silicon Controlled Rectifier (SCR)

The *silicon controlled rectifier* (SCR) is the most popular electrical power controller due to its fast switching action, small size, and high current and high voltage ratings.

4.2.1 Description

The structure of an SCR is shown in Figure 4.1(a), and Figure 4.1(b) shows the electrical symbol. It has three terminals: the anode (A) and the cathode (K) are its power terminals, and the gate (G) is the control terminal. When the SCR is forward-biased, that is, when the anode is made positive with respect to the cathode, a positive voltage on the gate with respect to the cathode turns on (triggers) the SCR. However, the current through the SCR cannot be turned off using the gate. It is turned off by interrupting the anode current. In a manner similar to that of a diode, the SCR blocks current in the reverse direction.

Figure 4.1
The SCR (a) PNPN structure
(b) symbol

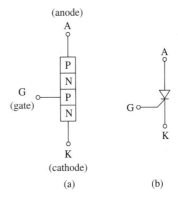

(a)   (b)

4.2.2 Two-Transistor Model of the SCR

The most common explanation of the positive feedback action that results when an SCR switches can be illustrated using the two-transistor model shown in Figure 4.2. The SCR can be represented as two separate complimentary transistors, an NPN (Q_1) and a PNP (Q_2) transistor. The collector of Q_1 is the base of Q_2, and the base of Q_1 is the collector of Q_2. A positive voltage on the gate forward-biases the base-emitter junction of transistor Q_1, turning it on. This

Figure 4.2
The SCR two-transistor
model (a) PNP-NPN
transistor analogy (b) SCR
transistor analogy

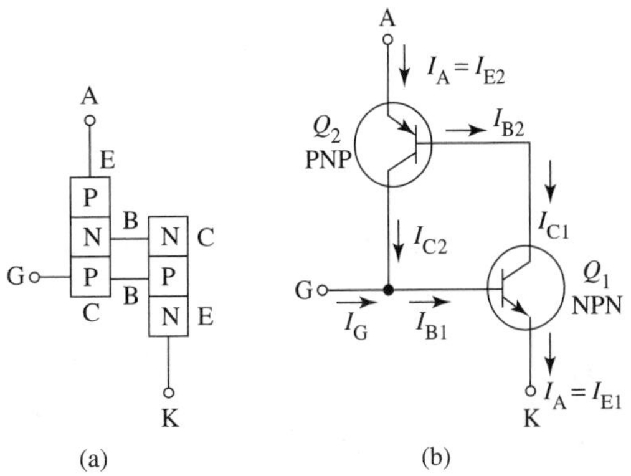

(a)   (b)

allows current through the NPN collector (the PNP base). If the SCR anode is positive, the PNP emitter-base junction is forward-biased, turning it on. After the PNP transistor is turned on, it in turn supplies the NPN with base current. This regenerative process, called *latching,* continues until both transistors are driven into saturation. Removal of the gate voltage will not turn the SCR off. Q_1 supplies Q_2 with base current and Q_2 supplies Q_1 with base current. The SCR remains on until its principal current (current from anode to cathode) is interrupted.

It is important to note that to turn the thyristor on, the gate requires a small-magnitude positive pulse for only a short period of time. Once the device is on, the gate signal serves no useful purpose and can be removed.

4.3   SCR Characteristic Curves

The volt-ampere characteristic of an SCR is shown in Figure 4.3. When the SCR is forward-biased, a small forward current called the *off-state current* flows through the device. This region of the curve is known as the forward blocking region. However, if the forward bias is increased until the anode voltage reaches a critical limit called the *forward breakover voltage* (V_{FBO}), the SCR turns on. The voltage across the SCR then drops to a low value, the *on-state voltage* (1–3 V), and the current increases sharply, limited only by the components in series with the SCR.

As can be seen from the three characteristic curves, the value of V_{FBO} can be controlled by the level of the gate current. If the gate-cathode junction is

Figure 4.3
Characteristic V-I curves of an SCR

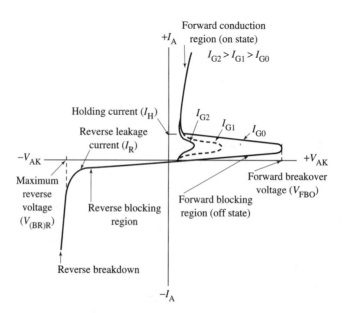

forward-biased, the SCR is turned on at a lower breakover voltage than with the gate open (I_G = 0). As the gate current is increased, the breakover voltage is decreased and the SCR characteristic closely resembles that of an ordinary diode. The only difference in the three curves is in the forward-blocking region. The I_G = 0 curve shows that the SCR can be turned on with no gate current, but this is not a desirable characteristic in SCRs. In practice, the SCR is made to switch on by applying a gate signal. At a low gate current (I_{G1}), the SCR turns on at a lower forward-anode voltage. At a higher gate current (I_{G2}), the SCR fires at a still lower value of forward-anode voltage.

The reverse characteristic is similar to that of a common PN junction diode. When the SCR is reverse-biased (that is, when the anode is made negative with respect to the cathode), there is a small *reverse leakage current* (I_R). If the reverse voltage is increased until the voltage reaches the *reverse breakdown voltage* ($V_{(BR)R}$), the reverse current will increase sharply. If the current is not limited to a safe value, the device can be destroyed. Care must be taken to make sure that the maximum reverse voltage across an SCR does not exceed its breakdown voltage.

Essentially, the SCR acts like a switch. When the applied voltage is below the breakover point, the switch is off. When the applied voltage reaches the breakover point or if a positive signal is applied to the gate, the switch is on. The SCR remains on as long as its anode current I_A stays above a certain value called the *holding current* (I_H).

After the SCR has been turned on, the gate loses control, that is, even if the gate current is reduced to zero, the SCR remains on. The SCR can only be turned off by removing the supply voltage or reducing the anode current to a level below the holding current. If the source is AC, the SCR is reverse-biased during the negative half-cycle and therefore turns off "naturally."

4.3.1 The Ideal Characteristic

The SCR can be represented by the idealized characteristic shown in Figure 4.4. It has three basic operating states. The *forward blocking* (off) state, the *forward conduction* (on) state, and the *reverse blocking* (off) state. The gate signal

Figure 4.4
Idealized characteristic
of an SCR

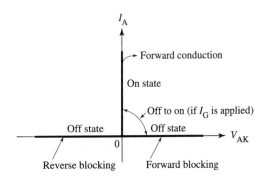

switches the SCR from the forward blocking state to the forward conducting state. The ideal SCR behaves like a diode after it has been turned on.

4.4 Testing SCRs

An SCR can be tested by forward-biasing it, supplying the gate with a trigger voltage, and noting whether the SCR stays on after the gate-voltage is removed. Small SCRs that have low holding currents can be tested easily with a digital multimeter.

Figure 4.5 shows how to test an SCR with an ohmmeter. To test the SCR, connect the positive ohmmeter lead to the anode and the negative lead to the cathode. This forward-biases the SCR. However, the ohmmeter should give a very high reading since the SCR is off. Then short the gate to the anode. This supplies the trigger signal to the gate, and the SCR should turn on. The resistance should decrease and should stay low even after the gate lead is removed from the anode. During the test, the anode and cathode must remain connected to the ohmmeter.

Figure 4.5
Testing an SCR with an ohmmeter

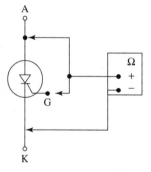

If the SCR turns on before the gate voltage is applied, it is shorted from anode to cathode. If it does not turn on after the gate voltage is applied, it is open. If it turns on during a gate trigger but turns off immediately when the gate lead is removed, its condition is doubtful. The current supplied by the ohmmeter may not be sufficient to keep the SCR above its holding current level.

Reverse-bias the SCR by connecting the cathode to the positive ohmmeter lead and the anode to the negative lead. The ohmmeter should show an infinite reading. If the gate is shorted to the anode, the ohmmeter should still indicate infinity.

Care must be taken with this method of testing. If the voltage applied by the ohmmeter to the gate-cathode junction is too high, the device may be destroyed.

SCRs can be tested more reliably by using a curve tracer to give the exact values of certain voltages or currents. Nevertheless, for a quick check of an SCR, an ohmmeter works quite well.

4.5 SCR Ratings

The ratings of a semiconductor device give the values for the various conditions that a manufacturer recommends for reliable operation of the device. The use of a device beyond the limits defined by its ratings usually leads to destruction. Therefore, a device should never be used beyond its rating limits.

4.5.1 Subscripts for SCR Ratings

Most of the important SCR ratings are given in terms of voltage (V) or current (I). Each rating is accompanied by one, two, or three subscripts, whose meanings are listed in Table 4–1.

4.5.2 SCR Current Ratings

4.5.2.1 Maximum Repetitive RMS Current Rating

The forward (anode-to-cathode) current that an SCR can safely carry depends on the maximum junction temperature (T_j). A major requirement in SCR applications is that the junction temperature not exceed its maximum value. It is not

Table 4–1 Subscript notations

A	anode or ambient
AVE	average
c	case
D	forward blocking region (no gate voltage)
F	forward
G	gate
H	holding current
j	junction
K	cathode
L	latching
M	maximum
O	open third terminal (for example, gate open)
pk	peak
R	reverse (when used as the first subscript) repetitive (when used as the second subscript)
S	shorted (when used as the first subscript) surge (when used as a second subscript)
T	total or maximum (when used as the first subscript) triggered (when used as a second subscript)

easy to measure and control T_j directly. However, we can easily measure and control the SCR voltages and currents that contribute to T_j. The largest contributor to T_j is repetitive RMS on-state current $I_{T(RMS)}$.

The RMS (or effective) current is used to rate the device, since it determines the heat dissipation. However, the DC or average current delivered to the load is usually more important. Therefore, manufacturers give their data in terms of average current ($I_{T(AVE)}$), which is called rated average on-state current. $I_{T(AVE)}$ is the maximum average current value that can be carried by the SCR in its on state. Average current is equal to RMS current in a pure DC circuit. However, the average value of a pulse is much lower than its RMS value.

Determining the RMS value of a nonsinusoidal waveform like the one shown in Figure 4.6 is quite difficult. We can simplify the calculations by approximating the nonsinusoidal waveform by a rectangular waveform whose height is equal to the peak value and whose width is equal to the pulse duration. This approximation will give a higher RMS value, but it leaves a slight safety factor.

Figure 4.6
Approximation of the SCR waveform (a) actual waveform (b) approximation

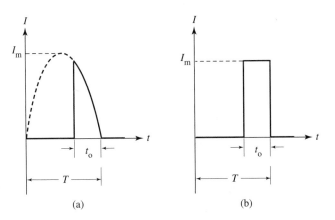

After approximating, the RMS value of the current can be found from

$$I_{RMS} = \sqrt{\frac{I_m^2 t_o}{T}} \qquad\qquad 4.1$$

where t_o is the pulse duration, T is the pulse-repetition time, and I_m is maximum current. The average value of the pulse is given by

$$I_{AVE} = \frac{I_m t_o}{T} \qquad\qquad 4.2$$

The *form factor* (f_o) is defined as the ratio of the RMS current to the average current

$$f_o = \frac{I_{RMS}}{I_{AVE}} \qquad\qquad 4.3$$

Table 4–2 Conduction angle versus form factor

Conduction angle (θ)	Form factor (f_o)
20°	5.0
40°	3.5
60°	2.7
80°	2.3
100°	2.0
120°	1.8
140°	1.6
160°	1.4
180°	1.3

If we know the form factor for a given waveform, the RMS current can be easily obtained from

$$I_{RMS} = f_o(I_{AVE}) \qquad\qquad \textbf{4.4}$$

The given current rating of a particular SCR is usually its maximum repetitive RMS current (I_{TRMS}). Care should be exercised if the current rating given is DC or average. The RMS rating I_{TRMS} is then

$$I_{T(RMS)} = f_o(I_{T(AVE)}) \qquad\qquad \textbf{4.5}$$

Table 4–2 gives the form factor as a function of conduction angle (θ). The conduction angle is the duration for which the SCR is on. It is measured as shown in Figure 4.7.

Figure 4.7
Conduction angle measurement

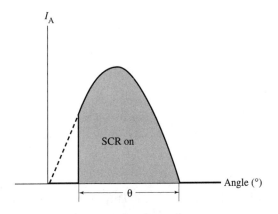

Example 4.1   Find the form factor of the waveform shown in Figure 4.8(a).

Figure 4.8
Approximating a
nonsinusoidal waveform
(a) actual (b) approximation

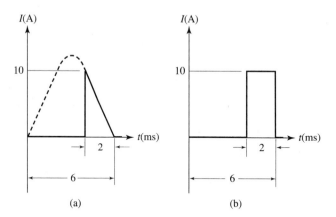

(a)

(b)

Solution   Approximate the waveform as shown in Figure 4.8(b).
Using Equation 4.1,

$$I_{RMS} = \sqrt{\frac{I_m{}^2 t_o}{T}}$$

$$= \sqrt{\frac{(10)^2 * 2}{6}}$$

$$= \sqrt{33.3}$$

$$= 5.8 \text{ A}$$

Using Equation 4.2

$$I_{AVE} = \frac{I_m t_o}{T}$$

$$= \frac{(10)2}{6}$$

$$= 3.3 \text{ A}$$

Using Equation 4.3,

$$f_o = \frac{I_{RMS}}{I_{AVE}}$$

$$= 1.7$$

Example 4.2   Find the RMS current in an SCR circuit when a DC ammeter reads 100 A with a
conduction angle of 60°.

Solution For a conduction angle of 60°, the form factor from Table 4–2 is

$$f_\mathrm{o} = 2.7$$

Using Equation 4.4,

$$I_\mathrm{RMS} = 2.7\,(100) = 270\ \mathrm{A}$$

4.5.2.2 Surge Current Rating (I_FM or I_TSM)

The surge current rating of an SCR (I_FM or I_TSM) is the peak anode current an SCR can handle for a brief duration. These durations must be isolated by enough time for the SCR junction to return to its rated operating temperature. The surge current rating may be five to twenty times as large as the repetitive RMS current rating ($I_\mathrm{T(RMS)}$). An SCR is designed to withstand a maximum of 100 surges during its operating life because it may be left with some kind of damage due to excessive exposure to high temperatures. To prevent damage to the SCR, the surge current rating and repetitive current rating should not be exceeded.

4.5.2.3 Latching Current (I_L)

A minimum anode current must flow through the SCR in order for it to stay on initially after the gate signal is removed. This current is called the *latching current* (I_L). If this current is not reached while the gate signal is being applied, the SCR may turn on, but it will turn off when the gate signal is removed.

4.5.2.4 Holding Current (I_H)

After the SCR is latched on, a certain minimum anode current is required to maintain conduction. If the anode current is reduced below this critical value, the SCR will turn off. The lowest value of current, just before the SCR turns off, is the holding current (I_H). Both holding and latching currents decrease when the temperature increases and vice versa. The ratio of the two currents (I_H and I_L) is approximately 2:1. These values are quite small compared to I_TRMS. I_H or I_L may pose a problem in circuits where the load current is much lower than I_TRMS or the SCR is operated at low temperatures.

Example 4.3 Find the maximum value of the load resistor that ensures SCR conduction in the circuit shown in Figure 4.9. The SCR has a holding current of 200 mA.

Figure 4.9
See Example 4.3

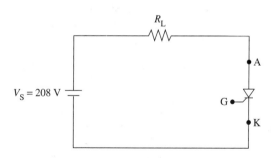

Solution In order for the SCR to remain on, the anode current must not fall below 200 mA. Therefore,

$$R_{\text{LMAX.}} = \frac{V_s}{I_h} = \frac{208}{0.2} = 1040 \ \Omega$$

4.5.3 SCR Voltage Ratings

The SCR voltage rating is a measure of the maximum voltage that can be applied across the device without causing a breakdown. The various ratings related to voltage are discussed next.

4.5.3.1 Peak Repetitive Forward Blocking Voltage (V_{DRM})

The peak repetitive forward blocking voltage (V_{DRM}) is one of the basic SCR ratings. It is the maximum instantaneous voltage that the SCR can block in the forward direction. V_{DRM} is selected to be less than the forward breakover voltage (V_{FBO}), see Figure 4.3. If the V_{DRM} rating is exceeded, the SCR will conduct even without a gate voltage.

4.5.3.2 Peak Repetitive Reverse Voltage (V_{RRM})

Like diodes, SCRs have a peak inverse voltage rating. It is called the peak repetitive reverse voltage rating (V_{RRM}). It is the maximum instantaneous voltage that an SCR can withstand, without breakdown, in the reverse direction.

The V_{DRM} of any SCR is lower than its V_{RRM}. Therefore, an SCR should not be subjected to a peak voltage greater than V_{DRM}. This will automatically satisfy the V_{RRM} rating. If the V_{RRM} rating is exceeded, the SCR is likely to be damaged. V_{DRM} and V_{RRM} both tend to increase as temperature increases.

4.5.3.3 Nonrepetitive Peak Reverse Voltage (V_{RSM})

The nonrepetitive peak reverse voltage rating (V_{RSM}) is the maximum transient reverse voltage that the SCR can withstand. SCRs can safely block nonrepetitive voltages, so this rating is generally 10–20% higher than the repetitive voltage rating. V_{RSM} can be increased by inserting a diode of equal current rating in series with the SCR.

4.5.4 SCR Classification According to Frequency and Switching Speed

SCRs are classified into two categories that reflect both frequency and switching speed ratings: *slow-switching* or *phase-control*–type SCRs and *fast-switching* or *inverter*-type SCRs. Phase-control SCRs have a large turnoff time and therefore have lower frequency capabilities. Inverter-type SCRs must be used in high-frequency operations, since the required circuit turnoff time becomes a significant portion of the total cycle time.

Manufacturers provide the ratings associated with speed in terms of the frequency (f_{max}) at which maximum anode current I_{max} or maximum power dissipation (P_{max}) decreases to zero, turn-on time (t_{ON}), and turnoff time (t_{OFF}).

4.5.5 SCR Rate-of-Change Ratings

4.5.5.1 The Critical Rate of Rise of the On-State Current (di/dt rating)

When the SCR initially conducts, the anode current is concentrated in a relatively small area beside the gate. A certain time is required for the conduction to spread out evenly to cover the whole body. However, if the rate at which the anode current increases is much greater than the rate at which the conduction area increases, the initial small area overheats, and this can destroy the SCR. Manufacturers set a safe value for the *rate of change of anode current* that their devices can withstand. This value is the *critical rate of rise of the on-state current,* commonly referred to as the *di/dt rating* of the device. It is expressed in amperes per microsecond (A/μs).

To prevent damage to SCRs by a high *di/dt* value, a small inductance (L) is added in series with the device. Inductance opposes any change in current, thus slowing down the rise of anode current. The required inductance (L) can be found from Equation 4.6:

$$L \geq \frac{V_p}{(di/dt)_{max}} \qquad\qquad \textbf{4.6}$$

where

L is inductance (in μH)

$(di/dt)_{max}$ is the rate-of-change-of-current rating of the SCR (in A/μs)

V_p is the peak value of the source voltage (in V)

Example 4.4   In Figure 4.10, the load resistance (R_L) is 10 Ω and the AC source voltage (V_s) is 208 V. Find the value of L required to limit the circuit *di/dt* to 20A/μs.

Figure 4.10
See Example 4.4

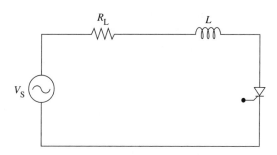

Solution

$$V_p = \sqrt{2}\,(208) = 294 \text{ V}$$

Using Equation 4.6,

$$L = \frac{294}{20} = 14.7 \;\mu\text{H}$$

4.5.5.2 The Critical Rate of Rise of the Off-State Voltage (*dv/dt* rating)

The application of a fast-rising forward voltage across an SCR under off-state conditions results in the flow of current across the junctions into the gate layer. This current, which is equivalent to an externally supplied gate current, will turn the SCR on if it exceeds the critical value. The steeper the waveform, the greater the likelihood that the device will trigger below its forward voltage rating. A high rate of rise of forward voltage (*dv/dt*) due to switching or surges can thus cause unscheduled firing. It also limits the maximum frequency that can be connected to the device.

The *dv/dt* rating gives the maximum rise time of a voltage pulse that can be applied to the SCR in the off state without causing it to fire. It is specified in volts/μS. A high *dv/dt* rating is desirable for many applications where the device is likely to see a fast-rising pulse.

The *RC snubber circuit* shown in Figure 4.11 is used to prevent unscheduled firing in circuits with high values of *dv/dt*. Because capacitance is opposition to a change in voltage, a small capacitance across the SCR reduces the rate at which the voltage across the SCR can change. An approximate value for the capacitance (*C*) can be obtained by finding the time constant (*T*) and dividing it by the load resistance (R_L):

$$T = \frac{V_{DRM}}{(dv/dt)_{max}} \qquad\qquad \textbf{4.7}$$

Then

$$C \geq \frac{T}{R_L} \geq \frac{V_{DRM}}{R_L (dv/dt)_{max}} \qquad\qquad \textbf{4.8}$$

Figure 4.11
A snubber circuit to reduce *dv/dt*

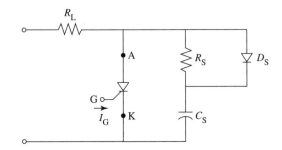

When the SCR is off, the capacitor charges up in a positive direction up to the instant of SCR turn-on. When the SCR is triggered on, the capacitor discharges and adds to the di/dt presented by the original circuit. Therefore, a small resistor (R_s) is added in series with the capacitor to slow the capacitor discharge and limit the current transient at turn-on.

An approximate value of R_s can be obtained from Equation 4.9. This equation actually gives the minimum resistance required, so a value greater than this is usually used. A suitable value would be 100 Ω.

$$R_s \geq \sqrt{\frac{V_{DRM}}{(di/dt)_{max}}} \qquad\qquad \textbf{4.9}$$

Although the addition of R_s protects the SCR from high values of di/dt, it lowers the dv/dt capability of the SCR. To extend dv/dt, a small diode D_s is connected across R_s. During times when dv/dt is high, the diode shorts R_s, but during times of high di/dt, the diode is off.

Example 4.5   An SCR has $V_{DRM} = 600$ V, $(dv/dt)_{max} = 25$ V/μS, and $(di/dt)_{max} = 30$ A/μS. It is used to energize a 100-Ω resistive load. Find the minimum values for an RC snubber circuit to avoid unintentional triggering.

Solution   Using Equation 4.7,

$$T = \frac{600 \text{ V}}{25 \text{ V/}\mu s} = 24 \ \mu s$$

Using Equation 4.8,

$$C = \frac{24 \times 10^{-6}}{100} = 0.24 \ \mu F$$

Using Equation 4.9,

$$R_s = \sqrt{\frac{600 \text{ V}}{30 \text{ A/}\mu s}} = 4.5 \text{ m}\Omega$$

Example 4.6   In Figure 4.12, the source voltage is 120 V and the load resistance is 10 Ω. The SCR can withstand a dv/dt value of 40 V/μS. If the snubber discharge current must be limited to 3 A, find the value of the snubber resistor and capacitor.

Figure 4.12
See Example 4.6

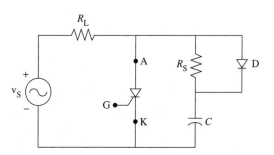

Solution   The peak value of the source voltage is:

$$V_p = \sqrt{2} * 120 = 169.7 \text{ V}$$

The minimum time is

$$T = \frac{169.7 \text{ V}}{40 \text{ V/}\mu\text{s}} = 4.24 \text{ }\mu\text{s}$$

Therefore,

$$C = \frac{4.24 \times 10^{-6}}{10} = 0.424 \text{ }\mu\text{F}$$

If the SCR is turned on at the instant the source voltage is at its peak value, the voltage across the capacitor is 169.7 V. To limit the snubber discharge current to 3 A, the value of R_s is

$$R_s = \frac{169.7}{3} = 42.4 \text{ }\Omega$$

4.5.6   Gate Parameters

There are six important gate ratings, classified in terms of current and voltage. The following four are maximum gate ratings. When any of these are exceeded, the SCR may be destroyed.

4.5.6.1   Maximum Gate Peak Inverse Voltage (V_{GRM})

V_{GRM} is the maximum value of negative DC voltage that can be applied without damaging the gate-cathode junction.

4.5.6.2   Maximum Gate Trigger Current (I_{GTM})

I_{GTM} is the maximum DC gate current allowed to turn on the device.

4.5.6.3   Maximum Gate Trigger Voltage (V_{GTM})

V_{GTM} is the DC voltage necessary to produce I_{GTM}.

4.5.6.4   Maximum Gate Power Dissipation (P_{GM})

P_{GM} is the maximum instantaneous product of gate current and gate voltage that can exist during forward-bias. If V_{GTM} and I_{GTM} are used at their extreme limits simultaneously, P_{GM} is certain to be exceeded.

The last two gate ratings are the smallest voltage and current needed to trigger the SCR. The gate trigger must exceed both to fire the SCR.

4.5.6.5   Minimum Gate Trigger Voltage (V_{GT})

V_{GT} is the minimum DC gate-to-cathode voltage required to trigger the SCR. The voltage applied between the gate and the cathode must exceed this value while providing adequate gate current to turn the SCR on.

4.5.6.6 Minimum Gate Trigger Current (I_{GT})

I_{GT} is the minimum DC gate current necessary to turn the SCR on.

Most SCRs require a gate current of 0.1–50 mA to fire. The magnitudes of the gate current and voltage required to trigger an SCR vary inversely with the temperature. At higher ambient temperatures, both gate-trigger requirements decrease. At lower temperatures, these requirements increase. Worst-case triggering conditions occur, therefore, at the minimum operating temperature.

Example 4.7   For the circuit in Figure 4.13 what is the minimum voltage (V_{IN}) that will fire the SCR, if the gate current needed to fire the SCR is 15 mA.

Figure 4.13
See Example 4.7

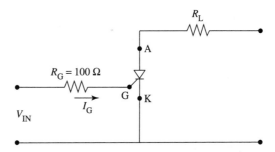

Solution   Since there is a regular PN junction between the gate and cathode, the gate-cathode voltage must be slightly greater than 0.6 V to forward-bias the junction. Let us assume a forward-bias voltage V_o of 0.7 V. Then

$$V_{IN(min)} = V_{RG} + V_o$$
$$= I_{GT} (100) + 0.7$$
$$= (15 * 10^{-3})(100) + 0.7$$
$$= 1.5 + 0.7 = 2.2 \text{ V}$$

Example 4.8   An SCR has the following ratings:

anode-cathode voltage drop = 1.5 V

gate-cathode junction voltage = 0.6V

gate current = 40 mA

If the SCR draws a current of 20 A, find:
a) the on-state power loss
b) the power loss in the gate

Solution

a) on-state power loss = 20 * 1.5 = 30 W

b) gate power loss

$$P_G = V_G * I_G$$
$$= 0.6 * 0.04$$
$$= 24 \text{ mW}$$

4.6   Junction Temperature Rating

In all semiconductor devices, the most important consideration is junction temperature (T_j). It not only defines the maximum and minimum limits but also determines whether the device can withstand prolonged operation. If the junction temperature in an SCR exceeds its maximum rating, the breakover voltage drops noticeably and the off-state current and reverse leakage current increase rapidly. The turnoff time also increases significantly. On the other hand, if the junction temperature falls below its minimum limit, the SCR may not trigger at all.

4.7   Increasing SCR Ratings

To ensure that SCR maximum ratings are not exceeded, an SCR with sufficient ratings must be chosen. We can increase SCR ratings by using external cooling to remove the heat produced by losses in the SCR. Adding external circuitry can also increase the voltage- and current-handling capabilities, and ratings can be extended by connecting SCRs in series and in parallel.

The reliability and life of semiconductor devices often depend on how well they are cooled. The power wasted as heat in the device also lowers efficiency. Usually SCRs dissipate about 1% of the total power. SCRs should be located in well-ventilated, cool places away from other heat-generating devices.

In most situations, the SCR case is not effective in carrying away the heat from the junction, so suitable heat dissipation devices like heat sinks should be provided. A heat sink is made of a metal, normally copper or aluminum, that is a good conductor of heat. It is quite thick where it contacts the SCR and thin where it contacts the air. This design provides a large surface area from which the heat can pass by convection and radiation to the surrounding air. To help heat conduction, silicon grease infused with metallic oxides is normally used between the adjoining surfaces of the SCR and the heat sink. The outside surface normally has parallel fins to let convection currents of air flow freely. For cooling larger SCRs, the convection flow can be further improved by using a fan or forced air. Water cooling is also used with very high power dissipating devices.

SCR ratings can be extended by adding external circuitry. For example, using an RC snubber circuit to extend the *dv/dt* capability and series inductance to extend the *di/dt* rating have already been mentioned. The V_{DRM} of an SCR can be increased by placing a resistor across the gate and cathode; this also decreases I_H and I_L; however, the gate drive requirement increases. V_{RSM} can be increased by inserting a diode of equal current rating in series with the SCR.

4.8   Series and Parallel SCR Connections

The maximum power that can be controlled by a single SCR is determined by its rated forward current and rated forward blocking voltage. To maximize one of these two ratings, the other has to be reduced. Although SCRs are currently available with very high voltage ratings, in many applications, such as transmission lines, the required voltage rating exceeds the voltage that can be provided by a single SCR. Then it is necessary to connect two or more SCRs in series. Similarly, for very high current applications, SCRs have to be connected in parallel. For high-voltage, high-current applications, series-parallel combinations of SCRs are used.

4.8.1   SCRs in Series

If the input voltage is higher than the voltage rating of a single SCR, two or more SCRs can be connected in series to increase their forward blocking capability. As in any other device, the characteristics of two SCRs (even two SCRs of the same make and rating) are different. Two SCRs in series divide the voltage between them in inverse proportion to their leakage currents. This leads to an unequal distribution of voltage across the two series SCRs. Figure 4.14 shows the leakage currents of two identical SCRs, SCR_1 and SCR_2, each having a forward blocking voltage V_{BO}. When two such SCRs are connected in series, the same current flows through the two devices; however, the voltage across SCR_1

Figure 4.14
The sharing of voltages between two series-connected SCRs, SCR_1 and SCR_2

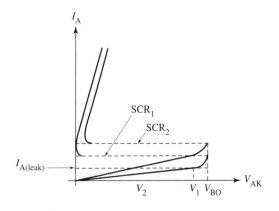

(V_1) is higher than that across SCR$_2$ (V_2) because the leakage current for SCR$_1$ is smaller than that of SCR$_2$ for the same voltage. Therefore, the two SCRs do not share the supply voltage equally. The maximum voltage that the SCRs can block is only $V_1 + V_2$, not $2V_{BO}$. To use the full forward blocking capabilities of each SCR, the forward blocking voltage must be equally distributed.

A nearly equal distribution of voltages during blocking is easily accomplished by connecting voltage-equalizing resistors R$_1$ and R$_2$ (Figure 4.15(a)) in parallel with each SCR such that each parallel combination has the same resistance. However, when a number of SCRs are connected in series, this method becomes uneconomical. A second approach, which permits a reasonably uniform distribution of voltages, is to use the same value resistance in parallel with each SCR. This allows a different but fixed voltage to appear across each SCR. In this arrangement (Figure 4.15(b)), the SCR with the lower leakage current will have a greater portion of the blocking voltage than the SCR with the higher leakage current.

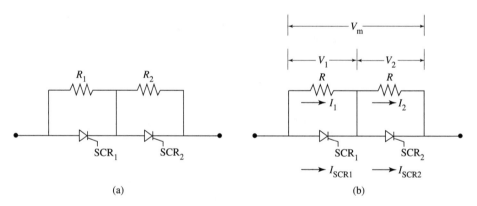

Figure 4.15
(a) Resistance equalization (b) voltage equalization

Let us assume that the leakage current of SCR$_1$(I_{SCR1}) is greater than the leakage current of SCR$_2$ (I_{SCR2}). SCR$_2$ will be required to have the larger voltage (V_2).

$$V_2 = I_2 R$$

The voltage across the series combination is

$$V_m = V_2 + I_1 R$$

Applying Kirchoff's current law (KCL) to the middle node,

$$I_{SCR1} + I_1 = I_{SCR2} + I_2$$
$$I_{SCR1} + I_{SCR2} = I_2 - I_2 = \Delta I$$

or

$$I_1 = I_2 - \Delta I$$
$$\begin{aligned}
V_m &= V_2 + (I_2 - \Delta I)R \\
&= V_2 + I_2 R - \Delta I R \\
&= V_2 + V_2 - \Delta I R \\
&= 2\,V_2 - \Delta I R
\end{aligned}$$
$$R = \frac{2\,V_2 - V_m}{\Delta I} \qquad\qquad\qquad\qquad \textbf{4.10}$$

Unequal voltage distribution among SCRs in series also occurs during turn-ons and turnoffs. One SCR may turn on or turn off before the other. The off SCR will then be subjected to the full source voltage. Shunt capacitors are very effective in equalizing voltages during switching. The capacitor also forces voltage sharing during sudden changes in supply voltage. A resistor is added in series with the capacitor to limit the current and di/dt (due to discharge of the capacitor through SCR) during turn-on. This C_s and R_s combination, shown in Figure 4.16, is essentially a snubber circuit. A diode (D) connected across R_s shunts it for forward voltages.

Figure 4.16
RC equalization for SCRs
connected in series

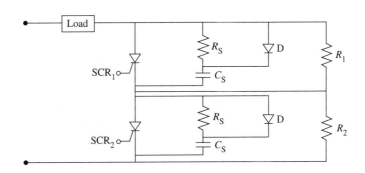

Example 4.9   The voltages across two SCRs connected in series are 200 V and 180 V. Calculate the value of the required equalizing resistor if the SCRs have a maximum difference of 1 mA in latching current. Also find the power dissipated by the blocking resistors.

Solution

$$R = \frac{2\,V_2 - V_m}{\Delta I} = \frac{2(200) - 380}{1(10^{-3})} = 20\ \text{k}\Omega$$

$$P_R = \frac{V_1^2}{R} + \frac{V_2^2}{R} = \frac{180^2 + 200^2}{20\,(10^3)} = 3.62\ \text{W}$$

4.8.2   SCRs in Parallel

When the load current exceeds the rating of a single SCR, SCRs are connected in parallel to increase their common current capability. If SCRs are not perfectly matched, this results in an unequal sharing of current between them. Figure 4.17 shows the *V-I* characteristics of two SCRs, SCR_1 and SCR_2. The ratings of the SCRs are the same. When these SCRs are connected in parallel, they will have equal voltage drops V_{SCR} across them. However, due to their mismatch in characteristics, SCR_2 is carrying the rated current (I_2), while SCR_1 is carrying a current I_1, which is much less than its rated value. The total rated current of the parallel connection is only $I_1 + I_2$ instead of $2I_2$.

Figure 4.17
Current sharing between
two parallel SCRs (a) two
SCRs in parallel (b) on-state
characteristics

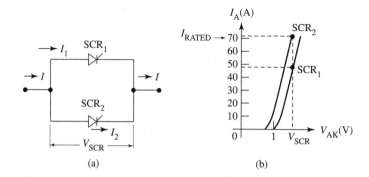

(a)

(b)

Matched-pair SCRs are generally available for parallel connection, but they are very expensive. With unmatched SCRs, equal current sharing is enforced by adding a low-value resistor or inductor in series with each SCR. Forced current sharing using equal-value resistors is shown in Figure 4.18. The basic requirement is to make current I_1 close to I_2; a maximum difference of 20% is acceptable. If we assume the voltage V_1 across SCR_1 to be greater than the voltage V_2 across SCR_2, the value of R can be obtained from

$$I_1R + V_1 = I_2R + V_2$$

$$R = \frac{V_1 - V_2}{I_2 - I_1}$$

4.11

Figure 4.18
Forced current sharing using
resistors

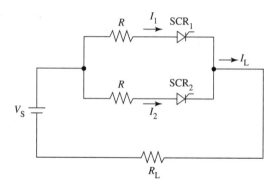

Example 4.10   In the circuit in Figure 4.18, the source voltage is 500 V and the load resistance is 5 Ω. Two SCRs with the characteristic shown in Figure 4.17(b), each rated at 70 A, are connected in parallel to share the load current. Find the value of the resistor that will give the proper current sharing. Also find the voltage drop across the parallel combination and the power dissipated by the current-sharing resistors.

Solution   The load current is

$$I_L = 500/5 = 100 \text{ A}$$

Let us choose a current difference of 15% between the two SCRs, so

$$I_2 = 50 + 15 = 65 \text{ A}$$
$$I_1 = 50 - 15 = 35 \text{ A}$$

From the characteristic curve,

$$V_1 = 1.6 \text{ V}$$
$$V_2 = 1.5 \text{ V}$$

Therefore,

$$R = \frac{V_1 - V_2}{I_2 - I_1} = \frac{1.6 - 1.5}{65 - 35} = 3.3 \text{ m } \Omega$$

The voltage drops across the two resistors are:

$$I_1 R = 35 \, (0.0033) = 0.12 \text{ V}$$
$$I_2 R = 65 \, (0.0033) = 0.22 \text{ V}$$

The voltage drop across the parallel branch is

$$V_1 + I_1 R = V_2 + I_2 R$$
$$= 1.6 + 0.12 = 1.5 + 0.22 = 1.72 \text{ V}$$

The power loss in the resistor is

$$I_1^2 R = 35^2 \, (0.0033) = 4.1 \text{ W}$$
$$I_2^2 R = 65^2 \, (0.0033) = 14.1 \text{ W}$$

Equalization using resistors is inefficient due to the extra power loss in the resistor. Moreover, resistors do not compensate for unequal SCR turn-on or turnoff times. One of the SCRs may turn on or turn off before the other. In either case, the on SCR must carry the full-load current momentarily until both are switched, and it can easily be damaged due to overloading. Figure 4.19 shows a center-tapped reactor in which the SCR carrying the greater current will induce a voltage proportional to the imbalance in current and with the polarity shown. Voltage in reactor L_1 opposes the flow of current, and voltage in L_2 causes an increase of current flow through the SCR that originally carried the lower current. As a result, a balanced current distribution is achieved. The reactors, although expensive, are more efficient.

Figure 4.19
Current sharing in SCRs with parallel reactors

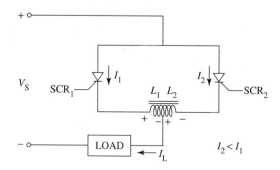

Since the main problem in high-current applications is excessive junction temperature, SCRs used in parallel must be *derated by* at least 15%. For example, two 50-A SCRs connected in parallel may carry only 85 (47.5 + 47.5) A. In addition, the SCRs are normally mounted on a common heat sink to equalize temperatures.

Example 4.11   For the circuit in Figure 4.18, find the value of the current-sharing resistors if $V_s = 1000$ V, $R_L = 1$ Ω, and the rated current of each SCR is 700 A. Also find the power loss in the resistors.

Solution   Choose

$$V_1 = 1.55 \text{ V}$$
$$V_2 \approx V_1 = 1.45 \text{ V}$$
$$I_L = 1000/1 = 1000 \text{ A}$$

Let one thyristor carry

$$I_1 = 400 \text{ A}$$

Therefore

$$I_2 = 1000 - 400 = 600 \text{ A}$$
$$R = \frac{1.55 - 1.45}{200} = 0.5 \text{ m}\Omega$$
$$P_{\text{LOSS}} = (600)^2(0.0005) + (400)^2(0.0005) = 260 \text{ W}$$

4.9 Power Loss

During a switching cycle, an SCR experiences power losses from the following sources:

1. on-state power loss
2. off-state power loss
3. switching power loss
4. gate-trigger loss

The off-state power loss when the SCR is blocking in either direction and triggering losses is small enough to be ignored under normal operating conditions. For low operating frequencies (below 400 Hz), the switching loss is also small. Therefore, the main source of power loss is the on-state power loss. This loss can be calculated by multiplying the on-stage voltage by the on-state current.

Switching power losses increase in high-frequency operations, particularly at turn on. Switching loss can be found by multiplying the instantaneous voltage by the instantaneous current.

Example 4.12   An SCR has the following ratings:

RMS current	$I_{T(RMS)}$ = 50 A
blocking voltage	V_{RRM} = 500 V
forward voltage drop at rated current	V_{SCR} = 1.5 V
leakage current at rated voltage	I_{RRM} = 5 mA
turn-on time	t_{ON} = 5 μs
turn-off time	t_{OFF} = 25 μs

If the SCR controls a 25-kW load at rated voltage and current, find the total power losses for a switching frequency of 100 Hz.

Solution   on-state power loss = 50 * 1.5 = 75 W

off-state power loss = 500 * 5 (10^{-3}) = 0.25 W

switching power loss

average power loss during switching = $\frac{1}{6}$ (500 * 50) = 1.04 kW

energy dissipated during turn-on interval = 1.04 (10^3) * 5 (10^{-6}) = 5.2 mJ

energy dissipated during turnoff interval = 1.04 (10^3) * 25 (10^{-6}) = 26 mJ

time period $T = 1/f = 1/100 = 10$ ms

assuming a duty cycle of 50%, on-time = off-time = 5 ms

on-state energy dissipation = 75 * 5 (10^{-3}) = 375 mJ

off-state energy dissipation = 0.25 * 5 (10^{-3}) = 1.25 mJ total energy dissipation per cycle = (0.0052 + 0.026 + 0.375 + 0.00125) J = 0.409 J

average power dissipation = 0.409/10 (10^{-3}) = 40.9 W

Example 4.13   Repeat Example 4.12 for a switching frequency of 50 kHz.

Solution   The on-state power loss is the same, 75 W. The off-state power loss is also the same, 0.25 W. The switching power loss is

average power loss during switching (same) = 1.04 kW

energy dissipated during turn-on interval (same) = 5.2 mJ

energy dissipated during turnoff interval (same) = 26 mJ

time period $T = 1/f = 1/50$ (10^3) = 20 μs

with a duty cycle of 50%, on-time = off-time = 10 μs

on-state energy dissipation = 75 * 10 (10^{-6}) = 0.75 mJ

off-state energy dissipation = 0.25 * 10 (10^{-6}) = 2.5 μJ

total energy dissipation per cycle = (5.2 + 26 + 0.75 + 0.0025) mJ = 31.95 mJ

average power dissipation = 31.95 * $10^{-3}/20$ (10^{-6}) = 1.6 kW

At the higher switching frequency, the total power dissipation increases to 1.6 kW, the switching power loss being the dominant factor. The 50-A SCR obviously cannot dissipate the enormous amount of heat that is generated. The higher the switching frequency, the higher the power dissipation during switching—this power dissipation limits the maximum frequency of operation. Furthermore, the current rating of the device must be increased at higher frequencies.

The limiting frequency above which the switching power loss becomes a dominant factor can usually be estimated from:

$$f \approx \frac{\text{on-state + off-state power loss averaged over a cycle}}{\text{energy dissipated during switching per cycle}} \qquad \textbf{4.12}$$

Example 4.14   Find the limiting switching frequency of the SCR in Example 4.12.

Solution   With a 50% duty cycle, the on-state + off-state power loss averaged over a cycle is

$$\frac{75 + 0.25}{2} = 37.625 \text{ W}$$

energy dissipated during switching per cycle is: (5.2 + 26) mJ = 0.03 J

limiting frequency ≈ 37.625/0.03 ≈ 1250 Hz

4.10   SCR Protection

To obtain satisfactory and reliable SCR operation, it is necessary to provide protection against overvoltages and overcurrents. Under overload or other abnormal operating conditions, an SCR may carry current many times greater than its rated value and can be destroyed permanently. Overvoltages, which arise from nonideal operation of the devices and from transients, may exceed the rated voltage of the SCR.

The SCR should be protected against any unusual transient conditions. Selecting an SCR rating based on abnormal operating conditions is not economical. Usually some safety margin is provided by choosing devices with ratings larger than those satisfactory for normal operation.

4.10.1   Overvoltage Protection

Overvoltage is one of the most important causes of SCR failure. An overvoltage is a voltage that exceeds the rated peak value of the supply voltage. Overvoltages are caused mainly by switching disturbances, usually due to the energy stored in the inductive components. The effect is to produce a transient voltage whose peak may exceed the rated forward blocking voltage of the SCR. Depending on the intensity of the overvoltage and the energy that it represents, the SCR may experience false triggering or may be permanently destroyed due to reverse breakdown.

To protect an SCR against overvoltages, a diode is connected in series with the SCR. The reverse voltage will be shared between the diode and the SCR. However, this method is inefficient since the voltage drop across the diode when conducting will introduce power loss. Another simple method is to choose an SCR with a higher voltage rating by introducing a voltage safety factor (V_F):

$$V_F = \frac{V_{RRM} \ (PIV) \ of \ the \ SCR}{peak \ value \ of \ the \ voltage \ applied \ to \ the \ SCR} \qquad \textbf{4.13}$$

Semiconductor devices are normally rated, for economic consideration, with a voltage safety factor of 1.5. Any overvoltages must be limited to less than this value. Protection against overvoltages can also be achieved by using an RC snubber circuit across the overvoltage source. The RC circuit will suppress the transient overvoltage by absorbing its energy. Another method is to connect a nonlinear resistor called a *varistor* across the SCR. The varistor provides a low-resistance path to the transient voltage and thus bypasses the device.

Example 4.15   Find the PIV rating of an SCR connected across a 220-V AC source. Use a safety factor of 1.5.

Solution

$$V_{peak} = \sqrt{2} * 220 = 311 \text{ V}$$

$$V_F = \frac{PIV}{311} = 1.5$$

$$PIV = 1.5 * 311 = 466.5 \text{ V}$$

4.10.2 Overcurrent Protection

An overcurrent can occur due to source failure during inversion, overload, or a short circuit. The SCR is usually protected by using conventional protective devices such as overcurrent relays, fast-acting fuses, and high-speed circuit breakers. The protective device is required to carry continuously the normal current that the SCR is rated to carry but to open the circuit in case of a short circuit or fault before the SCR is overheated and damaged.

4.10.3 I^2t Rating

The energy that must be dissipated by a fuse comes from either the power source or the energy stored in the inductive components of the circuit. The heat energy to be dissipated by the fuse is equal to I^2t. Therefore, the I^2t rating defines the thermal capacity of fuses and is used in the protection of SCRs. During overloads, faults, and short circuits, the SCR must withstand conditions leading to high junction temperatures. The I^2t rating for the SCR allows selection of the correct protection to avoid overheating the junction.

Manufacturers of fuses and SCRs specify in their data the I^2t ratings of their devices. The I^2t rating of the SCR is based on the device operating at maximum rated current and maximum junction temperature. Because fuses and SCRs are rated on a common basis, it is only necessary to select a fuse with an I^2t rating less than the I^2t rating of the SCR for proper protection. I is the RMS value of the current, and t is the time in seconds.

4.11 Gate Circuit Protection

The gate-pulse generating circuit must be protected against induced voltage transients. It is also necessary to separate it from the power wiring as widely as possible. Electrical isolation between the SCR and the gate circuit is usually provided by a pulse transformer or an optocoupler.

The pulse transformer, which has a ferrite core, has a primary winding and one or more secondary windings, which permit simultaneous gate pulses to be applied to SCRs in series or parallel connections. The most commonly used optocouplers consist of a combination of a light-emitting diode (LED) and a photo transistor assembled in a single package. This arrangement allows coupling of a signal from one circuit to another, while providing almost complete electrical isolation between the two circuits.

4.12   SCR Gate-Triggering Circuits

For proper operation of circuits using SCRs, the trigger circuits should supply the firing signal at precisely the correct time to assure turn-on when required. In general, the firing circuit used to trigger an SCR must meet the following criteria:

1. produce a gate signal of suitable magnitude and sufficiently short rise time
2. produce a gate signal of adequate duration
3. provide accurate firing control over the required range
4. ensure that triggering does not occur from false signals or noise
5. in AC applications, ensure that the gate signal is applied when the SCR is forward-biased
6. in three-phase circuits, provide gate pulses that are 120° apart with respect to the reference point
7. ensure simultaneous triggering of SCRs connected in series or in parallel

Three basic types of gate-firing signals are normally used: DC signals, pulse signals, and AC signals.

Triggering requirements are normally provided in terms of DC voltage and current. Since pulse signals are ordinarily used for firing SCRs, it is also necessary to consider the duration of the firing pulse. A trigger pulse that has a magnitude just equal to the DC requirements must have a pulse width that is long enough to ensure that the gate signal is provided during the full turn-on time of the SCR. As the magnitude of the gate signal increases, the turn-on time of the SCR decreases and the width of the gate pulse may be reduced. For highly inductive loads, the pulse width must be made long enough to ensure that the anode current rises to a value greater than the latching current of the SCR.

4.12.1   DC Signals

Figure 4.20(a) shows a simple circuit that applies a DC signal from an external trigger circuit. The switch S is closed to turn the SCR on. Closing the switch applies a DC current to the gate of the SCR, which is forward-biased by the source (V_S). Once the SCR is conducting, the switch can be opened to remove the gate signal. Diode D limits the magnitude of a negative gate signal to ≈ 1 V, and the resistor R_G is used to limit the gate current. Figure 4.20(b) shows an alternative circuit that provides the gate signal internally from the main power source. The two circuits operate in essentially the same way.

Applying a constant DC gate signal is not desirable because of the gate power dissipation, which would be present at all times. Also, DC gate signals are not used for triggering SCRs in AC applications, because the presence of a positive signal at the gate during the negative half-cycle would increase the reverse anode current and possibly destroy the device.

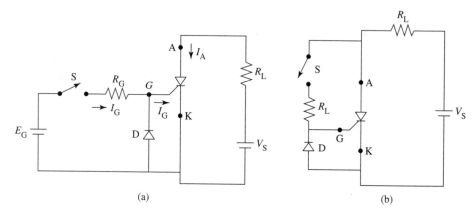

Figure 4.20
DC gating signals (a) from a separate source (b) from the same source

---

Example 4.16 The SCR in Figure 4.20(a) has a maximum gate current of 100 mA and a maximum V_{GK} of 2 V. If E_G is 15 V, find the value of R_G that will provide sufficient current for turn-on. Also find the power dissipated by the gate.

Solution Applying KVL around the input loop,

$$E_G - I_G * R_G - V_{GK} = 0$$
$$I_G * R_G = E_G - V_{GK}$$
$$R_G = \frac{E_G - V_{GK}}{I_G} = \frac{15 - 2}{100(10^{-3})} = 130 \ \Omega$$

Any value of R_G greater than 130 Ω will satisfy the gate current requirement for turn-on. However, R_G cannot be made infinitely large since the minimum gate current requirement must also be satisfied.

The power dissipated by the gate is

$$P_G = V_{GK} * I_G$$
$$= 2 * 100(10^{-3})$$
$$= 0.200 \ W$$

Example 4.17 The SCR in Figure 4.21 has a minimum gate current of 100 mA and a minimum V_{GK} of 2 V. If $R_S = 20 \ \Omega$ and $R_G = 30 \ \Omega$, find the value of the trigger voltage (V_{TRIG}) that will provide sufficient current for turn-on.

Solution Applying KVL around the input loop,

$$V_{TRIG} - I_S * R_S - V_{GK} = 0$$
$$V_{TRIG} = I_S * R_S + V_{GK}$$

Figure 4.21
See Example 4.17

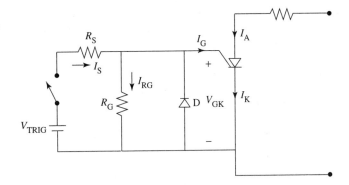

Applying KCL at the gate node,

$$I_S = I_{RG} + I_G$$

$$= \frac{V_{GK}}{R_G} + I_G$$

Therefore,

$$V_{TRIG} = \left(\frac{V_{GK}}{R_G} + I_G\right)R_s + V_{GK}$$

$$= \left(\frac{2}{30} + 0.1\right)20 + 2$$

$$= 5.33 \text{ V}$$

4.12.2   Pulse Signals

To reduce gate power dissipation, SCR firing circuits generate a single pulse or a train of pulses instead of a continuous DC gate signal. This allows precise control of the point at which the SCR is fired. In addition, it is easy to provide electrical isolation between the SCR and the gate-trigger circuit. Electrical isolation by means of a pulse transformer or optical coupler is important if several SCRs are gated from the same source. Isolation also reduces unwanted signals, such as transient noise signals, that could inadvertently trigger a sensitive SCR.

Figure 4.22(a) shows the most common method of producing pulses—using a unijunction transistor (UJT) oscillator. This circuit is ideal for triggering an SCR. It provides a train of narrow pulses at B_1. When the capacitor is charged to the peak voltage (V_P) of the UJT, the UJT turns on. This places a low resistance across the emitter–base 1 junction, and emitter current flows through the primary of the pulse transformer, applying a gate signal to the SCR. The pulse width of the output signal can be increased by increasing the value of C. One difficulty with this circuit is that due to the narrow pulse width, a latching current may not be attained before the gate signal is removed. An RC snubber circuit can be used to remove this problem.

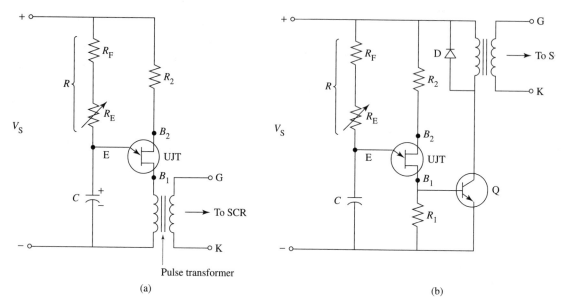

Figure 4.22
SCR trigger circuits using a UJT oscillator

The operation of the circuit shown in Figure 4.22(b) is similar. The width and rise time of the pulse can be improved by using the output across R_1 to drive a transistor Q connected in series with the transformer primary. When the pulse from the UJT is applied to the base of Q, the transistor saturates and the supply voltage V_s is applied across the primary. This induces a voltage pulse at the secondary of the pulse transformer, which is applied to the SCR. When the pulse to the base of Q is removed, it turns off. The current caused by the collapsing magnetic field in the transformer induces a voltage of opposite polarity across the primary winding. Diode D provides a path for current flow during this time.

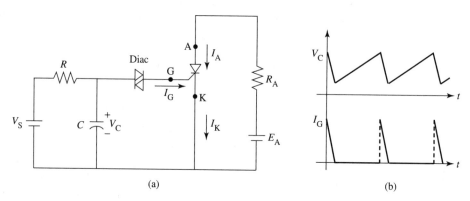

Figure 4.23
(a) An SCR trigger circuit using a diac (b) waveforms

A similar circuit using a diac (Figure 4.23 (a)) charges a capacitor slowly over a period of time determined by the *RC* time constant. After the capacitor has been charged to a voltage equal to the breakover voltage of the diac, it switches the diac into conduction. The capacitor is then rapidly discharged into the gate terminal of the SCR. After a short interval, the diac turns off and the cycle repeats. This arrangement requires a relatively low power to charge the capacitor from the DC source, but it supplies a large power for a short time for reliable SCR turn-on. The waveforms are shown in Figure 4.23 (b).

The trigger circuit in Figure 4.24 uses an optocoupler to obtain electrical isolation between the control circuitry and the load. Triggering via optocoupler also prevents false triggering from noise or transients. This triggering technique is especially popular in solid-state relays.

Figure 4.24
An SCR trigger circuit using an optocoupler

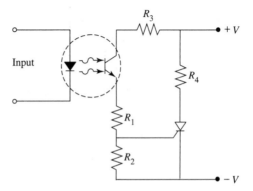

4.12.3 AC Signals

The most common method of controlling SCRs in AC applications is to derive the firing signal from the same AC source and to control its point of application to the SCR during the positive half-cycle. A simple resistive trigger circuit is shown in Figure 4.25(a). During the positive half-cycle, the SCR is in the for-

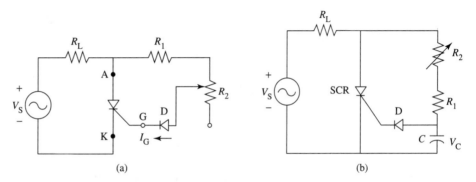

(a)   (b)

Figure 4.25
AC signals (a) resistive phase control (b) *RC* phase control

ward blocking state. At some value of V_S, the gate current is high enough to turn the SCR on. The exact moment of firing of the SCR is controlled by rheostat R_2. Diode D ensures that only positive current is applied to the gate. In Figure 4.25(b), an *RC* circuit produces the gating signal. The voltage across *C* lags the supply voltage by an amount that depends on the value of $(R_1 + R_2)$ and *C*. Increasing R_2 increases the time it takes for the voltage V_C to reach a level at which there is sufficient gate current to turn on the SCR.

Example 4.18   A gate-triggering circuit for an SCR provides a train of pulses with a frequency of 100 Hz and a pulse width of 2 ms. If the pulse has a peak power of 2 W, find the average power dissipated by the gate.

Solution   pulse period $T = 1/f = 1/100 = 10$ ms

duty cycle $d = T_{ON}/T = 2/10 = 0.2$

$P_{G(avg.)} = 0.2 * 2 = 0.4$ W

4.13   Triggering SCRs in Series and in Parallel

SCRs connected in series or parallel should be triggered from the same source and at the same instant. This can be achieved by using a relatively high gate-trigger voltage that fires the SCRs faster resulting in a uniform turn-on time. A pulse transformer is used to ensure that all gates are triggered simultaneously. Figure 4.26 shows a *gate-trigger pulse transformer* with properly insulated multiple secondary windings. The transformer also provides electrical isolation so

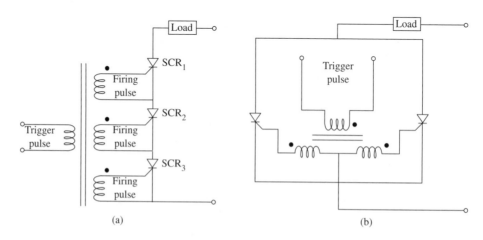

(a)   (b)

Figure 4.26
Simultaneous triggering of SCRs (a) triggering SCRs in series (b) triggering SCRs in parallel

that the trigger source is not loaded heavily, thus preventing other SCRs in the group from firing.

4.14   SCR Turnoff (Commutation) Circuits

If an SCR is forward-biased and a gate signal is applied, the device turns on. However, once the anode current I_A is above the holding current, the gate loses control. The only way to turn the SCR off is to reduce the anode current below the holding current value or to make the anode negative with respect to the cathode. The process of turnoff is known as *commutation*. In AC applications, the required condition for turnoff is achieved when the source reverses during the negative half-cycle. This method is called *natural* or *line commutation*. For DC applications, additional circuitry must be used to turn the SCR off. These circuits first force a reverse current through the SCR for a short period to reduce the anode current to zero. They then maintain the reverse bias for the necessary time to complete the turnoff. This process is called *forced commutation*.

It should be noted that if a forward voltage is applied instantly after the anode current is decreased to zero, the SCR will not block the forward voltage and will start conducting again, even though it is not triggered by a gate pulse. It is therefore important to keep the device reverse-biased for a finite time, called the *turnoff time* (t_{OFF}), before a forward anode voltage can be applied. The turnoff time of an SCR is specified as the minimum period between the instant the anode current becomes zero and the instant the device is able to block the forward voltage.

SCR turnoff can be accomplished in the following ways:

1. diverting the anode current to an alternate path
2. shorting the SCR from anode to cathode
3. applying a reverse voltage *(by making the cathode positive with respect to the anode)* across the SCR
4. forcing the anode current to zero for a brief period
5. opening the external path from its anode supply voltage
6. momentarily reducing the supply voltage to zero

The most common methods of achieving commutation are discussed briefly in the following subsections.

4.14.1   Capacitor Commutation

In DC circuits, the SCR can be turned off by switching the anode current to an alternate path for sufficient time to allow the SCR to recover its blocking ability. A simple circuit using a transistor switch Q is shown in Figure 4.27. When the

SCR is on, the transistor is in the off state. To turn the SCR off, a positive pulse is applied to the base of Q, turning it on. The anode current is diverted to the transistor. When the anode current falls below the holding current, the SCR turns off. The transistor is held on just long enough to turn off the SCR. This is not a useful method for repetitive on-off operations since the SCR is not actually reverse-biased and turnoff is therefore slow.

Figure 4.27
SCR turnoff circuits using a transistor switch

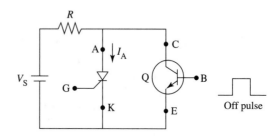

In another method, the SCR can be turned off by applying a reverse bias for enough time to allow the SCR to recover its forward blocking ability. A typical commutation circuit includes a commutation capacitor C and an auxiliary SCR_2, as shown in Figure 4.28. When the main SCR_1 is conducting, capacitor C charges to the source voltage V_S through R_L, initially with the polarity shown. At this instant, SCR_2 is off. To turn SCR_1 off, SCR_2 is triggered. When SCR_2 turns on, the capacitor is switched across SCR_1, applying a reverse voltage across it. If SCR_1 is reverse-biased long enough, it will turn off.

Figure 4.28
SCR turnoff circuits using a commutation capacitor

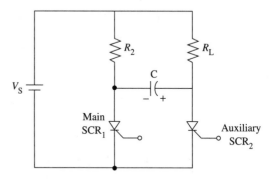

To ensure successful commutation, the value of the capacitance C can be determined by:

$$C \geq \frac{t_{OFF}}{0.693 \, R_L}$$

4.14

where

C = commutation capacitor in μF

R_L = load resistance in Ω

t_{OFF} = turnoff time in μS

Example 4.19 In the circuit in Figure 4.28, the source voltage is 220 V and the load resistance is 10 Ω. If the turnoff time for the SCR is 10 μs, find the minimum value of capacitance that will ensure commutation.

Solution The minimum value of C is:

$$C = \frac{t_{OFF}}{0.0693 \ R_L} = \frac{10 \ (10^{-6})}{0.693 * 10} = 1.44 \ \mu F$$

A suitable value of C would be 1.5 μF.

4.14.2 Commutation by External Source

In this type of commutation circuit, the commutation energy is obtained from an external source in the form of a pulse. A simple circuit is shown in Figure 4.29. The pulse generator reverse-biases the SCR and thus turns it off. The pulse width must be such that the SCR is reverse-biased for a period greater than the turnoff time of the SCR.

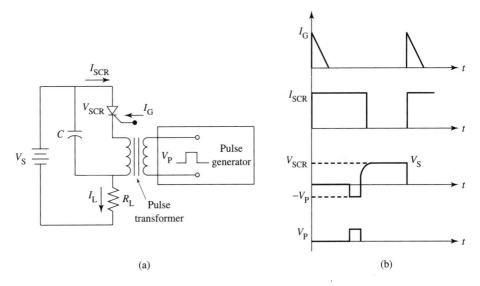

(a)   (b)

Figure 4.29
Commutation by external source (a) circuit (b) waveforms

When the SCR is triggered by applying a gate signal, current flows through the SCR, the secondary of the pulse transformer, and the load. To turn the SCR off, a positive pulse from the pulse transformer is applied to the cathode of the SCR. The capacitor is charged to only about 1 V and can be considered a short circuit for the duration of commutation.

4.14.3 Commutation by Resonance

The natural resonance set up in an LC circuit can be used directly to turn off an SCR, eliminating the need for an external source. Figure 4.30 shows a simple series resonant turnoff circuit. The underdamped LC resonating circuit in series with the load applies a reverse voltage to the SCR to turn it off.

Figure 4.30
Series resonant turnoff circuit

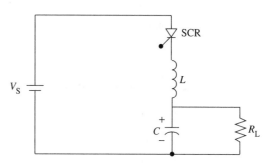

The parallel resonance commutation circuit shown in Figure 4.31 can also be used to turn an SCR off. The capacitor C is initially charged during the SCR off period to the source voltage with the polarity indicated. When the SCR is turned on, the capacitor discharges through the LC resonant circuit and applies a reverse voltage across the SCR to turn it off. Once the SCR is turned on, it conducts till the capacitor charges again to V_S and starts discharging through the SCR and L. Then it automatically turns off.

Figure 4.31
Parallel resonant turnoff circuit

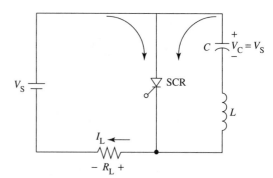

4.14.4 AC Line Commutation

This commutation method is used in circuits with an AC source. Figure 4.32 shows a typical line commutated circuit and its associated waveforms. Load current flows in the circuit during the positive half-cycle. The SCR is reverse-biased during the negative half-cycle of the input voltage. With zero gate signal, the SCR will turn off if the turnoff time of the SCR is less than the duration of the half-cycle, that is, for a period $T/2$. The maximum frequency at which this circuit can operate depends on the turnoff time of the SCR.

Figure 4.32
AC line commutation

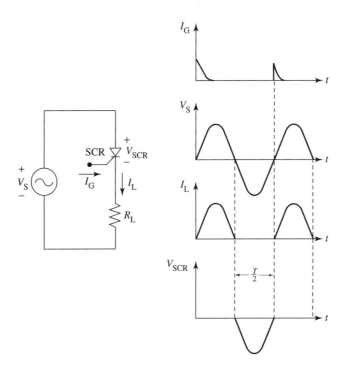

4.15   Other Types of Thyristors

We have now discussed the standard thyristor, the silicon controlled rectifier, in some detail. The other important members of the thyristor family are the gate turnoff GTO thyristor and the triac. There are also several low-power devices belonging to this family that are mainly used in trigger circuits to turn the thyristor on. The principal ones are the silicon controlled switch (SCS) and the diac. The MOS-controlled thyristor (MCT) is a new but very promising switching device.

4.15.1   The Silicon Controlled Switch (SCS)

The silicon controlled switch (SCS) is a four-layer PNPN device. Figure 4.33 shows the structure and the symbol for the device. The SCS has two gates, an anode gate (AG) and a cathode gate (KG).

Figure 4.33
The SCS (a) structure
(b) symbol

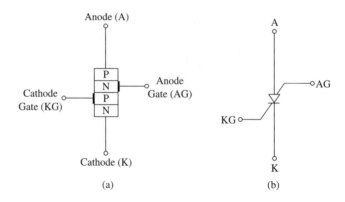

Figure 4.34 shows the electrical equivalent circuit. As shown, both transistor bases are accessible for applying gate pulses. Like an SCR, an SCS can be turned on by applying a positive pulse at the cathode gate. The device can also be turned on by applying a negative gate pulse at the anode gate. If the SCS is on, a positive pulse at the anode gate or a negative pulse at the cathode gate is required to turn it off. Usually, the turn-on anode gate current is larger in magnitude than the required cathode gate current.

Figure 4.34
Equivalent circuit for an SCS

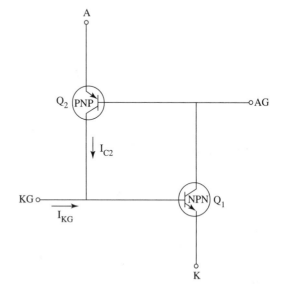

4.15.2   The Gate Turnoff Thyristor (GTO)

The gate turn off thyristor is a power semiconductor switch that turns on like a normal SCR, that is, with a positive gate signal. In addition, it can be turned off by means of a negative gate current. Both on-state and off-state operations of the device are therefore controlled by the gate current. A second very important characteristic of the GTO is its improved switching characteristics. The turn-on time is similar to that of the SCR, but the turnoff time is~~is~~ much smaller. This allows the use of GTOs in high-speed applications. However, the voltage and current ratings of available GTOs are lower than those of SCRs. GTOs also have higher on-state voltage drops and less leakage current. GTOs are used in motor drives, uninterruptible power supplies (UPS), static volt amperes reactive (VAR) compensators, choppers, and inverters at high power levels.

The structure of a GTO, shown in Figure 4.35(a), is essentially the same as that of an SCR. The symbol for the GTO is shown in Figure 4.35(b). If the anode (A) is made positive with respect to the cathode (K) and a positive gate signal is applied to the gate, the GTO turns on. It remains on until the anode current falls to a value below the holding current. To turn the device off, a negative gate signal is applied to the gate.

Figure 4.35
Gate turnoff thyristor
(a) structure (b) symbol

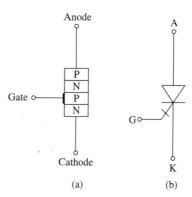

The *ideal V-I* characteristic of the GTO is shown in Figure 4.36. If there is no gate signal, the device remains off for either polarity of anode-cathode voltage. In the forward direction, if a positive signal I_G is applied to the gate, the

Figure 4.36
GTO ideal *V-I* characteristic

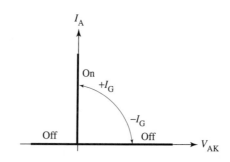

GTO turns on. It remains on, even if the gate signal is removed. If a negative signal $-I_G$ is applied to the gate, the GTO turns off. It stays off until a positive gate signal is reapplied. Like an SCR, the GTO can still be turned off by reducing the anode current to a value below the holding current.

The major disadvantage of the GTO is the increase in the magnitude of the gate current required to turn on or turn off the device, compared with the SCR. Usually, more gate current is needed to turn off the device than to turn it on. The GTO also has reduced reverse voltage blocking capability. Therefore, an inverse diode must be used across the SCR (as shown in Figure 4.37) if there is likelihood that a high reverse voltage may appear across the device. The power losses are also somewhat higher than those of the SCR due to increased conduction loss. However, these disadvantages far outweigh its convenience when used in DC applications that would require additional components to turn off the device if an SCR were used.

Figure 4.37
GTO snubber circuit

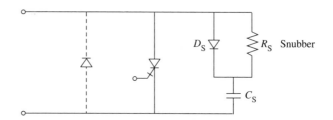

GTOs are subjected to large power losses when a sharp rise voltage is applied. To prevent this, a polarized snubber consisting of capacitors, resistors, and a diode is connected in parallel with the GTO (see Figure 4.37). The snubber circuit also limits dv/dt across the GTO during turnoff.

4.15.3 The Diac

A diac is a three-layer, two-terminal semiconductor switch. It operates like two diodes connected back-to-back in series. The only way to turn the device on is by exceeding the breakover voltage. It can be switched from the off to the on

Figure 4.38
Diac (a) structure (b) symbol

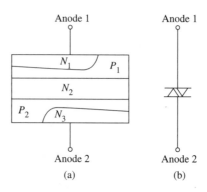

state for either polarity of applied voltage, making it useful in AC applications. Figure 4.38 shows the structure and the symbol for the device. Its terminals are called anode 1 and anode 2.

Figure 4.39 shows the *V-I* characteristic of a diac. When anode 1 is at a higher potential than anode 2, a small leakage current flows up to the breakover voltage V_{BO}, beyond which the diac conducts. The voltage across the diac drops to a low value and stays relatively constant. However, the current can increase rapidly to a high value, limited only by the external circuit. When anode 2 is at a higher potential than anode 1, similar behavior takes place. The breakover voltages are very close in magnitude in either direction, varying only about 10% from each other. Diacs are extensively used to trigger larger devices such as SCRs and triacs.

Figure 4.39
V-I characteristic of a diac

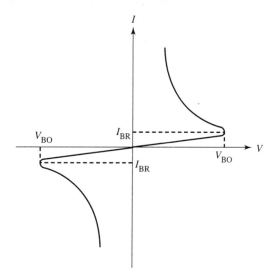

4.15.4   The Triac

The triac is simply a diac with a gate terminal added to control the turn-on. It is able to conduct current in both the forward and the reverse directions and can be controlled by a positive or a negative gate signal. This makes it useful for AC power control. Like an SCR, a triac has three terminals, but they are called *main terminal 1* (MT$_1$), *main terminal 2* (MT$_2$), and *gate*. The structure and symbol for the device are shown in Figure 4.40(a) and (b), respectively. A triac is also called a bidirectional SCR, since it can be considered an integration of two SCRs connected back-to-back in parallel, as shown in Figure 4.40(c).

Figure 4.41 shows the voltage-current characteristic of the triac. The characteristic is identical to that of an SCR for either polarity of applied voltage to the terminals MT$_1$ and MT$_2$. The breakover voltage of the triac can also be controlled by application of a positive or negative signal to the gate. As the mag-

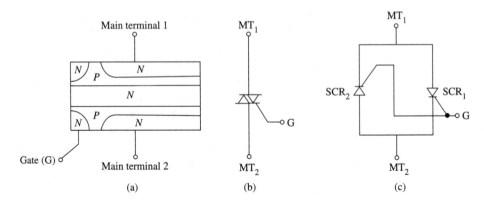

Figure 4.40
The Triac (a) structure (b) symbol (c) SCR equivalent circuit

nitude of the gate signal increases, the breakover point of the triac decreases. Once the triac turns on, the gate signal can be removed, and just like an SCR, the triac remains on until the main current falls below the holding current. The four operating modes for turn-on are summarized in Table 4–3.

Figure 4.41
Triac *V-I* characteristic

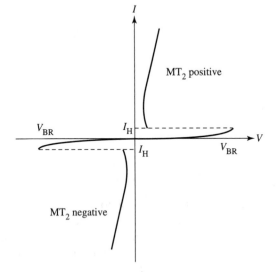

A triac is more economical and easy to control, but if the power to be regulated is greater than the triac's rating, a pair of SCRs can be used. In fact, due to the higher *dv/dt* capabilities and lower turnoff times of the SCR, high-power circuits normally employ two SCRs to achieve the function of a triac.

One limitation of the triac is its slow speed, which limits the operating frequency to a few hundred hertz. Triacs are therefore exclusively used to regulate

Table 4–3 Operating modes of the triac

MT_2-to-MT_1 voltage	Gate-to-MT_1 voltage
Positive	Positive
Positive	Negative
Negative	Positive
Negative	Negative

60-Hz AC voltage in applications such as light dimming, speed control of motors, and heating control and in solid-state AC relays.

4.15.5 The MOS-Controlled Thyristor (MCT)

The MOS-controlled thyristor (MCT) is a new device that combines the characteristics of a MOSFET and an SCR. It has a low forward voltage drop in the on state and a low turnoff time. It has high di/dt and dv/dt capabilities. It is similar in functionality to the GTO but has a lower turnoff gate current requirement. Its main disadvantage is that its reverse voltage blocking ability is very low.

Figure 4.42 shows the symbol and equivalent circuit for the MCT. In an MCT, an SCR and two MOSFETs are combined into a single device. The two MOSFETs have the same gate terminal, which is the gate of the MCT. They also have the same source terminal, which is the anode of the MCT. The N-channel MOSFET Q_{OFF}, which is connected between the anode and one of its internal layers, turns the SCR off while the P-channel MOSFET Q_{ON}, connected between the gate and anode, turns it on.

Figure 4.42
The MCT (a) symbol
(b) equivalent circuit.

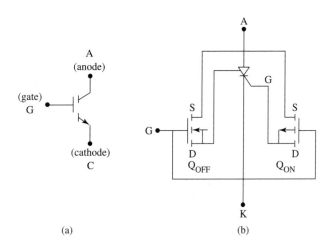

(a)   (b)

Unlike a GTO, which is turned on and off by establishing a gate current, the MCT is switched by establishing the proper voltage from gate to anode. When the gate-to-anode voltage is approximately −5V, Q_{ON} turns on and supplies gate current to the SCR. This turns on the SCR. The MCT is turned off by applying a gate-to-anode voltage of approximately +10V, which turns Q_{OFF} on. This shunts current away from the SCR and turns it off.

4.15.5.1   The MCT *V-I* Characteristic

Figure 4.43 shows the *V-I* characteristic of an MCT. If the anode (A) is made positive with respect to the cathode (K), when no voltage is applied to the gate, the MCT remains in the blocking state, allowing only a small leakage current (I_{LEAK}). The MCT remains in the off state until a breakover voltage V_{BO} is reached, at which point the MCT breaks down. However, the MCT is not turned on this way.

Figure 4.43
MCT *V-I* characteristic

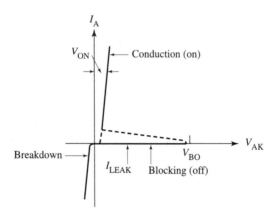

If the cathode (K) is made positive with respect to the anode (A) with either positive or negative voltage applied to the gate, the MCT breaks down at a low voltage. This situation is to be avoided.

The normal way to turn the device on is to forward-bias the MCT by making the anode positive with respect to the cathode and applying a negative voltage to the gate and anode. When on, the voltage drop across the MCT (V_{ON}) is very small (about 1 V) and the anode current is limited only by the load resistance. Once the MCT turns on, removal of the gate voltage will not turn it off. If the MCT is on, the application of a positive voltage to the gate turns the device off until a negative voltage is applied to the gate again.

4.16 Problems

4.1 What is the main difference between a diode and an SCR?

4.2 What two conditions must be met to cause an SCR to turn on?

4.3 After an SCR is turned on, what effect does the gate current have on the SCR?

4.4 How does the breakover voltage vary with the gate current?

4.5 How can an SCR be turned on with its gate terminal open?

4.6 What requirements should be satisfied by a gate trigger?

4.7 Explain the operation of an SCR using the two-transistor model.

4.8 Find the average and RMS values of the current waveform shown in Figure 4.6, using the approximation method. I_m is 80 A, t_o is 4 ms, and T is 20 ms.

4.9 Find the RMS current of an SCR if the average DC current is 80 A, with a conduction angle θ of 20°.

4.10 Find the RMS current of an SCR if the average DC current is 120 A, with a conduction angle θ of 40°.

4.11 Sketch the output-current waveform of an SCR controlling a load when V_{DRM} is slightly exceeded, causing breakover with no gate trigger.

4.12 When is the holding current of an SCR greater, at −30°C or +30°C?

4.13 Define I_h. When does the holding current pose a problem?

4.14 Explain the meaning of the *di/dt* rating of an SCR. What is likely to happen to an SCR when $(di/dt)_{max}$ is exceeded? How can the effects be reduced?

4.15 Explain the meaning of the *dv/dt* rating of an SCR. What is likely to happen to an SCR when $(dv/dt)_{max}$ is exceeded? How can the effects be reduced?

4.16 Find the minimum value of the series inductance L required to protect an SCR from excessive *di/dt*. The SCR has a *di/dt* rating of 10 A/μs, and the AC source voltage is 220 V.

4.17 Why is a snubber required across a thyristor? Sketch a snubber circuit and explain how it works.

4.18 Find values of an RC snubber circuit, if the SCR has the following ratings:

$V_{DRM} = 200$ V

$(dv/dt)_{max} = 200$ V/μs

$(di/dt)_{max} = 100$ A/μs

$R_L = 10\ \Omega$

4.19 What is the difference between an inverter-type SCR and a phase-control-type SCR?

4.20 Draw four SCRs in series and external circuitry to equalize voltages.

4.21 Describe the various methods used to turn on an SCR.

4.22 Describe the various methods used to turn off SCRs in DC circuits.

4.23 Draw the symbol of a triac and identify its terminals.

4.24 Sketch the *V-I* characteristic of a triac.

4.25 List the four operating modes of a triac.

4.26 Draw the symbol of a diac and identify its terminals.

4.27 Sketch the *V-I* characteristic of a diac.

4.28 Describe the operation of a GTO.

4.29 What is the main advantage of a GTO over a conventional SCR?

4.30 Draw the symbol of an MCT and identify its terminals.

4.31 Describe the operation of an MCT.

4.32 Sketch the *V-I* characteristic of an MCT.

4.17   Equations

$$I_{RMS} = \sqrt{\frac{I_m^2 t_o}{T}} \tag{4.1}$$

$$I_{AVE} = \frac{I_m t_o}{T} \tag{4.2}$$

$$f_o = \frac{I_{RMS}}{I_{AVE}} \tag{4.3}$$

$$I_{RMS} = f_o (I_{AVE}) \tag{4.4}$$

$$I_{TRMS} = f_o (I_{T(AVE)}) \tag{4.5}$$

$$L \geq \frac{V_P}{(di/dt)_{max}} \tag{4.6}$$

$$T = \frac{V_{DRM}}{(dv/dt)_{max}} \tag{4.7}$$

$$C \geq \frac{T}{R_L} \geq \frac{V_{DRM}}{R_L (dv/dt)_{max}} \tag{4.8}$$

$$R_S \geq \sqrt{\frac{V_{DRM}}{(di/dt)_{max}}} \tag{4.9}$$

$$R = \frac{2V_2 - V_m}{\Delta I} \tag{4.10}$$

$$R = \frac{V_1 - V_2}{I_2 - I_1} \tag{4.11}$$

$$f \approx \frac{\text{on-state + off-state power loss averaged over a cycle}}{\text{energy dissipated during switching per cycle}}$$

4.12

$$V_F = \frac{V_{RRM} \text{ (PIV) of the SCR}}{\text{peak value of the voltage applied to the SCR}}$$

4.13

$$C \geq \frac{t_{off}}{0.693 \, R_L}$$

4.14

Single-Phase Uncontrolled Rectifiers

5

Learning Objectives

After completing this chapter, the student should be able to

- describe with the help of waveforms the operation of a half-wave uncontrolled rectifier with resistive and inductive loads

- explain what is meant by a freewheeling diode

- explain the difference between a half-wave and a full-wave rectifier

- describe with the help of waveforms the operation of a full-wave uncontrolled rectifier with resistive and inductive loads

- discuss the advantages of a full-wave rectifier over a half-wave rectifier

5.1   Introduction

Rectification is the process of converting alternating current or voltage into direct current or voltage. An uncontrolled rectifier uses only diodes as the rectifying elements. The DC output voltage is fixed in magnitude by the amplitude of the AC supply voltage. However, the DC output is not *pure*—it contains significant AC components called *ripple*. To eliminate this ripple, a filter is inserted after the rectifier.

In this chapter, we will study single-phase, uncontrolled rectifiers ranging from a simple half-wave rectifier using a single diode to more complex full-wave bridge rectifiers using several diodes. We will assume that the diodes used have the same ideal characteristics as those in Figure 2.3.

5.2   The Half-Wave Rectifier (One-Pulse Rectifier)

5.2.1   With a Resistive Load

A simple half-wave rectifier circuit supplying a purely resistive load is shown in Figure 5.1(a). The source voltage is a sine wave with a maximum value V_m and period T. During the positive half-cycle, when the voltage at the anode is positive with respect to the cathode, the diode turns on. This allows current through the load resistor R. Thus, the load voltage (v_o) follows the positive half-sine wave. During the negative half-cycle, the voltage at the anode becomes negative with respect to the cathode and the diode turns off. Then no current flows through R. The output voltage (v_o) is shown in Figure 5.1(b), which also shows the load current.

The half-wave rectifier thus changes AC power to DC. The output voltage is pulsating DC containing a large ripple. Therefore, one-pulse circuits have limited practical value for high-power applications.

The quantities of interest in this circuit are average or DC load voltage and load current.

Average load voltage is given by:

$$V_{o(avg.)} = \frac{V_S \sqrt{2}}{\pi}$$

$$= V_m/\pi$$

$$= 0.318\ V_m$$ **5.1**

where

V_S = the RMS value of supply voltage

V_m = the maximum value of supply voltage.

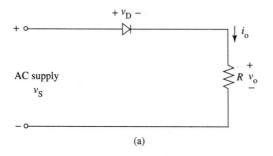

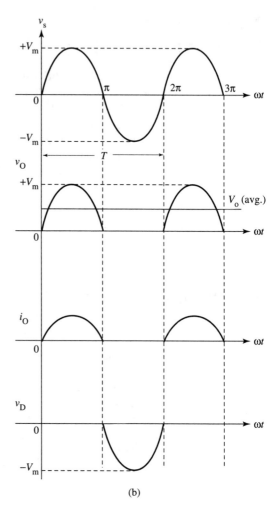

Figure 5.1
Half-wave rectifier with resistive load (a) circuit (b) waveforms

The current waveform has the same shape as the voltage waveform, so a relation similar to Equation 5.1 holds for the *average load current*:

$$I_{o(avg.)} = I_m/\pi$$
$$= 0.318\ I_m \qquad\qquad \textbf{5.2}$$

where $I_m = V_m/R$.

The *root-mean square (RMS) current* is given by:

$$I_{RMS} = I_m/2 \qquad\qquad \textbf{5.3}$$

The result in Equation 5.3 is characteristic of the half-wave circuit waveform.

The waveform of the diode voltage (Figure 5.1 (b)) shows that the diode must be able to withstand a reverse voltage that is equal to the peak source voltage. This voltage is used to select the proper diode in a given circuit. The *PIV (or PRV) rating* for the diode is therefore given by:

$$\text{PIV rating:} \geq V_m \qquad \textbf{5.4}$$

The purpose of a rectifier is to convert AC power into DC power. Since we are assuming *ideal* devices, with no power loss in the rectifier, the net power flow at the AC input terminal averaged over a full cycle must equal the DC power output. The *DC (average) power output* to the load is given by:

$$P_{o(avg.)} = V_{o(avg.)} * I_{o(avg.)}$$
$$= \frac{V_m}{\pi} * \frac{I_m}{\pi} = \frac{V_m * V_m}{\pi^2 * R} \qquad \textbf{5.5}$$
$$= \frac{V_m^2}{\pi^2 * R}$$

The *AC power input* is given by:

$$P_{AC} = V_{RMS} * I_{RMS}$$

V_{RMS} for a period from 0 to π (half-cycle) is $V_m/2$. Therefore,

$$P_{AC} = \frac{V_m}{2} * \frac{I_m}{2} = \frac{V_m * V_m}{4 * R} = \frac{V_m^2}{4R} \qquad \textbf{5.6}$$

The *rectifier efficiency* is defined as the ratio of DC output power to AC input power:

$$\eta = \frac{P_{o(avg.)}}{P_{AC}} \qquad \textbf{5.7}$$

The *form factor* is a measure of the goodness of the shape of the output voltage. Ideally, the DC output voltage of a rectifier should be constant. In practice, rectifiers provide outputs that are incomplete sine waves. The form factor is defined as the ratio of the RMS output voltage to the average value of output voltage:

$$\text{form factor (FF)} = \frac{V_{oRMS}}{V_{o(avg.)}} \qquad \textbf{5.8}$$

The ideal value of FF is unity. FF is unity if the output voltage is a constant DC value, for which $V_{oRMS} = V_{o(avg.)}$.

The output voltage of a rectifier contains both DC and AC (ripple) components. The frequency and magnitude of the ripple voltage is an important factor in the choice of rectifiers. The higher the frequency and the smaller the magnitude of the ripple, the easier it is to filter the ripple to within acceptable limits.

The *pulse number* is the ratio of the fundamental ripple frequency of the DC voltage to the frequency of the AC supply voltage.

$$\text{pulse number} = \frac{\text{fundamental ripple frequency}}{\text{AC source frequency}} \qquad \textbf{5.9}$$

The *ripple factor* is the ratio of the RMS value of the AC component to the DC component.

$$\text{ripple factor} = \frac{I_{\text{oAC}}}{I_{\text{oDC}}}$$

Ideally the ripple factor should be zero. The ripple factor can be determined by first finding the power dissipated in the load resistor R:

$$P_L = I_{\text{RMS}}^2 R = I_{\text{DC}}^2 R + I_{\text{AC}}^2 R$$

or

$$I_{\text{AC}}^2 = I_{\text{RMS}}^2 - I_{\text{DC}}^2$$

$$\frac{I_{\text{AC}}^2}{I_{\text{DC}}^2} = \frac{I_{\text{RMS}}^2}{I_{\text{DC}}^2} - \frac{I_{\text{DC}}^2}{I_{\text{DC}}^2}$$

$$RF^2 = \frac{I_{\text{RMS}}^2}{I_{\text{DC}}^2} - 1$$

$$RF = \sqrt{\frac{I_{\text{RMS}}^2}{I_{\text{DC}}^2} - 1} \qquad\qquad \textbf{5.10}$$

Example 5.1   A half-wave rectifier shown in Figure 5.1(a) is connected to a 50 V AC source. If the load resistance is 100 Ω, find
a) the maximum load voltage
b) the average load voltage
c) the maximum load current
d) the average load current
e) the RMs load current
f) the PIV rating for the diode
g) the DC (average) power output
h) the AC power input
i) the rectifier efficiency
j) the form factor
k) the pulse number
l) the ripple factor
m) the conduction angle

Solution   a) maximum load voltage

$$V_m = \sqrt{2}\, 50 = 70.7 \text{ V}$$

b) average load voltage

$$V_{o(\text{avg.})} = 0.318 \; V_m$$
$$= 0.318 * 70.7$$
$$= 22.5 \text{ V}$$

c) maximum load current

$$I_m = V_m/R = 70.7/100 = 707 \text{ mA}$$

d) average load current

$$I_{o(avg.)} = 0.318 * 0.707 = 225 \text{ mA}$$

e) RMS load current

$$I_{oRMS} = I_m/2 = 0.707/2 = 353.5 \text{ mA}$$

f) PIV rating for the diode

$$PIV \geq V_m = 70.7 \text{ V}$$

g) DC power output

$$P_{DC} = \frac{V_m^2}{\pi^2 * R} = \frac{70.7^2}{\pi^2 * 100} = 5.1 \text{ W}$$

h) AC power input

$$P_{AC} = \frac{V_m^2}{4 \, R} = \frac{70.7^2}{4 * 100} = 12.5 \text{ W}$$

i) rectifier efficiency

$$\eta = \frac{P_{DC}}{P_{AC}}$$

$$= \frac{4}{\pi^2}$$

$$= 0.405 \text{ or } 40.5\%$$

j) form factor

$$FF = \frac{V_{oRMS}}{V_{o(avg.)}} = \frac{V_m/2}{V_m/\pi} = \pi/2 = 1.57$$

k) pulse number

$$p = \frac{\text{fundamental ripple frequency}}{\text{AC source frequency}} = \frac{60}{60}$$

or

$$p = \frac{360°}{360°} = 1$$

l) ripple factor

$$RF = \sqrt{\frac{I_{RMS}^2}{I_{DC}^2} - 1}$$

$$= \sqrt{\frac{(I_m/2)^2}{(I_m/\pi)^2} - 1}$$

$$= \sqrt{(\pi/2)^2 - 1}$$

$$= 1.21$$

m) conduction angle

$$\theta = 180°$$

Example 5.2   In Figure 5.1(a), the source voltage is 120 V with a resistive load of 10 Ω. Find
a) the maximum load current
b) the average load voltage
c) the average load current
d) the RMS load current
e) the power absorbed by the load
f) the power factor
g) the PIV rating for the diode

Solution   The peak value of the source voltage is

$$V_m = \sqrt{2}\, V_{RMS} = (1.414)(120) = 169.7 \text{ V}$$

a) maximum load current $I_m = V_m/R = 169.7/10 = 16.97$ A

b) average load voltage $V_{o(avg.)} = 0.318\, V_m = 0.318*169.7 = 54.0$ V

c) average load current $I_{o(avg.)} = 0.318\, I_m = 0.318*16.97 = 5.4$ A

d) RMS load current $I_{RMS} = I_m/2 = 16.97/2 = 8.49$ A

e) power absorbed by the load $P = I_{RMS}^2\, R = 8.49^2 * 10 = 720$ W

f) apparent power $S = V_{RMS}*I_{RMS} = 120*8.49 = 1018.8$ VA

power factor $\dfrac{P}{S} = \dfrac{720}{1018.8} = 0.707$

g) PIV rating for diode: ≥ 169.7 V

5.2.2   With an Inductive (*RL*) Load

The half-wave circuit with an inductive (*RL*) load, which is the practical case, is shown in Figure 5.2(a). Let us analyze the operation of this circuit.

a) As in the case of a resistive load, the diode turns on when its anode becomes positive with respect to the cathode. The voltage across the load is therefore the same as the positive half-cycle of the AC source.

b) During this time, energy is transferred from the AC source and is stored in the magnetic field surrounding the inductor.

c) The current through an inductor cannot change instantaneously. Therefore the current increases gradually until it reaches its maximum value. Note that the current does not reach its peak when the voltage is at its maximum. This is consistent with the fact that the current through an inductor lags the voltage.

d) When the source voltage decreases, the current starts decreasing gradually, becoming zero when all the energy stored by the inductor is released to the circuit. The load current therefore exists for a little more than half the entire period.

e) At the same time, the collapsing magnetic field links with the inductor and induces a voltage that opposes the decrease in the applied voltage.

f) As soon as the current is zero, the diode is reverse-biased. The diode then remains off for the rest of the negative cycle. Figure 5.2(b) shows the waveforms.

During the interval from 0 to $\pi/2$, the source voltage v_S increases from zero to its positive maximum, while the voltage induced across the inductor v_L opposes the change of current through the load. In the time interval $\pi/2$ to π, the source voltage decreases from its positive maximum value to zero. At the same time, the induced voltage has reversed polarity and opposes the decrease in current; that is, it now aids the diode forward current.

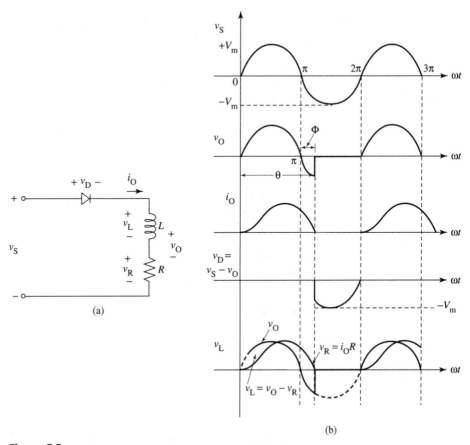

Figure 5.2
Half-wave rectifier with inductive load (a) circuit diagram (b) voltage and current waveforms

At π, the source voltage reverses and starts to increase to its negative maximum value. However, the voltage induced across the inductor is still positive and will sustain forward conduction of the diode until the induced voltage decreases to zero, at which point the diode blocks. Note that although the diode is reverse-biased, there is a current flow through the diode until the angle $\theta = \pi + \Phi$ is reached. This is the result of the energy stored in the magnetic field being returned to the source.

The waveform of voltage across the resistor is the same as the waveform of i_o($v_R = i_o R$). The difference between v_R and v_o is the voltage across the inductor (see Figure 5.2(b)).

g) If we increase the load inductance L, the diode will conduct current for more time during the cycle.

If we assume L to be infinite, the current that flows through the diode would be completely smooth and therefore *continuous*. In this situation, the diode would be on for the full cycle, the voltage across the diode would be zero, and the values of v_s and v_o would be equal. The circuit would no longer rectify, so the current would be alternating. This is not possible, therefore the output current must be zero for some duration that is less than half a cycle. The average output voltage therefore depends on the relative inductance and resistance of the load. The waveforms shown in Figure 5.2(b) are drawn for the case in which the average output voltage is not zero and the current is therefore not continuous.

The average value of the load voltage is given by:

$$V_{o(avg.)} = \frac{V_m}{2\pi} (1\text{-}\cos \theta) \qquad\qquad \textbf{5.11}$$

where $\theta = \pi + \Phi$ is the conduction angle, (Φ depends on the values of L and R). It should be clear from the above that as:

Φ approaches 180°, $V_{o(avg.)}$ approaches zero

θ approaches 360°, $V_{o(avg.)}$ approaches zero.

In other words, the inductive load reduces the average load voltage.

Note that if $\theta = 180°$, then $V_{o(avg.)} = V_m/\pi$ (Equation 5.1).

$$I_{o(avg.)} = \frac{V_m}{2\pi R}(1\text{-}\cos \theta) \qquad\qquad \textbf{5.12}$$

For high-power applications, one-pulse circuits have limited practical use due to their low output voltage and the large ripple that is contained in the DC output voltage.

5.2.3 With an Inductive Load and a Freewheeling Diode (FWD)

The circuit in Figure 5.2 can be modified to produce the circuit shown in Figure 5.3, which is of practical use in low-power applications. It is still considered a half-wave rectifier, although the load current can flow for the entire cycle.

The second diode D_2, which is added in parallel with the load, is known as a freewheeling-diode (FWD). This diode prevents negative voltage from appearing across the load, and as a result, it increases the average output voltage ($V_{o(avg.)}$) across the load as well as the average current $I_{o(avg.)}$. During the negative half-cycle of the supply voltage, the FWD conducts and provides an alternate path for the load current. During this interval of FWD conduction, the main diode D_1 is reverse-biased and stops conducting, and thus the source current is zero for this half-period. The FWD helps to prevent the load current from ever going to zero and thus reduces the ripple.

The load voltage waveform is the same as in the half-wave circuit with a resistive load. The average load voltage is similarly V_m/π, as given by Equation 5.1.

The voltage and current waveforms shown in this figure assume a large load inductance. The larger the load inductance, the smoother the load current becomes.

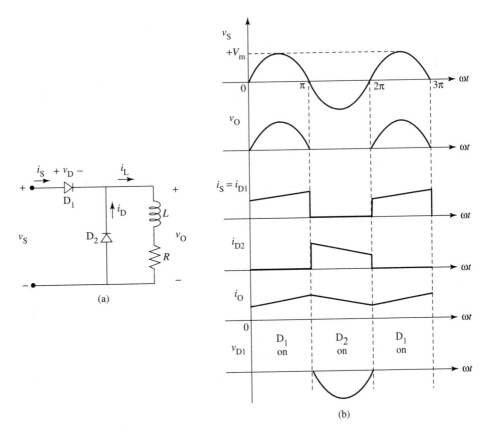

Figure 5.3
Inductive load with a FWD (a) circuit diagram (b) typical waveforms with large inductive load

---

Example 5.3 In the circuit shown in Figure 5.3, $V_s = 240$ V and $R = 10$ Ω. If the load inductance is large, find
a) the average load voltage
b) the average load current
c) the RMS value of load current
d) the power supplied to the load
e) the ripple factor
f) the power factor

Solution

$$V_m = \sqrt{2} \, (240) = 339.4 \text{ V}$$

a) average load voltage

$$V_{o(avg.)} = 0.318 * 339.4 = 108 \text{ V}$$

b) average load current

$$I_{o(avg.)} = 108/10 = 10.8 \text{ A}$$

c) Due to the large value of the load inductance, the load current is nearly constant. The RMS value of the load current therefore equals its average value.

d) Power supplied to the load

$$P_L = I_{RMS}^2 R = (10.8)^2 * 10 = 1166.4 \text{ W}$$

e) ripple factor

$$RF = \sqrt{\left[\frac{V_{RMS}}{V_{DC}}\right]^2 - 1} = \sqrt{\left[\frac{(339.4/2)}{108}\right]^2 - 1} = 1.211$$

f) Power factor

$$\frac{P}{S} = \frac{1166.4}{240 * 10.8} = 0.45$$

5.3 Full-Wave Center-Tapped Transformer Rectifier (Two-Pulse Rectifier)

A single-phase, half-wave rectifier is not very practical due to its low average output voltage, poor efficiency, and high ripple factor. These limitations can be overcome by full-wave rectification. Full-wave rectifiers are more commonly used than half-wave rectifiers, due to their higher average voltages and currents, higher efficiency, and reduced ripple factor.

5.3.1 With a Resistive Load

Figure 5.4(a) shows the schematic diagram of the full-wave rectifier using a transformer with a center-tapped secondary. The source voltage and load resistor are the same as in the half-wave case. During the positive half-cycle (Figure 5.4(b)), diode D_1 conducts and D_2 is reverse-biased. Current flows through the

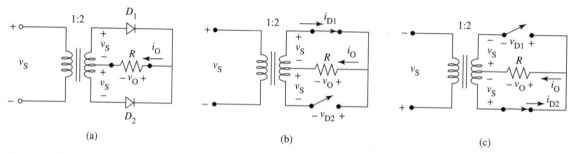

(a)   (b)   (c)

Figure 5.4
Full-wave center-tap rectifier (a) circuit (b) equivalent circuit during the positive half-cycle (c) equivalent circuit during the negative half-cycle

Figure 5.4
(d) Voltage and current
waveforms

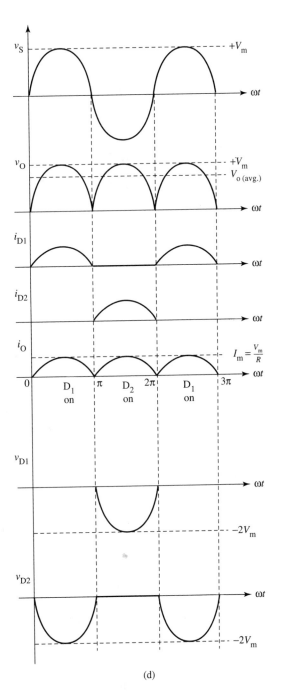

(d)

load, causing a positive drop. During the negative half-cycle (Figure 5.4(c)), diode D_2 conducts and D_1 turns off. Current flows through R, maintaining the same polarity for the voltage across the load (see Figure 5.4(d)). Therefore, the load voltage waveform consists of successive half-cycles of a sine wave, resulting in a higher average value and higher ripple frequency.

Average and RMS values are similar tot hose for the half-wave case:

$$V_{o(avg.)} = (2 \ V_m)/\pi = 0.636 \ V_m \qquad\qquad \textbf{5.13}$$

Note that the full-wave average is twice the half-wave average—this is obvious by inspecting the two graphs of voltage versus time. Similarly, the average load current is given by the same factor, so

$$I_{o(avg.)} = (2 \ I_m)/\pi = 0.636 \ I_m = \frac{0.636 \ V_m}{R} \qquad\qquad \textbf{5.14}$$

The RMS output current is given by

$$I_{oRMS} = \frac{I_m}{\sqrt{2}} = 0.707 \ I_m \qquad\qquad \textbf{5.15}$$

The graph of the voltage across the diode in Figure 5.4(d) shows that each diode must withstand a reverse voltage equal to $2V_m$. The PIV rating for the diodes used in this circuit is therefore given by:

$$\text{PIV rating for diodes:} \geq 2 \ V_m \qquad\qquad \textbf{5.16}$$

The average diode current is

$$I_{D1(avg.)} = I_{D2(avg.)} = I_{o(avg.)}/2 = I_m/\pi \qquad\qquad \textbf{5.17}$$

The RMS diode current is

$$I_{DRMS} = I_m/2 \qquad\qquad \textbf{5.18}$$

The average or DC power delivered to the load is given by

$$
\begin{aligned}
P_{o(avg.)} &= V_{o(avg.)} * I_{o(avg.)} \\
&= \frac{2 \ V_m}{\pi} * \frac{2 \ I_m}{\pi} = \frac{4 \ V_m * V_m}{\pi^2 * R} \\
&= \frac{4 \ V_m^2}{\pi^2 * R} \qquad\qquad \textbf{5.19}
\end{aligned}
$$

The AC power input is given by

$$
\begin{aligned}
P_{AC} &= V_{RMS} * I_{RMS} \\
&= \frac{V_m}{\sqrt{2}} * \frac{I_m}{\sqrt{2}} \\
&= \frac{V_m * V_m}{2 * R} \\
&= \frac{V_m^2}{2 \ R} \qquad\qquad \textbf{5.20}
\end{aligned}
$$

Example 5.4   The full-wave rectifier shown in Figure 5.4(a) is supplied from a 120 V, 60 Hz source. If the load resistance is 10 Ω, find

a) the maximum load current
b) the average load voltage
c) the average load current
d) the average diode current
e) the RMS load current
f) the power to the load
g) the PIV rating for the diode
h) the ripple frequency

Solution   Peak voltage:

$$V_m = \sqrt{2} \ V_{RMS} = (1.414)(120) = 169.7 \text{ V}$$

a) maximum load current

$$I_m = V_m/R = 169.7/10 = 16.97 \text{ A}$$

b) average load voltage

$$V_{o(avg.)} = 0.636 \ V_m = 0.636 * 169.7 = 108 \text{ V}$$

c) average load current

$$I_{o(avg.)} = 0.636 \ I_m = 0.636 * 16.97 = 10.8 \text{ A}$$

d) average diode current

$$I_{D1(avg.)} = I_{D2(avg.)} = I_{o(avg.)}/2 = 10.8/2 = 5.4 \text{A}$$

e) RMS load current

$$I_{oRMS} = I_m/\sqrt{2} = 16.97/\sqrt{2} = 12.0 \text{ A}$$

f) power to the load

$$I^2 R = 12^2 (10) = 1440 \text{ W}$$

g) PIV rating for diode = $2 \ V_m = 339.4 \ V$
h) Since two cycles of output occurs for every cycle of input,
ripple frequency = 2 * frequency of the ac input = 2 * 60 Hz = 120 Hz

---

Example 5.5   The full-wave rectifier shown in Figure 5.4(a) is supplied from a 50 V source. If the load resistance is 100 Ω, find

a) the average load voltage
b) the maximum load current
c) the average load current
d) the RMS value of load current
e) the PIV rating for the diode
f) the average power delivered to the load
g) the AC power input
h) the rectifier efficiency

    i) the form factor
    j) the pulse number
    k) the ripple factor
    l) the conduction angle
    m) the power factor

Solution   Peak voltage

$$V_m = \sqrt{2} \ (50) = 70.7 \text{ V}$$

a) average load voltage $V_{o(avg.)} = 0.636 \ V_m = 0.636 * 70.7 = 45 \text{ V}$

b) maximum load current $I_m = V_m/R = 70.7/100 = 707 \text{ mA}$

c) average load current $I_{o(avg.)} = 0.636 * 0.707 = 450 \text{ mA}$

d) RMS load current $I_{oRMS} = I_m/\sqrt{2} = 0.707/1.414 = 500 \text{ mA}$

e) PIV rating for the diode $\text{PIV} \geq 2 \ V_m = 141.4 \text{ V}$

f) average power delivered to the load

$$P_{o(avg.)} = \frac{4 \ V_m^2}{\pi^2 * R}$$

$$= \frac{4 \ (70.7)^2}{\pi^2 * 100}$$

$$= 20.25 \text{ W}$$

g) AC power input $P_{AC} = \dfrac{V_m^2}{2 \ R} = \dfrac{70.7^2}{2 * 100} = 25 \text{ W}$

h) rectifier efficiency $\eta = \dfrac{P_{(avg.)}}{P_{AC}}$

$$= 4 \ \frac{V_m^2}{\pi^2 * R} * \frac{2 \ R}{V_m^2}$$

$$= \frac{8}{\pi^2}$$

$$= 0.810 \text{ or } 81\%$$

i) form factor $FF = \dfrac{V_{o(RMS)}}{V_{o(avg.)}} = \dfrac{50}{45} = 1.11$

j) pulse number $p = \dfrac{\text{fundamental ripple frequency}}{\text{AC source frequency}} = 120/60$

    or $p = \dfrac{360°}{180°} = 2$

k) ripple factor $RF = \sqrt{\dfrac{I_{RMS}^2}{I_{DC}^2} - 1}$

$$= \sqrt{\frac{(0.5)^2}{(0.45)^2} - 1}$$

$$= 0.484$$

l) conduction angle $\theta = 180°$

m) power factor $PF = \dfrac{(V_m/\pi)(I_{oavg.}/2)}{(V_m/2\sqrt{2})(I_{oavg.}/\sqrt{2})} = 2/\pi = 0.64$

5.3.2 With an Inductive (RL) Load

Adding an inductance in series with the load resistance changes the voltage and current waveform, as shown in Section 5.2.2. Note that the load current continues to flow for a period of time after the diode is reverse-biased, and this results in a decrease in the magnitude of the average output voltage.

Figure 5.5 shows a center-tap full-wave rectifier with an inductive load and its associated voltage and current waveforms. As can be seen in Figure 5.5(b),

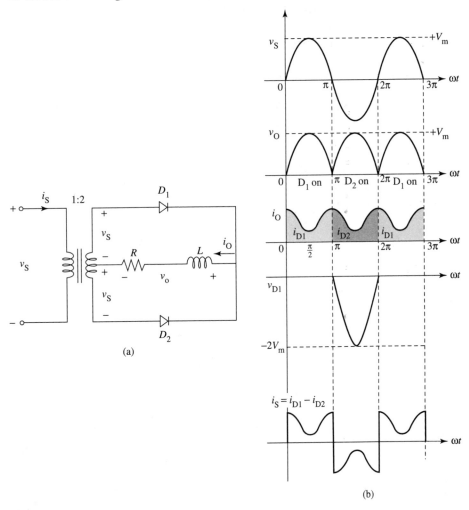

Figure 5.5
Center-tap rectifier with an inductive load (a) circuit diagram (b) voltage and current waveform

the load current is at its maximum when the source voltage (v_s) is zero. When v_s increases in magnitude during the interval from 0 to $\pi/2$, the inductor opposes the flow of current and stores energy in its magnetic fields. At $\pi/2$, when v_s has reached its maximum, the load current is at its minimum. In the interval between $\pi/2$ and π, where the source voltage decreases in magnitude, the induced voltage across the inductor opposes any decrease in the load current by aiding the source voltage. Therefore, the load current increases to a maximum value when $v_s = 0$. The process continues for every half-cycle of the rectified sine wave. The load current never reduces to zero since the energy stored in the magnetic field maintains the current flow.

The equations are similar to those for the center-tap rectifier with a resistive load. The average value of the load voltage is:

$$V_{o(avg.)} = (2\ V_m)/\pi = 0.636\ V_m \tag{5.21}$$

Figure 5.6
Current waveforms for a highly inductive load

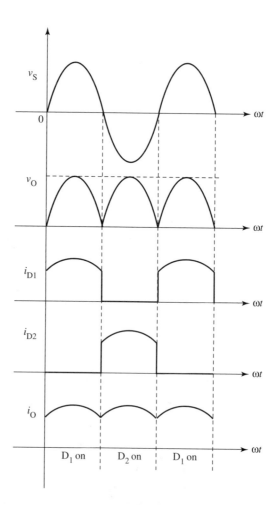

The average value of the load current is:

$$I_{o(avg.)} = \frac{2\ V_m}{\pi\ R} = 0.636\ \frac{V_m}{R} \qquad \textbf{5.22}$$

If the load inductance is sufficiently large, the load current is nearly constant, as shown in Figure 5.6.

The RMS value of the load current is:

$$I_{o(RMS)} = I_{o(avg.)} = V_{o(avg.)}/R \qquad \textbf{5.23}$$

$$I_{D(RMS)} = \frac{I_{o(avg.)}}{2} \qquad \textbf{5.24}$$

Example 5.6   The full-wave rectifier shown in Figure 5.5(a) is supplied from a 115 V source. If the load resistance is 100 Ω, find
a) the output DC voltage
b) the average load current
c) the power delivered to the load
d) the AC input power
e) the rectifier efficiency
f) the ripple factor
g) the form factor

Solution   The peak voltage is

$$V_m = \sqrt{2}\ (115) = 162.6\ V$$

a) The output DC voltage is

$$V_{o(avg.)} = 0.636\ V_m = 0.636 * 162.6 = 103.4\ V$$

b) The average load current is

$$I_{o(avg.)} = 103.4/100 = 1.03\ A$$

c) power delivered to the load

$$
\begin{aligned}
P_L &= V_{o(avg.)} * I_{o(avg.)}\\
&= 103.4 * 1.03\\
&= 107\ W
\end{aligned}
$$

d) AC input power

$$
\begin{aligned}
P_{AC} &= V_{RMS} * I_{RMS}\\
&= \frac{V_m^2}{2\ R}\\
&= \frac{(162.6)^2}{2\ (100)}\\
&= 132.2\ W
\end{aligned}
$$

e) rectifier efficiency $\qquad \eta = \dfrac{P_L}{P_{AC}} = \dfrac{107}{132.2} = 0.81\ \text{or}\ 81\%$

f) ripple factor

$$RF = \sqrt{\left[\frac{V_{RMS}}{V_{dc}}\right]^2 - 1}$$

$$= \sqrt{\left[\frac{(115)}{103.4}\right]^2 - 1}$$

$$= 0.48$$

g) form factor $$FF = \frac{V_{RMS}}{V_{o(avg.)}} = \frac{115}{103.4} = 1.11$$

5.4   The Full-Wave Bridge Rectifier

5.4.1   With a Resistive Load

Full-wave rectification can also be obtained by using a bridge rectifier like the one shown in Figure 5.7. This full-wave bridge rectifier uses four diodes. During the positive half-cycle of the source voltage (Figure 5.8(a), diodes D_2 and D_3 are forward-biased and can therefore be replaced by a closed switch. The load current flow during this period is through D_2 and the load R and then through D_3 and back to the source. This causes a positive drop across R.

Figure 5.7
A full-wave bridge rectifier circuit

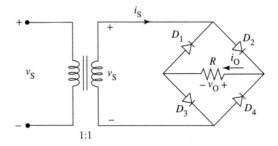

Figure 5.8(b) shows the full-wave bridge circuit during the negative half-cycle of the source voltage. Now diodes D_1 and D_4 are forward-biased and can therefore be replaced by closed switches. The load current path is now through D_4, through R, and then through D_1 to the source. The current path through R is in the same direction as before, so there is a positive drop across R during both half-cycles. Thus the full-wave bridge rectifier causes the load current to flow during both half-cycles. Figure 5.9 shows the appropriate waveforms.

The average and RMS values of voltage and current are similar to those for the full-wave center-tap case. However, the waveform of the voltage across the diode in Figure 5.9 shows that each diode must withstand a reverse voltage equal to V_m only.

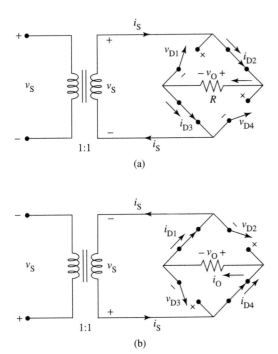

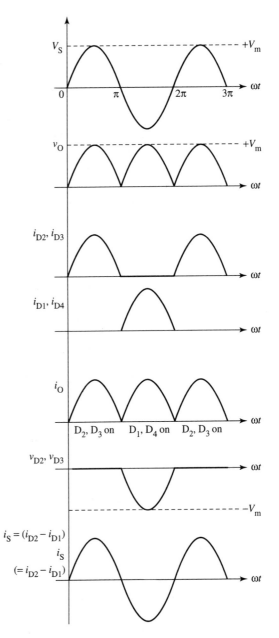

Figure 5.8
Bridge rectifier (a) positive half-cycle
(b) negative half-cycle

Figure 5.9
Bridge rectifier waveforms

PIV rating for diodes $\geq V_m$ **5.25**

Because there are two paths for the load current, the average diode current is just half of the average load current:

$$I_{D(avg.)} = I_{(avg.)}/2$$ **5.26**

Example 5.7   The full-wave bridge rectifier of Figure 5.7 is supplied by a 120 V source. If the load resistance is 10.8 Ω, find
a) the peak load voltage
b) the DC voltage across the load
c) the DC load current
d) the average current in each diode
e) the average output power
f) the rectifier efficiency
g) the ripple factor
h) the power factor

Solution   a) peak load voltage $V_m = \sqrt{2}\ V_{RMS} = (1.414)(120) = 170$ V

b) DC voltage across the load $V_{o(avg.)} = 0.636 * 170 = 108$ V

c) DC load current $I_{o(avg.)} = 108/10.8 = 10$ A

d) average current in each diode   Since the diodes carry the load current on alternate half-cycles,

$$I_{D(avg.)} = I_{o(avg.)}/2 = 10/2 = 5 \text{ A}$$

e) average output power $P_{o(avg.)} = V_{o(avg.)} * I_{o(avg.)} = 108 * 10 = 1080$ W

f) rectifier efficiency $\eta = 8/\pi^2 = 0.81$ or 81%

g) ripple factor RF $= \sqrt{\left[\dfrac{V_{RMS}}{V_{DC}}\right]^2 - 1} = \sqrt{\left[\dfrac{120}{108}\right]^2 - 1} = 0.482$

h) power factor   PF $= \dfrac{P}{S} = \dfrac{V_{o(avg.)} * I_{o(avg.)}}{V_{RMS} * I_{RMS}} = \dfrac{108 * 10}{120 * 10} = 0.9$

Example 5.8   The full-wave bridge rectifier of Figure 5.7 is supplied by a 120 V, 60 Hz source. If the load resistance is 10 Ω, find
a) the average load voltage
b) the maximum load current
c) the average load current
d) the RMS load current
e) the power to the load
f) the PIV rating for the diodes
g) the average diode current
h) the ripple frequency

Solution   The peak load voltage is

$$V_m = \sqrt{2}\ V_{RMS} = 1.414 * 120 = 170\ V$$

a) average load voltage $V_{o(avg.)} = 0.636 * 170 = 108\ V$

b) maximum load current $I_m = \dfrac{V_m}{R} = \dfrac{170}{10} = 17\ A$

c) average load current $I_{o(avg.)} = 0.636 * 17 = 10.8\ A$

d) RMS load current $= 17/\sqrt{2} = 12.0\ A$

e) power to the load $I_{RMS}^2\ R = 12^2 * 10 = 1440\ W$

f) PIV rating for diodes $V_m = 170\ V$

g) average diode current $I_{D(avg.)} = I_{o(avg.)}/2 = 10.8/2 = 5.4\ A$

h) Since two cycles of output occur for every cycle of input,
   ripple frequency = 2 * frequency of the AC input = 2 * 60 Hz = 120 Hz

---

Example 5.9   The full-wave bridge rectifier of Figure 5.7 is supplied by a 50 V source. If the
load resistance is 100 Ω, find
a) the average voltage across the load
b) the average load current
c) the RMS current
d) the pulse number
e) the conduction angle
f) the PIV rating for diodes
g) the form factor
h) the ripple factor

Solution   The peak load voltage is: $V_m = \sqrt{2}\ V_{RMS} = 1.414 * 50 = 70.7\ V$

a) average voltage across the load $V_{o(avg.)} = 0.636 * 70.7 = 45\ V$

b) average load current $I_{o(avg.)} = 45/100 = 0.45\ A$

c) RMS current $I_{RMS} = \dfrac{I_m}{\sqrt{2}} = \dfrac{V_m}{\sqrt{2}\ R} = 0.5\ A$

d) pulse number $p = 2$

e) conduction angle $\theta = 180°$

f) PIV rating $PIV = V_m = 70.7\ V$

g) form factor $FF = \dfrac{V_{RMS}}{V_{o(avg.)}} = \dfrac{50}{45} = 1.11$

h) ripple factor $RF = \sqrt{\left[\dfrac{V_{RMS}}{V_{DC}}\right]^2 - 1} = \sqrt{\left[\dfrac{50}{45}\right]^2 - 1} = 0.483$

5.4.2 With an Inductive (RL) Load

Adding an inductance in series with the load resistance changes the voltage and current waveform. Figure 5.10 shows a bridge rectifier with an inductive load. Let us assume the inductance L to be approximately equal to R. The load current no longer consists of half sine waves, but the average current is still the same as given by Equation 5.22:

$$I_{o(avg.)} = \frac{2\,I_m}{\pi} = \frac{2\,V_m}{\pi\,R}$$

The AC line current is no longer sinusoidal but is approximately a square wave. Figure 5.11 shows the voltage and current waveforms.

Figure 5.10
Bridge rectifier circuit (alternative drawing) with inductive load

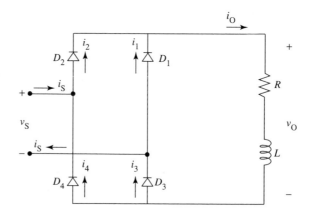

If we increase the load inductance until it is much larger than R, the ripple across the load becomes small. If we assume an infinite load inductance, the load current becomes constant and the circuit behavior is as shown by the waveforms in Figure 5.12. Diodes D_2 and D_3 conduct a constant load current on the positive half-cycle, while diodes D_1 and D_4 do the same on the negative half-cycle.

The source current is given by:

$$i_s = i_3 - i_1 = i_2 - i_4 \qquad\qquad \textbf{5.27}$$

Although not a sine wave, the AC source current is an alternating waveform of rectangular shape. The load is always connected to the source, but the connection is reversed on alternate half-cycles.

The output voltage (v_o) is a full-wave rectified waveform. Its average value can be determined from

$$V_{o(avg.)} = V_{L(avg.)} + V_{R(avg.)} \qquad\qquad \textbf{5.28}$$

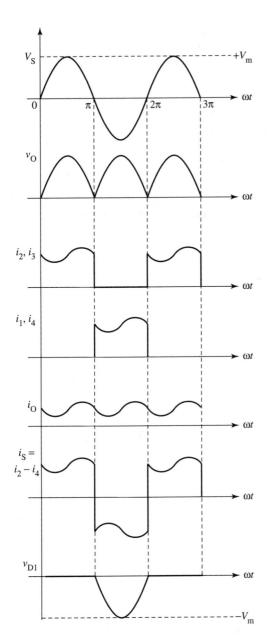

Figure 5.11
Waveforms for Figure 5.10 ($L \approx R$)

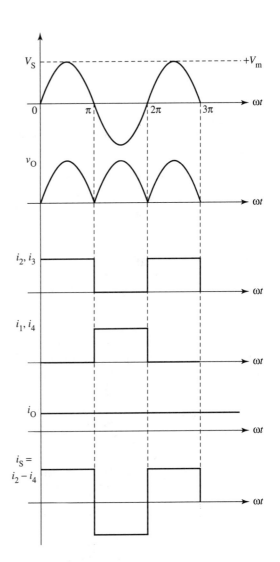

Figure 5.12
Waveforms for Figure 5.10 ($L \gg R$)

where v_R is the voltage across the resistive portion of the load and v_L is the induced voltage across the inductance. In periodic operation, $V_{L(avg.)}$, the average voltage across an inductor, must be zero. Therefore

$$V_{R(avg.)} = V_{o(avg.)} = (2\ V_m)/\pi$$
$$= 0.636\ V_m \hspace{4cm} \textbf{5.29}$$

The average load voltage is the same as for the resistive case. The average load current can be determined from

$$I_{o(avg.)} = V_{R(avg.)}/R$$
$$= 0.636\ (V_m/R) \hspace{3.5cm} \textbf{5.30}$$

Since the load current now is essentially constant, its RMS, maximum, and average values are the same:

$$I_{RMS} = I_{o(max.)} = I_{o(avg.)} \hspace{3cm} \textbf{5.31}$$

Since the diodes in the bridge conduct on alternate half-cycles, the average current in each diode is

$$I_{D(avg.)} = I_{o(avg.)}/2 \hspace{3.5cm} \textbf{5.32}$$

and the RMS current in each diode is

$$I_{D(RMS)} = \frac{I_{o(avg.)}}{\sqrt{2}} \hspace{3.5cm} \textbf{5.33}$$

Example 5.10 A full-wave bridge rectifier with an *RL* load is connected to a 120 V source. If the load resistance is 10 Ω and $L \gg R$, find
a) the average load voltage
b) the average load current
c) the maximum load current
d) the RMS value of load current
e) the average current in each diode
f) the RMS current in each diode
g) the power supplied to the load

Solution   The peak load voltage is

$$V_m = \sqrt{2}\ V_{RMS} = (1.414)(120) = 170\ V$$

a) average load voltage $V_{o(avg.)} = 0.636 * 170 = 108$ V

b) average load current $V_{o(avg.)}/R = 108/10 = 10.8$ A

c) maximum load current = average load current = 10.8 A

d) RMS value of load current = average load current = 10.8 A

e) average current in each diode $I_{D(avg.)} = I_{o(avg.)}/2 = 10.8/2 = 5.4$ A

f) RMS current in each diode $I_{D(RMS)} = \dfrac{I_{o(avg.)}}{\sqrt{2}} = \dfrac{10.8}{\sqrt{2}} = 7.6$ A

g) power supplied to the load $I_{RMS}^2 R = 10.8^2 * 10 = 1167$ W

5.5 Problems

5.1 Define a rectifier.

5.2 Sketch the schematic of a half-wave rectifier and explain its operation.

5.3 A half-wave rectifier has an average output voltage of 120 V. Find the PIV rating required for the diode.

5.4 A half-wave rectifier has a source voltage of 120 V. If the load resistance is 20 Ω, find
(a) the average load voltage
(b) the maximum diode current
(c) the average diode current
(d) the PIV rating for diode
(e) the average load power
(f) the ripple factor
(g) the form factor

5.5 A half-wave rectifier with an inductive load and a FWD has an input voltage of 120 V, a load resistance of 5 Ω, and a load inductance of 20 mH. Find
(a) the average load voltage
(b) the average load current
(c) the diode current and voltage ratings

5.6 What are the main advantages of a full-wave rectifier over a half-wave rectifier?

5.7 In a full-wave center-tap rectifier, $V_s = 208$ V and the load resistance $R = 100$ Ω. Find
(a) the maximum load voltage
(b) the average load voltage
(c) the maximum diode current
(d) the average diode current
(e) the PIV rating for diode
(f) the RMS value of the output voltage
(g) the ripple factor

5.8 A 2:1 step-down transformer supplies a full-wave center-tap rectifier from a 400 V source. If the load resistance $R = 500$ Ω and $L = 1$ H, find
(a) the average load voltage
(b) the average load current

(c) the maximum diode current

(d) the power delivered to the load

(e) the PIV rating for diode

5.9 For a single-phase full-wave rectifier with an *RL* load, the voltage across the entire secondary winding is 400 V. If the load resistance is 100 Ω, find

(a) the average load voltage

(b) the RMS load voltage

(c) the maximum diode current

(d) the average diode current

(e) the PIV rating for diode

5.10 Sketch the output waveform of Figure 5.7 with all diodes reversed.

5.11 In a full-wave bridge rectifier, V_s = 240 V and the load resistance R = 10 Ω. Find

(a) the average load voltage

(b) the average load current

(c) the RMs load current

(d) the average diode current

(e) the average output power

(f) the PIV rating for diode

5.12 Sketch the output waveform of Figure 5.13. Find the average output voltage and the PIV rating for the diodes.

Figure 5.13
See Problem 5.12

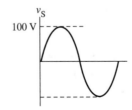

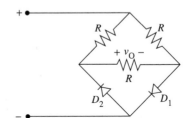

5.6 Equations

$$V_{o(avg.)} = 0.318 \ V_m \qquad \textbf{5.1}$$

$$I_{o(avg.)} = I_m/\pi \qquad \textbf{5.2}$$

$$I_{oRMS} = I_m/2 \qquad \textbf{5.3}$$

$$\text{PIV rating:} \geq V_m \qquad \textbf{5.4}$$

$$P_{o(avg.)} = \frac{V_m^2}{\pi^2 * R} \qquad \textbf{5.5}$$

$$P_{AC} = \frac{V_m^2}{4 \ R} \qquad \textbf{5.6}$$

$$\eta = \frac{P_{(avg.)}}{P_{AC}}$$ **5.7**

$$\text{form factor (FF)} = \frac{V_{oRMS}}{V_{o(avg.)}}$$ **5.8**

$$\text{pulse number} = \frac{\text{fundamental ripple frequency}}{\text{AC source frequency}}$$ **5.9**

$$RF = \sqrt{\frac{I_{RMS}^2}{I_{DC}^2 - 1}}$$ **5.10**

$$V_{o(avg.)} = \frac{V_m}{2\pi}(1\text{-cos }\theta)$$ **5.11**

$$I_{o(avg.)} = \frac{V_m}{2\pi R}(1\text{-cos }\theta)$$ **5.12**

$$V_{o(avg.)} = (2\ V_m)/\pi = 0.636\ V_m$$ **5.13**

$$I_{o(avg.)} = (2\ I_m)/\pi = 0.636\ I_m = \frac{0.636\ V_m}{R}$$ **5.14**

$$I_{oRMS} = \frac{I_m}{\sqrt{2}} = 0.707\ I_m$$ **5.15**

PIV rating for diodes: $\geq 2\ V_m$ **5.16**

$$I_{D1(ave.)} = I_{D2(ave.)} = I_{o(ave.)}/2 = I_m/\pi$$ **5.17**

$$I_{D(RMS)} = I_m/2$$ **5.18**

$$P_{o(avg.)} = \frac{4\ V_m^2}{\pi^2 * R}$$ **5.19**

$$P_{AC} = \frac{V_m^2}{2\ R}$$ **5.20**

$$V_{o(avg.)} = (2\ V_m)/\pi = 0.636\ V_m$$ **5.21**

$$I_{o(avg.)} = \frac{2\ V_m}{\pi\ R} = 0.636\ \frac{V_m}{R}$$ **5.22**

$$I_{(RMS)} = I_{o(avg.)} = V_{o(avg.)}/R$$ **5.23**

$$I_{D(RMS)} = I_{o(avg.)}/\sqrt{2}$$ **5.24**

PIV rating for diodes: $\geq V_m$ **5.25**

$$I_{D(avg.)} = I_{o(avg.)}/2$$ **5.26**

$$i_s = i_3 - i_1 = i_2 - i_4$$ **5.27**

$$V_{o(avg.)} = V_{L(avg.)} + V_{R(avg.)}$$ **5.28**

$$V_{R(avg.)} = V_{o(avg.)} = (2\ V_m)/\pi$$ **5.29**

$$I_{o(avg.)} = V_{R(avg.)}/R$$ **5.30**

$$I_{RMS} = I_{o(max.)} = I_{o(avg.)}$$ **5.31**

$$I_{D(avg.)} = I_{o(avg.)}/2$$ **5.32**

$$I_{D(RMS)} = \frac{I_{o(avg.)}}{\sqrt{2}}$$ **5.33**

6

Single-Phase Controlled Rectifiers

Chapter Outline

Learning Objectives

After completing this chapter, the student should be able to

- describe with the help of waveforms the operation of a half-wave controlled rectifier with resistive and inductive loads
- describe with the help of waveforms the operation of a full-wave controlled center-tap rectifier with resistive and inductive loads
- describe with the help of waveforms the operation of a full-wave controlled

bridge rectifier with resistive and inductive loads

- discuss the advantages and disadvantages of a center-tap converter versus a bridge converter
- describe the operation of a half-controlled bridge rectifier
- describe the operation of a dual converter

6.1   Introduction

To build a *controlled rectifier* or a *phase-controlled rectifier,* the diodes in the rectifier circuit in Chapter 5 are replaced by SCRs. These circuits produce a variable DC output voltage whose magnitude is varied by phase control, that is, by controlling the duration of the conduction period by varying the point at which a gate signal is applied to the SCR.

Unlike a diode, an SCR will not automatically conduct when the anode-to-cathode voltage becomes positive—a gate pulse must be provided. If we adjust the delay time of the gate pulse, and if this process is done repeatedly, then the rectifiers output can be controlled. This process is called *phase control.*

Controlled rectifiers, or *converters,* as they are generally called, are broadly classified into full-controlled and half-controlled types. The full-controlled or two-quadrant type uses SCRs as the rectifying devices. The DC current is unidirectional, but the DC voltage may have either polarity. With one polarity, the flow of power is from the AC source to the DC load—this is called *rectification.* With a reversal of the DC voltage by the load, the flow of power is from the DC source to the AC supply; this process is called *inversion.*

If we replace half of the SCRs with diodes, the circuit is classified as a half-controlled or semiconverter circuit. Such a circuit also allows the average value of the DC output voltage to be varied by phase control of the SCR. However, the polarity of the DC output voltage and the direction of current cannot change, that is, the flow of power is from the AC source to the DC load. Converters of this type are also called *one-quadrant converters.*

Controlled rectifiers provide DC power for various applications, such as DC motor speed control, battery charging, and high-voltage DC transmission. Phase control is suited for frequencies less than 400 Hz, typically 60 Hz. The main drawback of phase control is *radio frequency interference (RFI).* The chopped half-sine wave produces strong harmonics that interfere with radio, television, and other communication equipment.

In this chapter we will study controlled rectifiers, ranging from the simplest configuration, the half-wave rectifier (which is seldom used in power electronics applications because of the high ripple voltage content of its output), to the center-tap and the bridge rectifier circuit.

6.2   Half-Wave Controlled Rectifiers

6.2.1   With a Resistive Load

Figure 6.1(a) shows a half-wave controlled rectifier circuit with a resistive load. During the positive half-cycle of the supply voltage, the SCR is forward-biased and will conduct if a trigger pulse is applied to the gate. If the SCR turns on at t_o, load current flows and the output voltage v_o will be the same as the input voltage. At time $t = \pi$, the current falls *naturally* to zero, since the SCR is reverse-biased. During the negative half-cycle, the SCR blocks the flow of current, and no voltage is applied to the load. The SCR stays off until the gate signal is applied again at $(t_o + 2\pi)$. The period from 0 to t_o in Figure 6.1(b) represents the time in the positive half-cycle when the SCR is off. This angle (measured in degrees) is called the *firing angle* or *delay angle* (α). The SCR conducts from t_o to π; this angle is called the *conduction angle* (θ).

The average or DC value of the load voltage is given by

$$V_{o(avg.)} = \frac{V_m(1 + \cos \alpha)}{2\pi} \qquad\qquad \textbf{6.1}$$

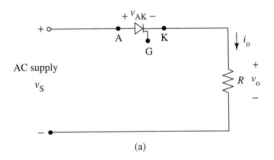

(a)

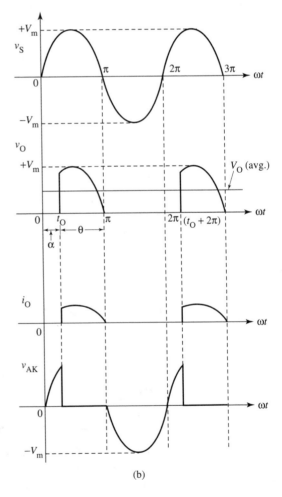

(b)

Figure 6.1
Half-wave controlled rectifier (a) circuit (b) voltage and current waveforms

where

V_m = maximum value of the AC source voltage = $\sqrt{2}\ V_s$

V_s = RMS value of the AC source voltage

Similarly, the average output current is

$$I_{o(avg.)} = \frac{I_m(1 + \cos \alpha)}{2\ \pi}$$

$$= \frac{V_m(1 + \cos \alpha)}{2\ \pi\ R} \qquad \textbf{6.2}$$

The RMS value of the load current is given by

$$I_{RMS} = \left(\frac{I_m}{2}\right)\left[1 - \frac{\alpha}{\pi} + \frac{\sin 2\ \alpha}{2\pi}\right]^{1/2} \qquad \textbf{6.3}$$

These equations tell us that the magnitude of the output voltage is controlled by the firing angle. Increasing α by firing the SCR later in the cycle lowers the voltage, and vice versa. The maximum output voltage, $V_{do} = V_m/\pi$, occurs when $\alpha = 0°$. This is the same voltage as for a half-wave diode circuit. Therefore, if the SCR is fired at $\alpha = 0°$, the circuit acts like a diode rectifier.

The normalized average voltage is

$$V_n = V_{o(avg.)}/V_{do} = \frac{V_m(1 + \cos \alpha)/2\ \pi}{V_m/\pi} = \frac{1 + \cos \alpha}{2} \qquad \textbf{6.4}$$

V_n as a function of α is known as the *control characteristic* of the rectifier and is shown in Figure 6.2.

Figure 6.2
Control characteristic for a half-wave rectifier

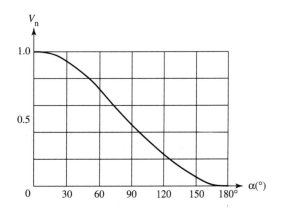

Example 6.1   A half-wave controlled rectifier is supplied from a 120 V source. If the load resistance is 10 Ω, find the load voltage and power to the load for the following delay angles:

a) $\alpha = 0°$

b) $\alpha = 45°$

c) $\alpha = 90°$
d) $\alpha = 135°$
e) $\alpha = 180°$

Solution   peak load voltage $= V_m = \sqrt{2}\ V_S = 1.414 * 120 = 170$ V

$$\text{average load voltage} = \frac{V_m\ (1 + \cos \alpha)}{2\pi}$$

a) For $\alpha = 0°$, we get

$$V_{o(avg.)} = \frac{170\ (1 + \cos 0°)}{2\pi} = 54.0\ \text{V}$$

$$P_L = V_{o(avg.)}^2/R = 54.0^2/10 = 293\ \text{W}$$

b) For $\alpha = 45°$, we get

$$V_{o(avg.)} = \frac{170\ (1 + \cos 45°)}{2\pi} = 46.2\ \text{V}$$

$$P_L = V_{o(avg.)}^2/R = 46.2^2/10 = 213\ \text{W}$$

c) For $\alpha = 90°$, we get

$$V_{o(avg.)} = 27.1\ \text{V}$$

$$P_L = 73.2\ \text{W}$$

d) For $\alpha = 135°$, we get

$$V_{o(avg.)} = 7.92\ \text{V}$$

$$P_L = 6.3\ \text{W}$$

e) $\alpha = 180°$, we get

$$V_{o(avg.)} = 0\ \text{V}$$

$$P_L = 0\ \text{W}$$

Therefore, the power can be varied from zero to a maximum of 293 W.

Example 6.2   A half-wave controlled rectifier connected to a 150 V, 60 Hz source is supplying a resistive load of 10 Ω. If the delay angle α is 30°, find:

a) the maximum load current
b) the average load voltage
c) the average load current
d) the RMS load current
e) the power supplied to the load
f) the conduction angle
g) the ripple frequency
h) the power factor

Solution   peak load voltage $= V_m = \sqrt{2}\ V_s = 1.414 * 150 = 212$ V

a) maximum load current

$$I_m = \frac{V_m}{R} = \frac{212}{10} = 21.2 \text{ A}$$

b) average load voltage $= \dfrac{V_m(1 + \cos \alpha)}{2\pi} = \dfrac{(212)\ (1 + \cos 30°)}{2\pi} = 63 \text{ V}$

c) average load current $= \dfrac{(I_m)(1 + \cos \alpha)}{2\pi} = \dfrac{(21.2)(1 + \cos 30°)}{2\pi} = 6.3 \text{ A}$

d) RMS load current

$$I_{RMS} = \left(\frac{I_m}{2}\right) \sqrt{\left[1 - \frac{\alpha}{\pi} + \frac{\sin 2\alpha}{2\pi}\right]}$$

$$= \frac{(21.2)}{2} \sqrt{\left[1 - \frac{30}{180} + \frac{\sin 60}{2\pi}\right]}$$

$$= 10.5 \text{ A}$$

e) power supplied to the load $= I_{RMS}^2\ R = 10.5^2(10) = 1094 \text{ W}$

f) conduction angle

$$\theta = 180° - \alpha = 180° - 30° = 150°$$

g) ripple frequency

$$f_r = \text{input supply frequency} = 60 \text{ Hz}$$

h) $S = V_S * I_{RMS} = 150 * 10.5 = 1575 \text{ VA}$

$$PF = \frac{P}{S} = \frac{1094}{1575} = 0.69$$

Example 6.3 A half-wave controlled rectifier is connected to a 120 V source. Calculate the firing angle necessary to deliver 150 W of power to a 10 Ω load.

Solution $V_{o(avg.)} = \dfrac{V_m(1 + \cos \alpha)}{2\pi}$

Rearranging,

$$V_m (1 + \cos \alpha) = 2\pi\ V_{o(avg.)}$$

$$1 + \cos \alpha = \frac{2\pi\ V_{o(avg.)}}{V_m}$$

$$\cos \alpha = \frac{2\pi\ V_{o(avg.)}}{V_m} - 1$$

$$\alpha = \cos^{-1}\left\{\frac{2\pi\ V_{o(avg.)}}{V_m} - 1\right\}$$

Now,

$$V_m = \sqrt{2} * 120 = 170 \text{ V}$$

and

$$P_{avg.} = \frac{V^2_{o(avg.)}}{R}$$

$$V^2_{o(avg.)} = P_{avg.} * R = 150 * 10 = 1500$$

$$V_{o(avg.)} = \sqrt{1500} = 38.7 \text{ V}$$

Therefore

$$\alpha = \cos^{-1} \left\{ \frac{2\pi\ 38.7}{170} - 1 \right\}$$

$$= 64.5°$$

6.2.2   With an Inductive (RL) Load

A half-wave rectifier with a load consisting of R and L is shown in Figure 6.3(a). If the SCR is triggered at a firing angle of α, the load current increases slowly, since the inductance in the load forces the current to lag the voltage. The voltage across the load (v_o) is positive, and the inductor is storing energy in its magnetic field. When the applied voltage becomes negative, the SCR is reverse-biased. However, the energy stored in the magnetic field of the inductor is returned and maintains a forward-decaying current through the load. The cur-

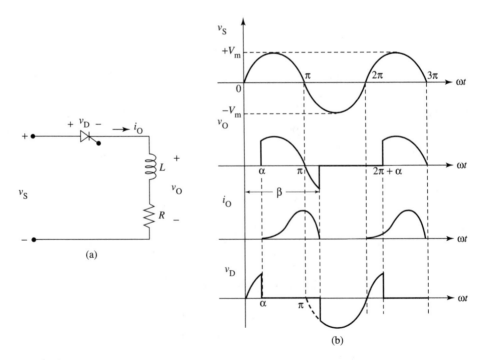

Figure 6.3
Half-wave rectifier with an RL load (a) circuit (b) waveforms for voltage and current

rent continues to flow until β (called the *advance angle*), when the SCR turns off. The voltage across the inductor then changes polarity, and the voltage across the load becomes negative. As a result, the average output voltage becomes less than it would be with a purely resistive load. The waveforms for output voltage and current are shown in Figure 6.3 (b); they contain a significant amount of ripples.

The average load voltage is given by:

$$V_{o(avg.)} = \frac{V_m(\cos \alpha - \cos \beta)}{2\pi}$$ **6.5**

6.2.3 With a Freewheeling Diode

To cut off the negative portion of the instantaneous output voltage and smooth the output current ripple, a freewheeling diode is used as shown in Figure 6.4. When the load voltage tends to reverse, the FWD becomes forward-biased and turns on. The SCR then becomes reverse-biased and turns off. Therefore, the

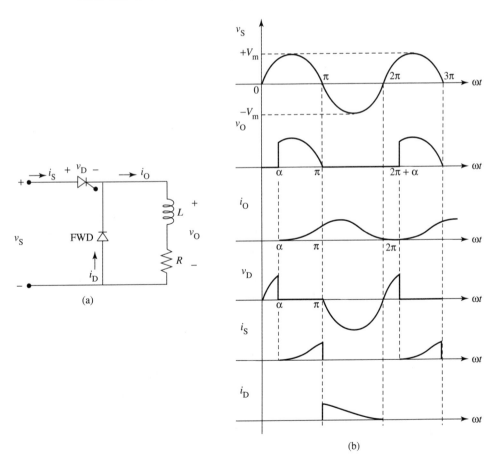

Figure 6.4
RL load with FWD (a) circuit (b) waveforms

current that was flowing from the source to the load through the SCR now free-wheels between the load and the diode. Note that the current continues to flow in the load after the SCR is turned off, due to the energy stored in the inductor. The output voltage is the same as in a circuit with a resistive load. The average value of the output voltage is given again by Equation 6.1.

6.3   Full-Wave Controlled Center-Tap Rectifiers

6.3.1   With a Resistive Load

Figure 6.5 shows the basic arrangement of a single-phase, center-tap controlled rectifier with a resistive load. Phase control of both the positive and the nega-tive halves of the AC supply is now possible, thus increasing the DC voltage and reducing the ripple compared to those of half-wave rectifiers.

Figure 6.5
Full-wave center-tap controlled rectifier circuit

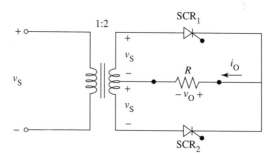

During the positive half-cycle of the input voltage, SCR$_1$ is forward-biased. If we apply the gate signal at α, SCR$_1$ turns on. The output voltage (v_o) follows the input voltage. The load current ($i_o = v_o/R$) has the same waveform as the load voltage. At π, when the current through SCR$_1$ becomes zero, it turns off naturally. During the negative half-cycle, SCR$_2$ is forward-biased. SCR$_2$ is fired at ($\pi + \alpha$). The output voltage again follows the input voltage. The current through SCR$_2$ becomes zero at 2π, and it turns off. SCR$_1$ is fired again at ($2\pi + \alpha$) and SCR$_2$ at ($3\pi + \alpha$), and the cycle repeats. Figure 6.6 shows the resulting voltage and current waveforms.

The average value of the load voltage is twice that given by Equa-tion 6.1:

$$V_{o(avg.)} = \frac{V_m(1 + \cos \alpha)}{\pi} \qquad\qquad \textbf{6.6}$$

and

$$I_{RMS} = I_m \left[1 - \frac{\alpha}{\pi} + \frac{\sin 2\alpha}{2\pi} \right]^{1/2} \qquad\qquad \textbf{6.7}$$

Figure 6.6
Full wave center-tap
rectifier waveforms with a
resistive load

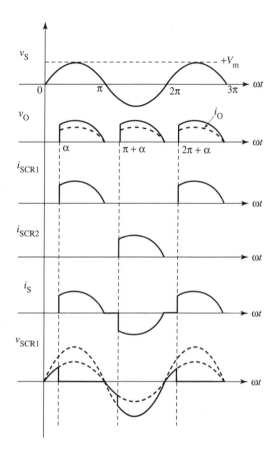

6.3.2   With an Inductive (*RL*) Load

Figure 6.7 shows the waveforms for voltage and current, assuming a highly inductive load so that the load current is continuous (i.e., present at all times). SCR_1 conducts for 180° from α to $(\pi + \alpha)$, and the load voltage follows the input voltage. At $(\pi + \alpha)$, SCR_2 is fired. SCR_1 now turns off, since the supply voltage immediately appears across it and applies a reverse bias. SCR_2 now conducts for 180° from $(\pi + \alpha)$ to $(2\pi + \alpha)$ and supplies power to the load.

The average value of the load voltage is given by

$$V_{o(avg.)} = \frac{2}{\pi} V_m \cos \alpha \qquad\qquad \textbf{6.8}$$

The output voltage is at its maximum when $\alpha = 0°$, zero when $\alpha = 90°$, and its negative maximum when $\alpha = 180°$. The normalized average output voltage is:

$$V_n = \frac{V_{o(avg.)}}{V_{do}} = \cos \alpha \qquad\qquad \textbf{6.9}$$

Figure 6.7
Full wave center-tap
rectifier waveforms with an
RL load

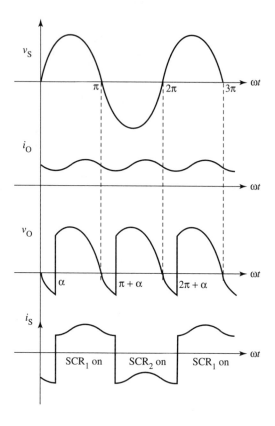

Figure 6.8
Control characteristic for a
center-tap rectifier

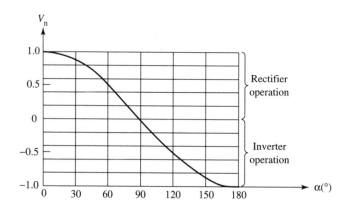

The *control characteristic* (the plot of V_n as a function of α) is shown in Figure 6.8.

The RMS output voltage is given by

$$V_{o(RMS)} = V_{s(RMS)} \qquad\qquad \textbf{6.10}$$

Example 6.4   Explain with the help of waveforms the operation of a full-wave center-tap rectifier with RL load for the following firing angles:

a) 0°
b) 45°
c) 90°
d) 135°
e) 180°

Solution   (a) Figure 6.9 shows the waveforms for voltage and current, assuming a highly inductive load, with a firing angle of 0°.

During the positive half-cycle of the source voltage, SCR$_1$ is forward-biased and SCR$_2$ is reverse-biased. The voltage across the load is v_s. During the negative half-cycle, SCR$_2$ is forward-biased and SCR$_1$ is reverse-biased. The voltage across the load is v_s. The application of a gate pulse with a zero firing-delay angle results in an output similar to that of an uncontrolled rectifier. Each SCR conducts for 180° and supplies current to the load for this period.

b) If we increase the firing angle, the average DC output voltage decreases as shown in Figure 6.10, which contains the voltage and current waveforms with $\alpha = 45°$. If SCR$_1$ is triggered at 45°, SCR$_2$ will conduct up to this point, even though the source voltage is negative, due to the highly inductive nature of the load. When SCR$_1$ turns on, SCR$_2$ is reverse-biased and turns off. The off SCR is subjected to twice the source voltage in the negative direction (see the v_{SCR1} waveform). The current to the load is supplied in turn by SCR$_1$ and SCR$_2$, each conducting for 180°.

c) If we increase the firing angle α to 90°, the two SCRs still remains in conduction for 180° as shown in Figure 6.11. However, the average DC voltage becomes zero, so there is no transfer of power from the AC source to the DC load.

In summary, as the firing angle is increased from 0 to 90°, the power supplied to the DC load decreases, becoming zero when $\alpha = 90°$.

d) If we increase the firing angle α beyond 90°, load current can flow only if there is a negative source of voltage at the DC load side. This is possible, for example, in a DC motor under regenerative conditions. When $\alpha = 135°$, the average DC voltage ($V_{o(avg.)}$), as shown in Figure 6.12, is negative. The load current still flow in each SCR for 180° in its original direction, but the load voltage has changed polarity. The power now flows from the DC load to the AC source. This circuit acts like an *inverter.*

e) If we increase the firing angle α to 180°, the average DC voltage reaches its maximum negative value. The SCRs remain in conduction for 180° (Figure 6.13).

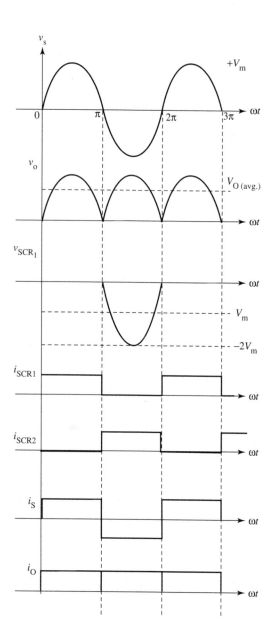

Figure 6.9
Voltage and current waveforms for $\alpha = 0°$

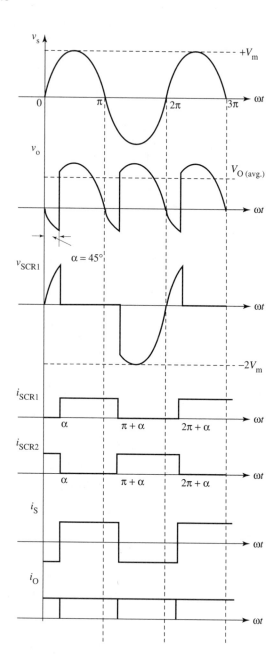

Figure 6.10
Voltage and current waveforms for $\alpha = 45°$

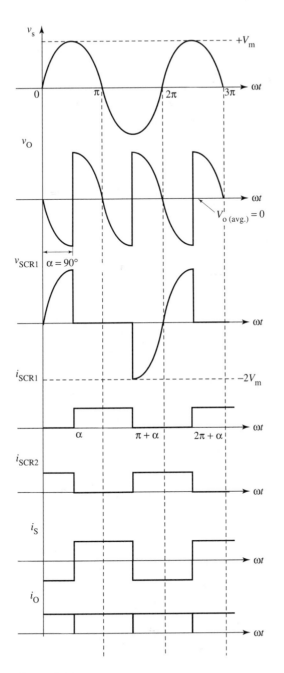

Figure 6.11
Voltage and current waveforms for $\alpha = 90°$

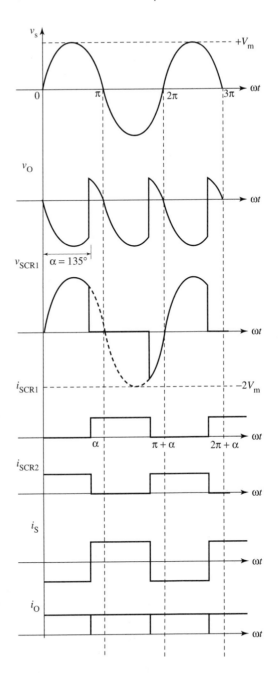

Figure 6.12
Voltage and current waveforms for $\alpha = 135°$

Figure 6.13
Voltage and current
waveforms for $\alpha = 180°$

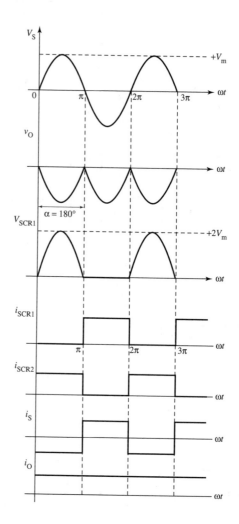

Example 6.5 Show the direction of power flow and the operating mode (rectifying or inversion) of the center-tap converter circuit with the following firing angles:
a) $\alpha > 0°$
b) $\alpha < 90°$
c) $\alpha > 90°$
d) $\alpha < 180°$

Solution a, b) For firing angles in the range $0° < \alpha < 90°$, the average output voltage is positive and the converter operates in the rectifying mode. In this mode, the power to the load is positive, that is, power flow is from the AC source to the DC load.

c, d) For firing angles in the range 90° < α < 180°, the output voltage is negative and the converter operates in the inversion mode. In this mode, the power to the load is negative, that is, power flow is from the DC load to the AC source.

6.3.3 With a Freewheeling Diode

A freewheeling diode connected across the inductive load (as shown in Figure 6.14) modifies the voltage and current waveforms of Figure 6.7. As the load voltage tends to go negative, the FWD becomes forward-biased and starts conducting. Thus, the load voltage is clamped to zero volts. A nearly constant load

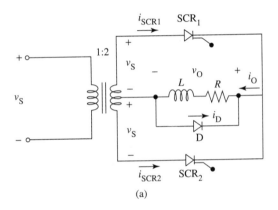

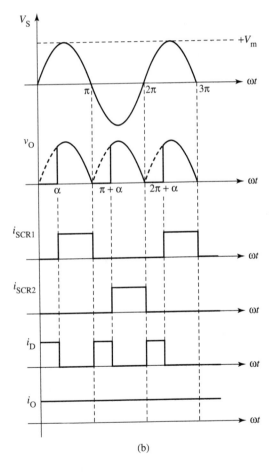

Figure 6.14
Center-tap rectifier with *RL* load and FWD (a) circuit (b) voltage and current waveforms

current is maintained by the freewheeling current through the diode. The average load voltage is given by:

$$V_{o(avg.)} = \frac{V_m (1 + \cos \alpha)}{\pi} \qquad \textbf{6.11}$$

The freewheeling diode carries the load current during the delay period α when the SCRs are off. Therefore, the current through the FWD(D) is given by:

$$I_D = I_{o(avg.)} \frac{\alpha}{\pi} = \frac{V_m(1 + \cos \alpha)}{\pi R} \frac{\alpha}{\pi}$$

$$= \frac{V_m (1 + \cos \alpha)\alpha}{\pi^2 R} \qquad \textbf{6.12}$$

6.4 Full-Wave Controlled Bridge Rectifiers

6.4.1 In Circuits With a Resistive Load

Figure 6.15 shows a full-wave controlled bridge rectifier circuit with a resistive load. In this circuit, diagonally opposite pairs of SCRs turn on and off together. The circuit operation is similar to that of the full-wave center-tap circuit discussed in Section 6.3. The average DC output voltage can be controlled from zero to its maximum positive value by varying the firing angle. The average value of the DC voltage is

$$V_{o(avg.)} = \frac{V_m(1 + \cos \alpha)}{\pi} \qquad \textbf{6.13}$$

and

$$I_{RMS} = \frac{I_m}{\sqrt{2}} \sqrt{1 - \frac{\alpha}{\pi} + \frac{\sin 2\alpha}{2\pi}} \qquad \textbf{6.14}$$

The SCRs are controlled and fire in pairs with a delay angle of α. The current and voltage waveforms become full-wave, as shown in Figure 6.16.

Figure 6.15
Full-wave bridge rectifier circuit

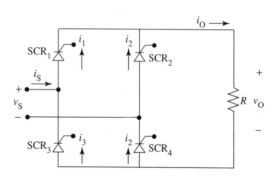

Figure 6.16
Waveforms of the bridge rectifier with a resistive load

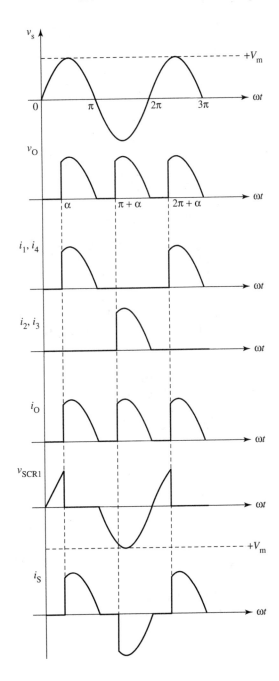

The average values for voltage and current are again twice those of the half-wave case. Therefore, from Equations 6.1 and 6.2,

$$V_{o(avg.)} = \frac{V_m(1 + \cos \alpha)}{\pi}$$

6.15

and

$$I_{o(avg.)} = \frac{I_m(1 + \cos \alpha)}{\pi}$$

$$I_{o(avg.)} = \frac{V_m(1 + \cos \alpha)}{\pi R}$$

6.16

The RMS value of the load current is given by

$$I_{RMS} = \frac{I_m}{\sqrt{2}} \sqrt{\left[1 - \frac{\alpha}{\pi} + \frac{\sin 2\alpha}{2\pi}\right]}$$

6.17

Example 6.6   The full-wave bridge rectifier shown in Figure 6.15 is supplied from a 150 V source with a load resistance of 10 Ω. If the firing angle α is 30°, find:
a) the average load voltage
b) the average load current
c) the maximum load current
d) the RMS load current
e) the power supplied to the load
f) the ripple frequency
g) the power factor

Solution

$$V_m = \sqrt{2}\,(150) = 212 \text{ V}$$

a) average load voltage $= \dfrac{V_m(1 + \cos \alpha)}{\pi} = \dfrac{(212)(1 + \cos 30°)}{\pi} = 126 \text{ V}$

b) average load current $= \dfrac{(V_m)(1 + \cos 30°)}{R\pi} = 12.6 \text{ A}$

c) maximum load current

$$I_m = \frac{V_m}{R} = \frac{212}{10} = 21.2 \text{ A}$$

d) RMS load current

$$I_{RMS} = \frac{I_m}{\sqrt{2}} \sqrt{\left[1 - \frac{\alpha}{\pi} + \frac{\sin 2\alpha}{2\pi}\right]}$$

$$= 14.8 \text{ A}$$

e) power supplied to the load $= I_{RMS}^2 R = 14.8^2(10) = 2182 \text{ W}$

f) ripple frequency

$f_r = 2 *$ input supply frequency $= 2 * 60 = 120$ Hz

g) $S = V_s * I_{RMS} = 150 * 14.8 = 2220$ VA

$$PF = \frac{P}{S} = \frac{2182}{2220} = 0.98$$

6.4.2 With an Inductive (*RL*) Load

Figure 6.17 shows the bridge rectifier with the addition of an inductive load. The load current tends to keep flowing since the inductor induces a voltage that acts to oppose an increase or decrease in current. Therefore, SCR keeps conducting even though the voltage may have fallen to zero. The current maintains conduction in the SCR even after the voltage across the SCR has reversed.

Figure 6.17
Bridge rectifier with an *RL* load

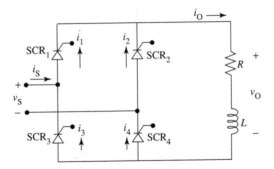

When the inductance is small or the delay angle α is kept large, the DC output current reaches zero every half cycle at $(\pi + \beta)$, as shown in Figure 6.18. During this period, neither pair of SCRs is on, and therefore the current is said to be *discontinuous*.

The average value of the output voltage is:

$$V_{o(avg.)} = \frac{V_m}{\pi} (\cos \alpha - \cos \beta)$$

If the load inductance is assumed to be large or α becomes small, the load current cannot reach zero and it flows continuously, as shown in Figure 6.19. Therefore, one pair of SCRs is conducting at all times. The current is said to be *continuous*.

During the positive half-cycle, SCR_1 and SCR_4 conduct. Applying KVL around the loop containing, V_s, SCR_1 and SCR_2 at an instant when $i_s > 0$ (between α and $(\pi + \alpha)$) gives

$v_s = v_{SCR1} - v_{SCR2}$

Here, $v_{SCR1} = 0$ since SCR_1 is conducting. Therefore $v_{SCR2} = -v_s$, which means that SCR_2 is reverse-biased.

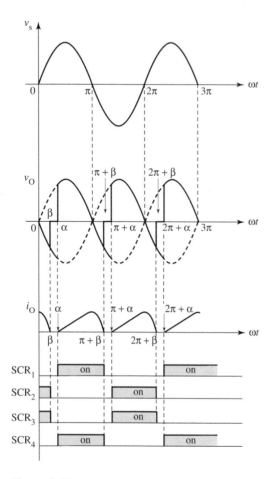

Figure 6.18
Waveform of bridge rectifier with small inductive load

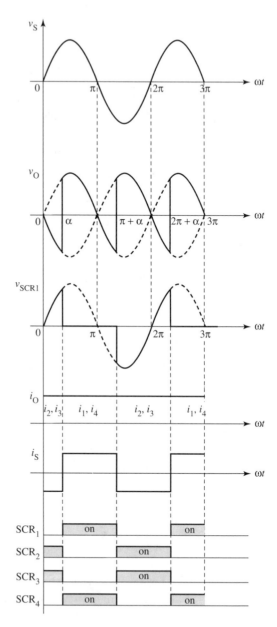

Figure 6.19
Waveform of bridge rectifier with $L >>> R$

Applying KVL around the loop containing v_s, SCR_1, the load, and SCR_4 gives

$$v_s = v_{SCR1} + v_o + v_{SCR4}$$

Again, SCR_1 and SCR_4 are conducting and have zero voltage across them. Therefore, $v_o = v_s$ (during the interval from α to π).

During the negative half-cycle, the source voltage $v_s < 0$. The preceding equations do not change form, although some of the quantities change sign. Now, since v_s is negative, SCR_2 and SCR_3 are forward-biased and will turn on when they receive a gate signal. Load current still flows in the same path through SCR_1 and SCR_4 until SCR_2 and SCR_3 are triggered. Therefore, from π to $(\pi + \alpha)$, the load voltage is negative since $v_s < 0$. At $(\pi + \alpha)$, SCR_2 and SCR_3 are triggered, which supplies a voltage $V_o = -v_s$ to the load.

The average value of this output voltage varies with α:

$$V_{o(avg.)} = \frac{2}{\pi} V_m \cos \alpha \qquad \textbf{6.19}$$

The RMS output voltage is constant, independent of the firing angle, and equals the RMS value of the supply voltage if the output current is continuous:

$$V_{o(RMS)} = \frac{V_{max}}{\sqrt{2}} = V_{S(RMS)} \qquad \textbf{6.20}$$

The average current is still $V_{o(avg.)}/R$, so

$$I_{o(avg.)} = \left(\frac{2}{\pi}\right)\left(\frac{V_m}{R}\right) \cos \alpha \qquad \textbf{6.21}$$

For $\alpha = 0°$ (no phase control), these equations reduce to those for the diode case:

$$I_{SCR(avg.)} = \frac{V_m}{\pi R} \qquad \textbf{6.22}$$

$$I_{SCR(RMS)} = \frac{I_{SCR(avg.)}}{\sqrt{2}} \qquad \textbf{6.23}$$

The normalized voltage $V_n = V_{o(avg.)}/V_{do} = \cos \alpha$. The *control characteristic (a plot of V_n as a function of α)* is shown in Figure 6.20.

Figure 6.20
Control characteristic for a bridge rectifier

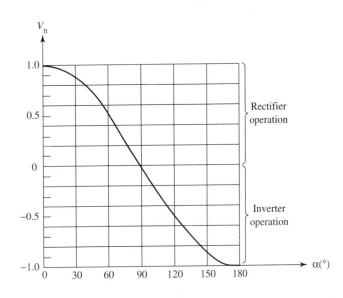

Note that when α becomes larger than 90°, the average value of output voltage becomes negative. This means that from 90 to 180°, power flows from the DC load side to the AC source side and the circuit operates as an inverter. When rectification and inversion are obtained from one converter, the process is called *two-quadrant operation* and the converter is called a *full converter*.

Example 6.7 A full-wave controlled rectifier with an inductive load is connected to a 230 V source. The resistive portion of the load is equal to 0.5 Ω. If the voltage across the load is 200 V, find the firing angle α.

Solution From Equation 6.19,

$$V_{o(avg.)} = \frac{2}{\pi} V_m \cos \alpha$$

$$200 = \frac{2}{\pi} (\sqrt{2} * 230) \cos \alpha$$

$$\cos \alpha = 0.96$$

$$\alpha = 15°$$

Example 6.8 A full-wave controlled rectifier with an inductive load is connected to a 120 V source. The resistive portion of the load is equal to 10 Ω. If the delay angle α is 30°, find
a) the average load voltage
b) the average load current
c) the maximum load current
d) the RMS load current
e) the average current in each SCR
f) the power supplied to the load
g) the form Factor
h) the ripple Factor
i) the rectifier efficiency

Solution

$$V_m = \sqrt{2} (120) = 208 \text{ V}$$

a) average load voltage

$$V_{o(avg.)} = \frac{2}{\pi} V_m \cos \alpha$$

$$= \frac{2}{\pi} (208)(\cos 30)$$

$$= 115 \text{ V}$$

b) average load current

$$I_{o(avg.)} = \frac{V_{o(avg.)}}{R} = \frac{115}{10} = 11.5 \text{ A}$$

c) maximum load current = average load current = 11.5 A

d) RMS load current = average load current = 11.5 A

e) Since the SCRs in bridge conduct on alternate half-cycles, the average SCR current is

$$\frac{1}{2} I_{o(\text{avg.})} = 5.75 \text{ A}$$

f) power supplied to the load = $I_{\text{RMS}}^2 R = 11.5^2(10) = 1323 \text{ W}$

g) form factor

$$\text{FF} = V_{o(\text{RMS})}/V_{o(\text{avg.})} = 120/115 = 1.04$$

h) ripple factor

$$\text{RF} = \sqrt{\text{FF}^2 - 1} = \sqrt{1.04^2 - 1} = 0.3$$

i) rectifier efficiency

$$\eta = V_{o(\text{avg.})}/V_{o(\text{RMS})} = 115/120 = 0.96$$

6.4.3 With a Freewheeling Diode

If a diode is connected across the load, the circuit can operate only as a rectifier because the diode prevents negative values of v_o from appearing across the load. Figure 6.21 shows the bridge rectifier circuit with the addition of a freewheeling diode (D). The diode provides an extra path for the flow of load current. Three paths are now possible: SCR_1 and SCR_4, SCR_2 and SCR_3, and the path through diode D.

Figure 6.21
Full-wave bridge rectifier
with FWD

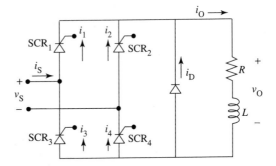

Negative values of v_o will forward-bias D and provide zero voltage across the load. Therefore, the negative portions of v_o in Figure 6.19 are now replaced by $v_o = 0$, as shown in Figure 6.22. During this interval, the load current freewheels through D and the SCR currents and source current are zero. To illustrate this, let us apply KVL around the path that contains v_s, SCR_1, v_o, and SCR_4:

$$v_s = v_{\text{SCR1}} + v_o + v_{\text{SCR4}}$$

Figure 6.22
Voltage and current
waveforms for Figure 6.21

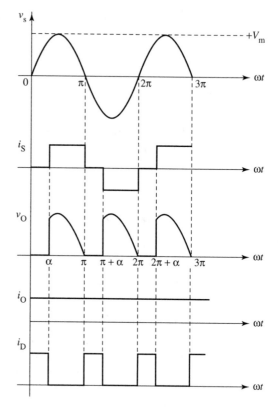

For the negative portion when $v_s < 0$, the FWD is on and $v_o = 0$. Therefore,

$$v_{SCR1} + v_{SCR4} < 0$$

This means that SCR_1 and SCR_4 in series are reverse-biased and turn off. SCR_2 and SCR_3 are already off since v_s is negative. The load current therefore transfers to the freewheeling diode.

The load voltage waveform is the same as that for the resistive load case. The average load voltage is the same as in Equation 6.13:

$$V_{o(avg.)} = \frac{V_m(1 + \cos \alpha)}{\pi} \qquad \textbf{6.24}$$

The average value of the output voltage can be varied from 0 to $2 V_m/\pi$ as α increases from 0 to π. This voltage is never negative.

The current in the FWD is the same as the load current from 0 to α and from π to $(\pi + \alpha)$, while it is zero for the remaining time. The average current in the FWD is given by:

$$I_{D(avg.)} = \frac{I_{o(avg.)}(\alpha)}{\pi} \qquad \textbf{6.25}$$

The freewheeling diode's maximum current is the same as the maximum value of the load current.

Example 6.9   A full-wave bridge rectifier with a freewheeling diode supplies an *RL* load. The source voltage is 120 V and the resistive portion of the load is 10 Ω. If the delay angle $\alpha = 30°$, find

a) the average load voltage
b) the average load current
c) the maximum load current
d) the RMS load current
e) the average current in each SCR
f) the power supplied to the load
g) the average current in the FWD

Solution

$$V_\text{m} = \sqrt{2}\ (120) = 208\ \text{V}$$

a) average load voltage

$$V_\text{o(avg.)} = \frac{V_\text{m}\ (1 + \cos \alpha)}{\pi}$$

$$= \frac{(208)}{\pi}\ (1 + \cos 30)$$

$$= 124\ \text{V}$$

b) average load current $= \dfrac{V_\text{o(avg.)}}{R} = \dfrac{124}{10} = 12.4\ \text{A}$

c) maximum load current = average load current = 12.4 A
d) RMS load current = average load current = 12.4 A
e) Since the SCRs in the bridge conduct on alternate half-cycles for $(180° - 30°) = 150°$, the average SCR current is

$$(12.4)\frac{(150)}{360} = 5.2\ \text{A}$$

f) power supplied to the load $= I_\text{RMS}^2 R = 12.4^2(10) = 1538$ W
g) From Equation 6.25,

$$I_\text{D(avg.)} = (12.4)\frac{(30)}{180} = 2.06\ \text{A}$$

6.5   Half-Controlled or Semicontrolled Bridge Rectifiers

Full or two-quadrant converters can operate with both positive and negative average DC load voltages. In the rectifying mode, they supply power from the AC source to the DC load. In the inversion mode, they remove power from the DC load and return it to the AC source.

Figure 6.23
Full-wave semicontrolled
bridge rectifier circuit

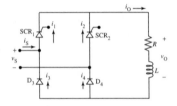

Figure 6.24
Waveform for a semi-
controlled bridge rectifier
with an inductive load

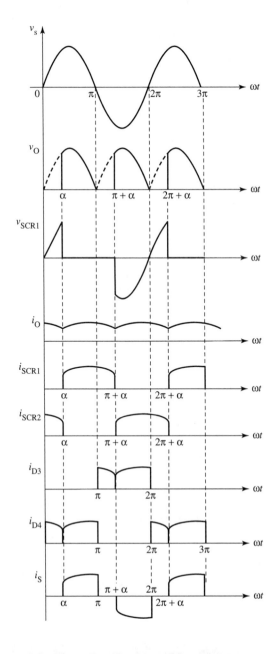

There are various applications that require power flow only from the AC source to the DC load and therefore are operated in only the rectifying mode. This is accomplished in bridge rectifiers by replacing half of the SCRs with diodes. These circuits are called *one-quadrant* or *semicontrolled* bridge rectifiers. An alternative method of obtaining one-quadrant operation in bridge rectifiers is to connect a freewheeling diode across the output terminals of the rectifier.

A basic semicontrolled bridge circuit is shown in Figure 6.23. Its operation is the same as that of a fully controlled bridge rectifier with a resistive load. When the source voltage is positive, SCR_1 and D_4 are forward-biased. If we trigger SCR_1 at α, current will flow through D_4, the load, and SCR_1. SCR_1 turns off at π when the source reverses. The load voltage is the same as the input voltage during this period (α to π). At ($\pi + \alpha$), SCR_2 is triggered, causing current to flow through D_3 and the load. At time 2π, SCR_2 turns off and the cycle repeats.

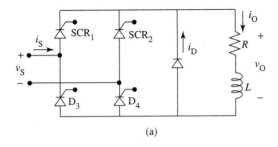

(a)

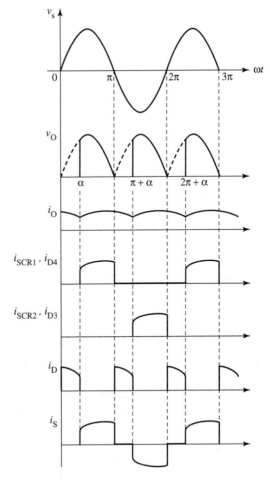

Figure 6.25
Semicontrolled bridge rectifier with FWD (a) circuit (b) waveform

With an inductive load, commutation (current transfer) occurs every half-cycle to bypass the load current through the diode, as shown in Figure 6.24. The current freewheels through SCR_1 and D_3 during the interval π to $(\pi + \alpha)$ and through SCR_2 and D_4 during the interval 2π to $(2\pi + \alpha)$. As a result, the negative portion of the output voltage is cut off, and the waveform of the output voltage becomes the same as with a pure resistive load.

The average output voltage can therefore be varied from its maximum positive value to zero as the firing angle is varied from 0 to 180°. The average value of the output voltage is given by Equation 6.24, that is, the voltage is the same as that of a full converter with FWD:

$$V_{o(avg.)} = \frac{V_m}{\pi}(1 + \cos \alpha) \qquad \qquad \textbf{6.26}$$

If the circuit in Figure 6.23 contains a highly inductive load the load current will flow throughout the entire negative half-cycle (even if the gate signal is removed) and therefore the circuit will lose control. If a freewheeling diode is used (Figure 6.25), the FWD becomes forward-biased and begins to conduct as the load voltage tends to reverse. The load current freewheels through the FWD. Therefore, during the internal π to $(\pi + \alpha)$, the output voltage becomes zero. Similarly, during the interval 2π to $(2\pi + \alpha)$, the FWD will clamp the negative voltage excursions to zero.

$$V_{o(avg.)} = \frac{V_m}{\pi}(1 + \cos \alpha) \qquad \qquad \textbf{6.27}$$

Example 6.10   Draw the output voltage waveform for a full-wave semicontrolled rectifier like the one shown in Figure 6.23 for the following delay angles:
a) $\alpha = 0°$
b) $\alpha = 45°$
c) $\alpha = 90°$

Solution   a) $\alpha = 0°$. During the positive half-cycle, SCR_1 and D_4 are conducting, and during the negative half-cycle, SCR_2 and D_3 are conducting. The output voltage is the same as that of a diode bridge rectifier (Figure 6.26(a)).

b) $\alpha = 45°$. During the positive half-cycle, the voltage across the load is zero until SCR_1 is turned on at 45°. The load current flows through SCR_1 and D_4, and the source voltage is applied to the load. During the negative half-cycle, SCR_1 becomes reverse-biased at π. If we assume an inductive load, SCR_1 maintain conduction until SCR_2 is turned on. The load current will freewheel through SCR_1 and D_3.

When SCR_2 is turned on at $(\pi + 45°)$, SCR_1 turns off and the load current flows through SCR_2 and D_3 until SCR_2 and D_3 are reverse-biased. At this point, SCR_2 remains in conduction and the load current freewheels through SCR_2 and D_4. During the period when the load current is freewheeling, no current is supplied from the AC source.

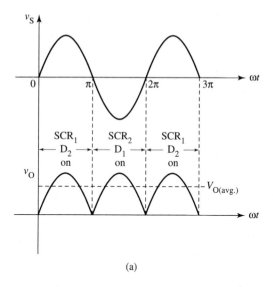

(a)

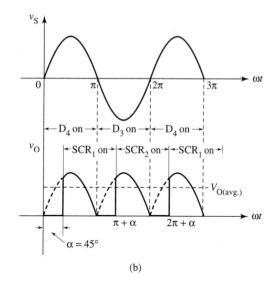

(b)

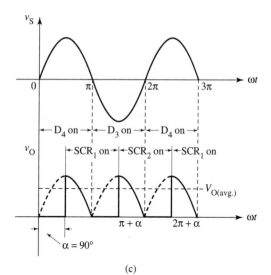

(c)

Figure 6.26
Output voltage waveforms for (a) $\alpha = 0°$, (b) $\alpha = 45°$,
(c) $\alpha = 90°$

Both SCRs remain in conduction for 180°. Figure 6.26(b) shows the output voltage waveform.

c) $\alpha = 90°$. The operation is exactly the same as in the case when $\alpha = 45°$ (see Figure 6.26(c)).

An alternative circuit configuration is shown in Figure 6.27. The freewheeling current is limited to the path that includes the two diodes, D_2 and D_4, in series. Therefore, the period of conduction for the diodes increases and that of the SCRs decreases. The SCRs now block for a full 180°. The waveforms are the same as in Figure 6.24.

Figure 6.27
Semicontrolled bridge converter, an alternative circuit configuration

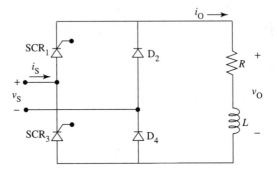

Another possible arrangement of the semicontrolled bridge converter is shown in Figure 6.28. Here the AC input voltage is rectified by the diode bridge to give full-wave voltage output. The output is then controlled by the SCR. The FWD allows the flow of current through the load during the time the SCR is off. The average output voltage is the same as in Equation 6.24.

Figure 6.28
Semicontrolled bridge converter, alternative arrangement

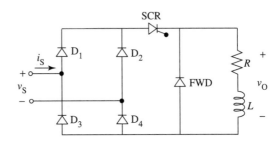

6.6 Dual Converters

It should be noted that the semiconverter or one-quadrant converter discussed in the previous section allows one polarity of output voltage and current. Power is therefore positive, that is, power flow is from the AC supply to the DC load. Thus a semiconverter can operate only in the rectifying mode. Full or two-quadrant converters can operate with positive polarity of output voltage and current but can also have a positive current and a reverse voltage. Full converters can therefore operate as rectifiers or as inverters with current flow in only one direction.

A dual converter or four-quadrant converter can operate as a rectifier or as an inverter with current flow in both directions. Two full converters can be connected back to back as shown in Figure 6.29 to form a dual converter. The dual converter provides virtually instantaneous reversal of current through the

load. One converter causes the load current to flow in one direction, while the
other converter causes the flow of load current in the reverse direction.

Figure 6.29
Dual converters

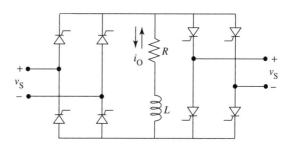

6.7 Problems

6.1 Draw the waveforms for load voltage, load current, and voltage across the
SCR for a half-wave controlled rectifier for the following cases:
(a) a resistive load with a firing angle of 45°
(b) a resistive load with a firing angle of 135°
(c) an inductive load with a firing angle of 0°
(d) an inductive load with a firing angle of 90°

6.2 A half-wave controlled rectifier (Figure 6.1(a)) supplied from a 120 V
source is used to control the power to a 15 Ω load. If the delay angle is
75°, find
(a) the power supplied to the load
(b) the maximum SCR current
(c) the average SCR current
(d) the SCR maximum reverse voltage

6.3 A half-wave controlled rectifier with a resistive load is fed from a 45 V
AC source. For the given delay angles, calculate $V_{o(avg.)}$ and tabulate
the results in Table 6.1. Also plot the control characteristic in Figure 6.30.
$$(V_n = \frac{V_{o(avg.)}}{V_{do}}; \ V_{do} = V_{o(avg.)} \text{ with } \alpha = 0°)$$

Table 6–1

α	0°	30°	60°	90°	120°	150°	180°
$V_{o(avg.)}$							
V_n							

Figure 6.30
Plot of V_n vs. α

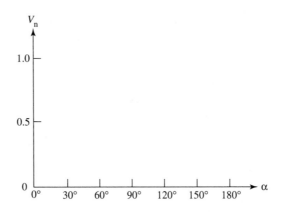

6.4 A full-wave center-tap rectifier is fed from a transformer with a secondary voltage of 120 V (center-tap to line). If it is used to charge a 12 V battery having an internal resistance of 0.1 Ω, find
(a) the firing angle necessary to produce a charging current of 10 A
(b) the average SCR current
(c) the PIV rating of the SCR

6.5 A full-wave center-tap rectifier is fed from a 120 V source. If the load inductance is much greater than the load resistance, find the average output voltage for the following firing angles:
(a) 45°
(b) 60°
(c) 90°
(d) 135°

6.6 Draw the waveforms for load voltage, load current, and voltage across the two SCRs for a full-wave center-tap rectifier for the following cases:
(a) a resistive load with a firing angle of 60°
(b) a resistive load with a firing angle of 120°
(c) an inductive load with a firing angle of 100°

6.7 A full-wave center-tap rectifier with a resistive load is fed from a 90 V (45 V on each secondary winding) AC source. For the given delay angles, calculate $V_{o(avg.)}$ and tabulate the results in Table 6.2. Also plot the control characteristic in Figure 6.31.

Table 6–2

α	0°	30°	60°	90°	120°	150°	180°
$V_{o(avg.)}$							
V_n							

Figure 6.31
Plot of V_n vs α

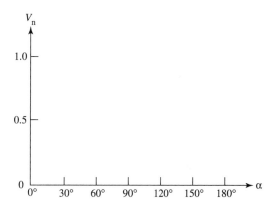

6.8 A full-wave rectifier (Figure 6.15) fed from a 200 V source supplies an average current of 10 A. If the firing angle is 30°, find
(a) the average power delivered to the load
(b) the average SCR current
(c) the maximum SCR current
(d) the SCR peak reverse voltage
(e) the value of the load resistance

6.9 Discuss the advantages and disadvantages of a center-tap converter versus a bridge converter.

6.10 A full-wave bridge rectifier (Figure 6.15) connected to a 120 V source provides a DC output voltage of 90 V. If the power supplied to the load is 1 KW, find
(a) the firing angle α
(b) the value of the load resistor R
(c) the SCR maximum current
(d) the SCR maximum voltage

6.11 A full-wave bridge rectifier with an FWD (Figure 6.21) is connected to a 220 V source. The load resistance is 20 Ω while the load inductance is large. If the firing angle α is 60°, find
(a) the SCR average current
(b) the SCR maximum voltage
(c) the power supplied to the load

6.12 Draw the waveforms for load voltage, load current, and voltage across the SCRs for a full-wave bridge rectifier for the following cases:
(a) a resistive load with a firing angle of 60°
(b) a resistive load with a firing angle of 120°

6.13 A full-wave bridge rectifier with a resistive load is fed from a 45 V AC source. For the given delay angles, calculate $V_{o(avg.)}$ and tabulate the results in Table 6–3. Also plot the control characteristic in Figure 6.32.

Table 6–3

α	0°	30°	60°	90°	120°	150°	180°
$V_{o(avg.)}$							
V_n							

Figure 6.32
Plot of V_n vs α

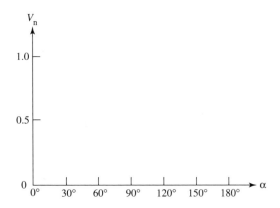

6.14 A full-wave bridge rectifier with an FWD (Figure 6.21) is connected to a 120 V source. The average current supplied to the load is 20 A. If the firing angle α is 30°, find
(a) the SCR average current
(b) the SCR maximum voltage
(c) the power supplied to the load

6.15 A full-wave bridge rectifier with an FWD (Figure 6.21) is connected to a 208 V source. The average voltage across the load is 100 V. If the load resistance is 10 Ω, find
(a) the SCR maximum current
(b) the SCR maximum voltage
(c) the firing angle α
(d) the power supplied to the load

6.16 In the circuit in Figure 6.17, v_S is a source of 230 V and R is 10 Ω. If the firing angle $\alpha = 120°$, find:
(a) the average load current
(b) the SCR average current
(c) the FWD average current
(d) the SCR maximum reverse voltage
(e) the average power supplied to the load

6.17 A bridge rectifier is supplied from a 120 V source with a resistive load R of 10 Ω. If the firing angle α is 30°, find the average value of the load voltage.

6.18 Repeat problem 6.17 with an RL load.

6.19 A bridge rectifier is connected to an RL load. The source voltage is 220 V and the load resistance R is 10 Ω. If the firing angle is 45°, find the average value of the load voltage.

6.20 Repeat problem 6.19, if an FWD is connected across the load.

6.21 Draw the waveforms for the single-phase, semicontrolled converters shown in Figure 6.28.

6.22 Draw the waveforms for load voltage and load current for a semicontrolled bridge rectifier for the following cases:
(a) a resistive load with a firing angle of 140°
(b) an inductive load with a firing angle of 140°

6.23 A semicontrolled bridge rectifier with a resistive load is fed from a 45 V AC source. For the given delay angles, calculate $V_{o(avg.)}$ and tabulate the results in Table 6–4. Also plot the control characteristic in Figure 6.33.

Table 6–4

α	0°	30°	60°	90°	120°	150°	180°
$V_{o(avg.)}$							
V_n							

Figure 6.33
Plot of V_n vs α

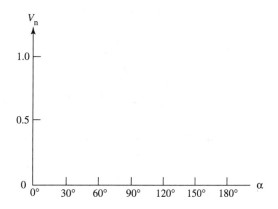

6.24 Briefly explain the operation of a one-quadrant, a two-quadrant, and a four-quadrant converter.

6.8 Equations

$$V_{o(avg.)} = \frac{V_m(1 + \cos \alpha)}{2\pi} \tag{6.1}$$

$$I_{o(avg.)} = \frac{V_m(1 + \cos \alpha)}{2\pi R} \tag{6.2}$$

$$I_{RMS} = \left(\frac{I_m}{2}\right)\left[1 - \frac{\alpha}{\pi} + \frac{\sin 2\alpha}{2\pi}\right]^{1/2} \tag{6.3}$$

$$V_n = V_{o(avg.)}/V_{do} = \frac{V_m(1 + \cos \alpha)/2\pi}{V_m/\pi} = \frac{1 + \cos \alpha}{2} \tag{6.4}$$

$$V_{o(avg.)} = \frac{V_m(\cos \alpha - \cos \beta)}{2\pi} \tag{6.5}$$

$$V_{o(avg.)} = \frac{V_m(1 + \cos \alpha)}{\pi} \tag{6.6}$$

$$I_{RMS} = I_m\left[1 - \frac{\alpha}{\pi} + \frac{\sin 2\alpha}{2\pi}\right]^{1/2} \tag{6.7}$$

$$V_{o(avg.)} = \frac{2}{\pi} V_m \cos \alpha \tag{6.8}$$

$$V_n = \frac{V_{o(ave)}}{V_{do}} = \cos \alpha \tag{6.9}$$

$$V_{o(RMS)} = V_{s(RMS)} \tag{6.10}$$

$$V_{o(avg.)} = \frac{V_m(1 + \cos \alpha)}{\pi} \tag{6.11}$$

$$I_D = \frac{V_m(1 + \cos \alpha)\alpha}{\pi^2 R} \tag{6.12}$$

$$V_{o(avg.)} = \frac{V_m(1 + \cos \alpha)}{\pi} \tag{6.13}$$

$$I_{RMS} = \frac{I_m}{\sqrt{2}}\left[1 - \frac{\alpha}{\pi} + \frac{\sin 2\alpha}{2\pi}\right]^{1/2} \tag{6.14}$$

$$V_{o(avg.)} = \frac{V_m(1 + \cos \alpha)}{\pi} \tag{6.15}$$

$$I_{o(avg.)} = \frac{V_m(1 + \cos \alpha)}{\pi R} \tag{6.16}$$

$$I_{RMS} = \frac{I_m}{\sqrt{2}} \sqrt{\left[1 - \frac{\alpha}{\pi} + \frac{\sin 2\alpha}{2\pi}\right]} \tag{6.17}$$

$$V_{o(avg.)} = \frac{V_m}{\pi} (\cos \alpha - \cos \beta) \qquad \textbf{6.18}$$

$$V_{o(avg.)} = \frac{2}{\pi} V_m \cos \alpha \qquad \textbf{6.19}$$

$$V_{o(RMS)} = \frac{V_{max}}{\sqrt{2}} = V_{S(RMS)} \qquad \textbf{6.20}$$

$$I_{o(avg.)} = \frac{(2)}{\pi} \frac{(V_m)}{R} \cos \alpha \qquad \textbf{6.21}$$

$$I_{SCR(avg.)} = \frac{V_m}{\pi R} \qquad \textbf{6.22}$$

$$I_{SCR(RMS)} = \frac{I_{SCR(avg.)}}{\sqrt{2}} \qquad \textbf{6.23}$$

$$V_{o(avg.)} = \frac{V_m(1 + \cos \alpha)}{\pi} \qquad \textbf{6.24}$$

$$I_{D(avg.)} = \frac{I_{o(avg.)}(\alpha)}{\pi} \qquad \textbf{6.25}$$

$$V_{o(avg.)} = \frac{V_m}{\pi}(1 + \cos \alpha) \qquad \textbf{6.26}$$

$$V_{o(avg.)} = \frac{V_m}{\pi}(1 + \cos \alpha) \qquad \textbf{6.27}$$

Three-Phase Uncontrolled Rectifiers

7

Chapter Outline

Learning Objectives

After completing this chapter, the student should be able to

- discuss the advantages of three-phase rectifiers over single-phase rectifiers

- describe with the help of waveforms the operation of a three-pulse uncontrolled rectifier with resistive and inductive loads

- describe with the help of waveforms the operation of a six-pulse uncontrolled rectifier with resistive and inductive loads

- explain the operation of a twelve-pulse uncontrolled rectifier

7.1 Introduction

Single-phase rectifiers are relatively simple in construction, but they have limited power-handling capabilities and generate significant ripples in the DC output voltage. Three-phase rectifiers provide a smoother DC output, so the output filtering is done more easily. The filter components of large power rectifiers are large and expensive, so a reduction or elimination of filters is important. Therefore, for high-power applications, using three-phase rectifiers is desirable. All the single-phase rectifier circuits discussed in Chapter 5 have corresponding three-phase versions. The circuits covered in this chapter can also be used with SCRs in three-phase controlled rectifier circuits (see Chapter 8).

Three-phase rectifiers have the following advantages compared with single-phase rectifiers:

1. higher output voltage for a given input voltage
2. lower amplitude ripples (although they never fall to zero), i.e., output voltage is smoother
3. higher frequency ripples, simplifying filtering
4. higher overall efficiency

Three-phase rectifiers use either three, six, or twelve diodes. Using more diodes reduces the cost by distributing the load, thus allowing the use of lower-rated devices.

7.2 Three-Phase Half-Wave (Three-Pulse) Rectifiers

7.2.1 With a Resistive Load

A basic three-phase half-wave rectifier consisting of three diodes and a resistive load is shown in Figure 7.1. The circuit can be analyzed by first determining the periods in which each diode is on and then applying the appropriate source voltage across the load resistor R. Each diode conducts for 120° intervals in the sequence D_1, D_2, D_3, . . . to give the combined output voltage v_o shown in Figure 7.2.

At any given time, the most positive instantaneous voltage turns its respective diode on. The on diode connects its most positive source terminal to the other two diode cathodes, keeping the other two diodes off. Therefore, only one diode is on at any time (ignoring the moment of switching). The sudden switchover from one diode to another is called *commutation*.

The input voltage waveform v_s in Figure 7.2 is used to find the periods when each diode is on. Consider the interval between 0° and 30°. During this time, the phase voltage v_{CN} is higher than both v_{AN} and v_{BN}. As a result, diode D_3 is forward-biased and the output voltage (v_o) becomes equal to v_{CN}. During

Figure 7.1
Three-phase half-wave
rectifier circuit diagram

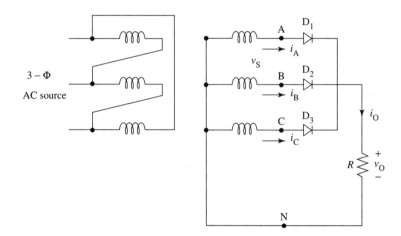

Figure 7.2
Load voltage waveforms

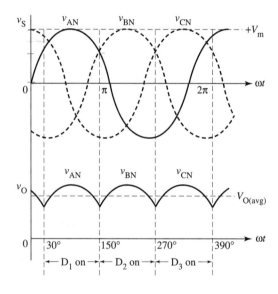

this interval, the voltage across D_1 is v_{AC} and that across D_2 is v_{BC}. Diodes D_2
and D_3 are therefore reverse-biased. From 30° to 150°, the most positive voltage is v_{AN}; it turns diode D_1 on and appears across R as v_o. At 150°, the instantaneous voltage of v_{BN} becomes greater than v_{AN}. Diode D_1 becomes reverse-
biased and turns off as diode D_2 becomes forward-biased and begin to conduct.
This applies v_{BN} across R from 150° to 270°. At 270°, v_{CN} again becomes the
most positive and D_3 turns on. Diode D_3 connects v_{CN} across R from 270°. The
cycle is then repeated.

The output voltage across the load v_o follows the peaks of the input supply voltage and pulsates between V_{max} and $0.5\ V_{max}$. This circuit is called a
three-pulse rectifier, since the output repeats itself three times in every cycle of
v_s. The ripple voltage is smaller than that produced by a single-phase rectifier.
The ripple frequency (f_r) of the output voltage is

$$f_r = n f_s$$ **7.1**

where

n = pulse number or number of diodes = 3

and

f_s = AC supply frequency

Therefore,

$f_r = 3 * 60 = 180$ Hz

Filtering is thus easier since the size of the filter is reduced as the ripple frequency increases.

A general expression for the average load voltage is

$$V_{o(avg.)} = \frac{n}{\pi} V_m \sin\left(\frac{\pi}{n}\right)$$ **7.2**

For the case of a three-pulse rectifier,

$$V_{o(avg.)} = 0.827 \ V_m$$ **7.3**

In terms of line voltage, average load voltage is given by

$$V_{o(avg.)} = 0.477 \ V_{L(m)}$$ **7.4**

where,

V_m = maximum value of phase voltage

$V_{L(m)}$ = maximum value of line voltage

Because the load is resistive, the load current has the same waveform as the load voltage. The individual diode currents are equal to the load current during the time when a particular diode conducts for its 120° interval. Each diode current is then zero for a 240° interval (see Figure 7.3).

In general, each diode conducts for a period of $\frac{2\pi}{n}$.

The average load current is given by

$$I_{o(avg.)} = \frac{n}{\pi} I_m \sin\left(\frac{\pi}{n}\right)$$ **7.5**

$$= 0.827 \ I_m$$ **7.6**

where

$I_m = V_m/R$

The average current in each diode is only one-third the load current:

$$I_{D(avg.)} = I_{o(avg.)}/n = I_{o(avg.)}/3$$ **7.7**

The maximum load current and maximum diode current are obviously the same, and because the load is resistive,

$$I_{o(m)} = \frac{V_m}{R}$$

$$= 1.21 \ I_{o(avg.)}$$ **7.8**

Figure 7.3
Current waveforms

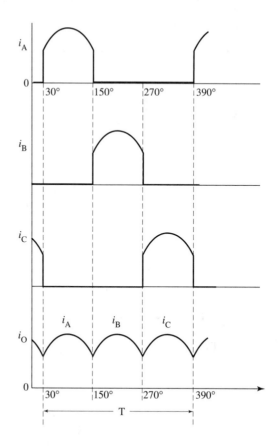

The RMS value of load current is

$$I_{o(RMS)} = I_m \left[\frac{1}{2\pi} \left(\frac{\pi}{n} + \frac{1}{2} \sin \frac{2\pi}{n} \right) \right]^{1/2}$$

$$= I_m \left[\frac{1}{2\pi} \left(\frac{\pi}{3} + \frac{1}{2} \sin \frac{2\pi}{3} \right) \right]^{1/2}$$

$$= 0.408 \ I_m$$ **7.9**

In general, the ripple factor is given by

$$RF = \frac{\sqrt{2}}{n^2 - 1}$$

$$= \frac{\sqrt{2}}{3^2 - 1}$$

$$= 0.177$$ **7.10**

and the form factor by

$$FF = \sqrt{n}$$

$$= \sqrt{3}$$

$$= 1.732$$ **7.11**

Table 7–1

Period	On diode	Off diodes	Diode voltages		
			v_{D1}	v_{D2}	v_{D3}
0 to 30°	D_3	D_1 and D_2	v_{AC}	v_{BC}	0
30 to 150°	D_1	D_2 and D_3	0	v_{BA}	v_{CA}
150 to 270°	D_2	D_3 and D_1	v_{AB}	0	v_{CB}
270 to 390°	D_3	D_1 and D_2	v_{AC}	v_{BC}	0

Table 7–1 shows the voltage across the diodes for various 120° intervals. The voltage across any diode can therefore be plotted by first plotting the waveform for the line voltages and then picking the appropriate voltage from Table 7–1. The waveforms are shown in Figure 7.4. Note that the line voltages are shown leading their respective phase voltages by 30°.

Figure 7.4
Voltage across the diodes
(a) v_{D1}

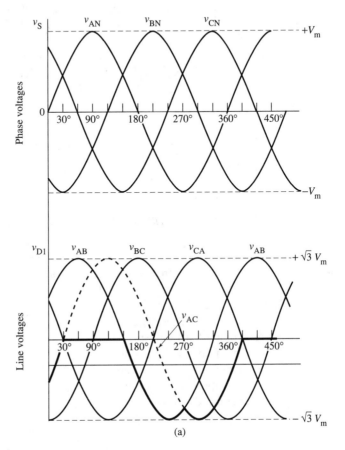

(a)

Figure 7.4
Voltage across the diodes
(b) v_{D2}

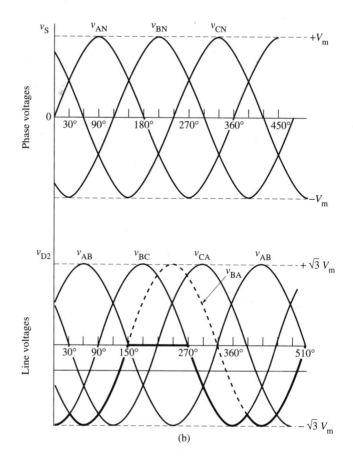

(b)

The PIV rating for the diodes should be

$$\text{PIV rating} \geq V_{L(m)} \text{ or } \sqrt{6}\ V_{s(m)}. \qquad \textbf{7.12}$$

Example 7.1   A three-pulse uncontrolled rectifier is connected to a 3 Φ, 4-wire, 220 V AC source. If the load resistance is 20 Ω, find

a) the maximum load voltage
b) the average load voltage
c) the average load current
d) the maximum load current
e) the maximum diode current
f) the PIV rating of the diode
g) the average diode current
h) the form factor

Figure 7.4
Voltage across the diodes
(c) v_{D3}

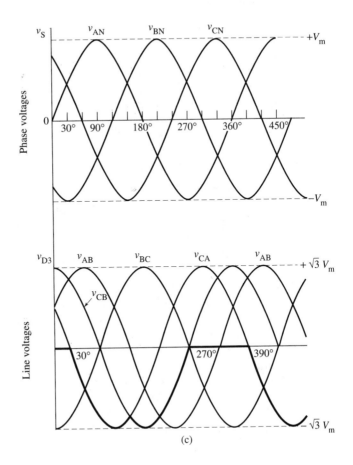

(c)

i)   the pulse number

j)   the conduction angle

Solution   a)   The maximum value of line voltage is

$$V_{L(m)} = \sqrt{2}\,(220) = 311 \text{ V}$$

The maximum value of the phase voltage is

$$V_m = 311\,/\,\sqrt{3} = 179.6 \text{ V}$$

b) $V_{o(avg.)} = 0.827 * 179.6 = 148.5$ V

c) $I_{o(avg.)} = V_{o(avg.)}/R = 148.5/5.20 = 7.4$ A

d) $I_{o(m)} = V_m/R = 179.6/20 = 9$ A

e) $I_{D(m)} = I_{o(max)} = 9$ A

f)   PIV $\geq V_{L(m)} = 311$ V

g) $I_{D(avg.)} = \dfrac{I_{o(avg.)}}{3} = 7.4/3 = 2.5$ A

h) FF $= \sqrt{n} = \sqrt{3} = 1.732$

i) $P = 3$

j) $\theta = 120°$

7.2.2 With an Inductive (*RL*) Load

In the practical case where the load contains an inductance in series with the resistance (see Figure 7.5), the load current is more constant and has a negligible ripple. The higher the inductance, the more the current tends to flatten out. Ideally, if L is infinite, the ripple will be zero.

The output voltage v_o still has ripples, but the ripple voltage will be zero across the resistive portion of the load. Ideally, the entire ripple voltage will be absorbed across the inductive portion of the load.

There is no change in the waveform of the output voltage, and the average output voltage remains the same as in Equation 7.2.

The average diode current is

$$I_{D(avg.)} = I_{o(avg.)}/3$$

If we assume the load current to be nearly constant,

$$I_{o(RMS)} = I_{o(avg.)} \qquad\qquad 7.13$$

Since the current waveform is rectangular in shape,

$$I_{D(m)} = I_{o(m)} = I_{D(avg.)} \qquad\qquad 7.14$$

Figure 7.5
The three-pulse rectifier with an *RL* load (a) circuit diagram

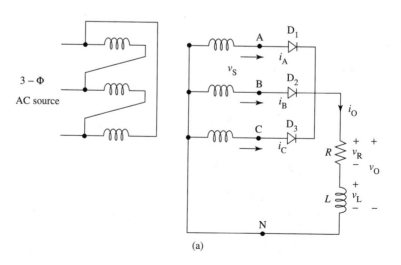

(a)

Figure 7.5
The three-pulse rectifier with an *RL* load (b) voltage and current waveforms

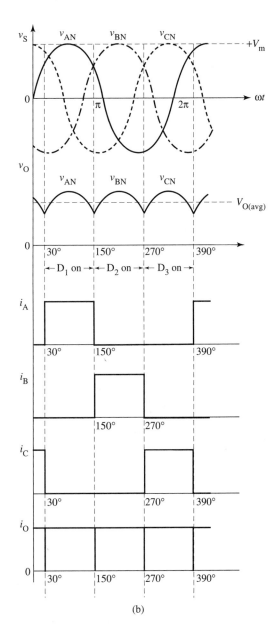

(b)

Example 7.2 If the load in Example 7.1 has a large inductance added to it, find
a) the maximum load voltage
b) the average load voltage
c) the average load current
d) the maximum load current

e) the RMS load current
f) the maximum diode current
g) the PIV rating of the diode
h) the average diode current

Solution   a) from Example 7.1(a),

$$V_m = 179.6 \text{ V}$$

b) from Example 7.1(b),

$$V_{o(avg.)} = 148.5 \text{ V}$$

c) from Example 7.1(c),

$$I_{o(avg.)} = 7.4 \text{ A}$$

d) since there is no ripple,

$$I_{o(m)} = I_{o(avg.)} = 7.4 \text{ A}$$

e) RMS load current

$$I_{o(RMS)} = I_{o(avg.)} = 7.4 \text{ A}$$

f) maximum diode current

$$I_{D(m)} = I_{o(avg.)} = 7.4 \text{A}$$

g) from Example 7.1(f)

$$\text{PIV} \geq 311 \text{ V}$$

h) from Example 7.1(g),

$$I_{D(avg.)} = 2.5 \text{ A}$$

■

7.3   Three-Phase Full-Wave (Six-Pulse) Bridge Rectifiers

7.3.1   With a Resistive Load

The three-phase full-wave (six-pulse) bridge rectifier is one of the most important circuits in high-power applications. The rectifier can be connected directly
to a three-phase source, or it can use a three-phase transformer connected in a
Δ-Y, Y-Δ, or Δ-Δ connection. A six-pulse rectifier provides an output that has
less ripple than that of the three-pulse rectifier. Figures 7.6 shows the circuit diagram; the diodes are numbered in the order in which they conduct. The bridge
rectifier uses both the positive and the negative half of the input voltage. The
negative peaks are turned over across the load resistor R. Therefore, the ripple
frequency is six times the AC source frequency.

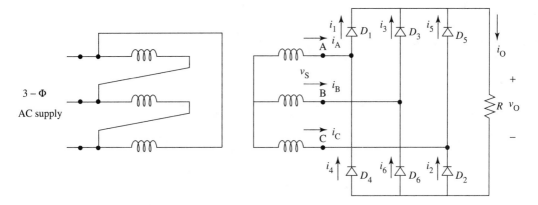

Figure 7.6
Full-wave bridge rectifier circuit diagram

7.3.1.1 Operation of the Full-Wave Bridge Rectifier

We will make the following assumptions in analyzing the operation of the bridge rectifier:

1. Two series diodes are always conducting, while the other four are blocking.

2. One of the conducting diodes is an odd numbered (D_1, D_3, or D_5), while the other is even-numbered (D_2, D_4, or D_6).

3. Each diode conducts for 120°, or one-third of a cycle.

4. Current flows out from the most positive source terminal, through an odd-numbered diode, through the load, through an even-numbered diode, and then back to the most negative source terminal.

The current flow at any given time can therefore be determined by finding the highest positive and the highest negative source terminals. The most positive terminal will forward-bias its respective odd-numbered diode and turn it on. The most negative terminal will forward-bias its respective even-numbered diode and turn it on.

To find out the most positive and the most negative source terminals, we can plot any two line voltages with respect to a common reference terminal. We arbitrarily choose terminal B as the reference. The two line voltages are v_{AB} and v_{CB}, as shown in Figure 7.7(b). v_{CB} is actually the inverse of v_{BC} (see Figure 7.7(a)).

The condition of the diodes can be found easily from Figure 7.7(b). During the interval 0° to 60°, the voltage at terminal C is the highest. Thus, from Figure 7.6, the anode of D_5 is at the most positive voltage in the circuit. This forward-biases D_5, turning it on. From 60° to 180°, terminal A becomes the most positive; therefore, D_1 is forward-biased and turns on. At 180°, terminal A voltage goes

Figure 7.7
Three-phase AC source voltage waveforms (a) line voltages with ABC phase sequence (b) line voltages v_{AB} and v_{CB}

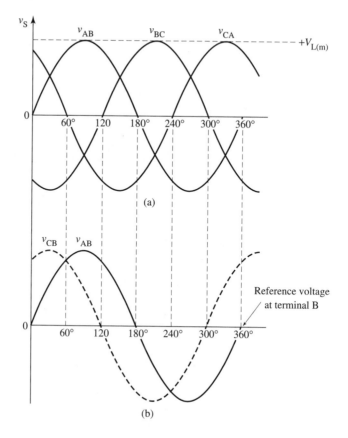

(a)

(b)

below the voltage at terminal B. Now the voltage at terminal B is the most positive, and this turn D_3 on. At 300°, the voltage at terminal C is again the most positive, turning D_5 on.

Similarly, the state of the even-numbered diodes can be determined from Figure 7.7(b) by finding the most negative terminal voltage. From 0° to 120°, terminal B is most negative; from 120° to 240°, terminal C is most negative; and from 240° to 360°, terminal A is most negative. The results are summarized in Table 7–2.

We can use Table 7–2 to draw the simplified equivalent circuits of the six-pulse rectifier shown in Figure 7.8. From these circuits, we can easily determine the output voltage for each 60° period. Figure 7.9(a) shows the three line voltages and their inverse voltages; for each 60° interval, a portion of the voltage from Figure 7.9(a) is redrawn as v_o in Figure 7.9(b) to obtain the complete output voltage waveform.

The output voltage fluctuates between 1.414 V_s and 1.225 V_s, where V_s is the RMS value of the line voltage. The average DC load voltage is twice that of a half-wave rectifier, and its value is given by

$$V_{o(avg.)} = 1.654\ V_m \qquad\qquad\qquad \textbf{7.15}$$

Table 7–2

Period	Highest positive voltage	Highest negative voltage	On diodes	
			Odd-numbered	Even-numbered
0 to 60°	C	B	D_5	D_6
60 to 120°	A	B	D_1	D_6
120 to 180°	A	C	D_1	D_2
180 to 240°	B	C	D_3	D_2
240 to 300°	B	A	D_3	D_4
300 to 360°	C	A	D_5	D_4

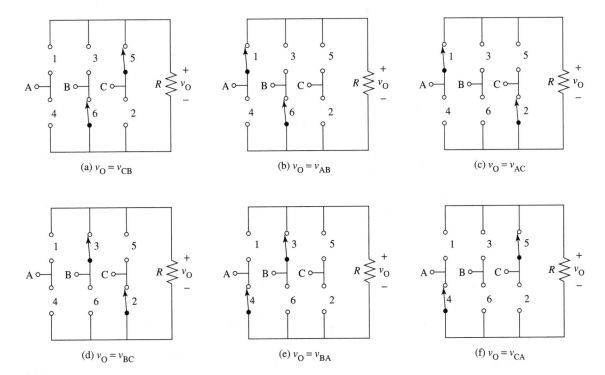

Figure 7.8
Equivalent circuits of the six-pulse rectifier

Figure 7.9
(a) Source voltages
including the three inverted
voltages (b) output voltage
waveforms of the bridge
rectifier

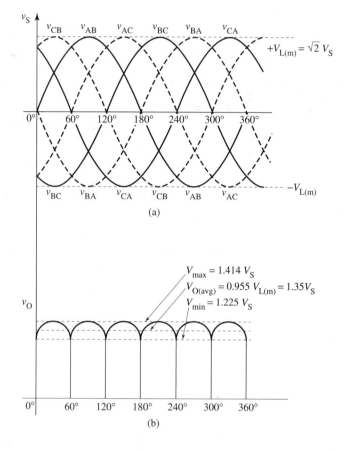

(a)

(b)

where V_m is the maximum value of the phase voltage. In terms of line voltage, the average load voltage is given by

$$V_{o(avg.)} = 0.955 \ V_{L(m)} \qquad \qquad \textbf{7.16}$$

The current through the diodes is shown in Figure 7.10. Each diode still conducts for 120°. The line currents i_A, i_B, i_C supplied by the transformer can be obtained from Kirchoff's current law:

$$i_A = i_1 - i_4$$
$$i_B = i_3 - i_6$$
$$i_C = i_5 - i_2$$

The line currents are also plotted in Figure 7.10. Note that the three line currents consist of identical waves that are 120° out of phase.

The average load current is given by

$$I_{o(avg.)} = \frac{3}{\pi} I_m = \frac{3}{\pi} \frac{V_m}{R} = 0.955 \ \frac{V_m}{R} \qquad \qquad \textbf{7.17}$$

Figure 7.10
The diode and line current waveforms

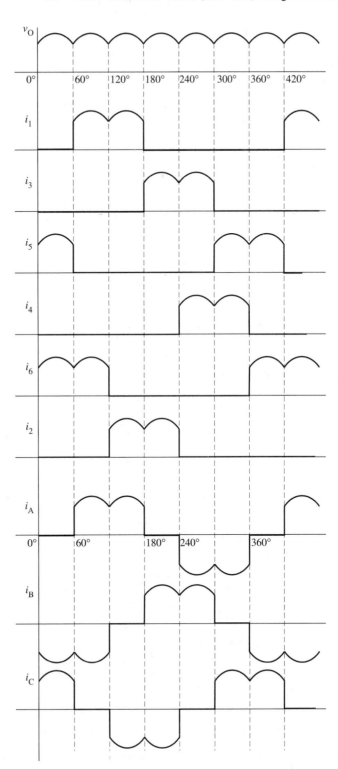

The average current in one of the diodes is only one-third the load current:

$$I_{D(avg.)} = I_{o(avg.)}/3 \qquad\qquad \textbf{7.18}$$

The RMS value of the diode current is

$$I_{D(RMS)} = \frac{1}{\sqrt{3}} I_{o(avg.)} \qquad\qquad \textbf{7.19}$$

The ripple factor is

$$RF = \frac{\sqrt{2}}{(n^2 - 1)} = 0.0404 \qquad\qquad \textbf{7.20}$$

and the ripple frequency is

$$f_r = 6 f_s \qquad\qquad \textbf{7.21}$$

The diode conduction periods is

$$\frac{2\pi}{3} = 120°. \qquad\qquad \textbf{7.22}$$

The maximum blocking voltage for diode

$$PIV \geq V_{L(m)} \qquad\qquad \textbf{7.23}$$

Table 7–3 shows the voltage across the diodes for various 60° intervals.

The voltage across any diode can therefore be plotted by first drawing the waveform for the line voltages (see Figure 7.11) and then picking the voltage from Table 7–3.

Table 7–3

Period	On diodes	Diode voltages					
		v_{D1}	v_{D2}	v_{D3}	v_{D4}	v_{D5}	v_{D6}
0 to 60°	D_5 and D_6	v_{AC}	v_{BC}	v_{BC}	v_{BA}	0	0
60 to 120°	D_6 and D_1	0	v_{BC}	v_{BA}	v_{BA}	v_{CA}	0
120 to 180°	D_1 and D_2	0	0	v_{BA}	v_{CA}	v_{CA}	v_{CB}
180 to 240°	D_2 and D_3	v_{AB}	0	0	v_{CA}	v_{CB}	v_{CB}
240 to 300°	D_3 and D_4	v_{AB}	v_{AC}	0	0	v_{CB}	v_{AB}
300 to 360°	D_4 and D_5	v_{AC}	v_{AC}	v_{BC}	0	0	v_{AB}

Figure 7.11
Voltage across the diodes
(a), (b), and (c)

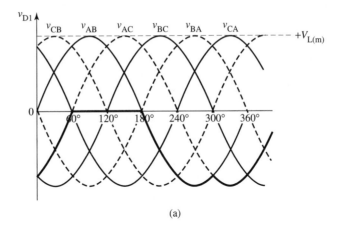

(a)

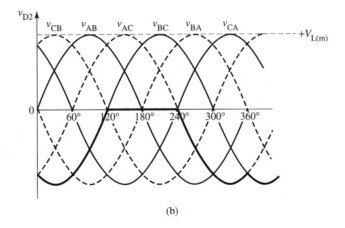

(b)

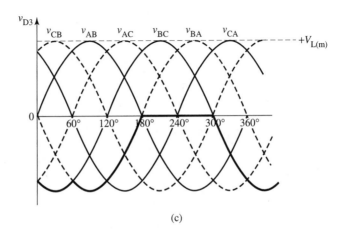

(c)

Figure 7.11
Voltage across the diodes
(d), (e), and (f)

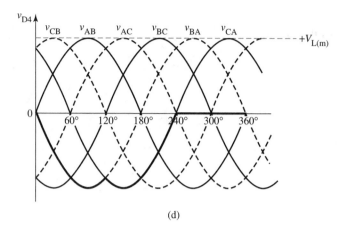

(d)

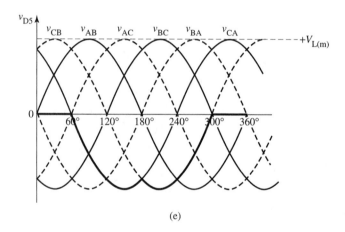

(e)

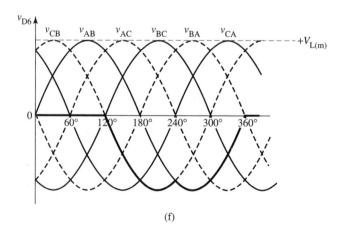

(f)

Example 7.3 A six-pulse uncontrolled rectifier is connected to a 3 Φ, 208 V, 60 Hz source. If the load resistance is 5 Ω, find
a) the average load voltage
b) the average load current
c) the average diode current
d) the maximum diode current
e) the PIV rating of the diode
f) the ripple frequency
g) the peak-to-peak ripple voltage
h) the form factor
i) the pulse number
j) the conduction angle

Solution a) $V_{L(m)} = \sqrt{2}\,(208) = 294$ V

$$V_{o(avg.)} = 0.955\ V_{L(m)} = 0.955 * 294 = 281 \text{ V}$$

b) $I_{o(avg.)} = V_{o(avg.)}/R = 281/5 = 56.2$ A

c) $I_{D(avg.)} = \dfrac{I_{o(avg.)}}{3} = 56.2/3 = 18.7$ A

d) $I_{D(max)} \approx I_{o(avg.)} = 56.2$ A
e) PIV across each diode $\geq V_{L(m)} = 294$ V
f) ripple frequency

$$f_r = 6\,f_s = 6 * 60 = 360 \text{ Hz}$$

g) The output voltage fluctuates between V_{min} and V_{max}:

$$V_{min} = 1.225\ V_s = 1.225 * 208 = 255 \text{ V}$$
$$V_{max} = 1.414\ V_s = 1.414 * 208 = 294 \text{ V}$$

The peak-to-peak ripple voltage is $294 - 255 = 39$ V.
h) form factor

$$\text{FF} = V_s/V_{o(avg.)} = 208/281 = 0.74$$

i) pulse number

$$P = 6$$

j) The conduction Angle is

$$\theta = 120°$$

7.3.2 With an Inductive (*RL*) Load

If the load is inductive, the output voltage waveform remains the same as with a resistive load, and the load current will have reduced ripples. The waveforms for current are shown in Figure 7.12, where a very large value of inductance is assumed.

Figure 7.12
Bridge rectifier current
waveforms with an
inductive load

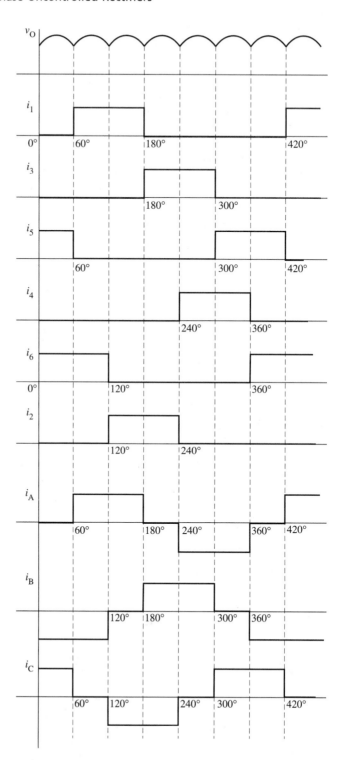

The average or DC load voltage is given by

$$V_{o(avg.)} = \frac{3}{\pi} \, V_{L(m)} = 0.955 \; V_{L(m)} \qquad\qquad \textbf{7.24}$$

The average load current is:

$$I_{o(avg.)} = \frac{V_{o(avg.)}}{R}$$

$$= \frac{0.955 \; V_m}{R} \qquad\qquad \textbf{7.25}$$

The diode conducts for one third of the cycle. Therefore, the average diode current is

$$I_{D(avg.)} = I_{o(avg.)}/3 \qquad\qquad \textbf{7.26}$$

The RMS value of the diode current is

$$I_{D(RMS)} = \frac{1}{\sqrt{3}} I_{o(avg.)} \qquad\qquad \textbf{7.27}$$

The RMS value of the line currents (i_A, i_B or i_C) is $\dfrac{\sqrt{2}}{\sqrt{3}} I_{o(avg.)} = 0.82 \; I_{o(avg.)}$

The RMS value of the load current is given by

$$I_{o(RMS)} = 0.956 \; I_m \qquad\qquad \textbf{7.28}$$

Note that there is little difference between the average and the RMS values of the load current because the current waveform has a very small ripple.

The PIV across each element is

$$V_{L(m)} = \sqrt{2} \; V_s \qquad\qquad \textbf{7.29}$$

The peak-to-peak ripple is only $(1.414 - 1.225) \; V_s = 0.189 \; V_s$.   **7.30**

The six-pulse rectifier is a big improvement over the three-pulse rectifier and it forms the basic building block of most larger rectifier installations.

Example 7.4   A six-pulse uncontrolled rectifier connected to a 3 Φ, 220 V source is supplying an RL load. If the inductance is very large and the load resistance is 50 Ω, find

a) the average load voltage
b) the average load current
c) the PIV rating of the diode
d) the average diode current
e) the RMS load current
f) the RMS diode current
g) the power to the load

Solution a) $V_{L(m)} = \sqrt{2}\,(220) = 311$ V

$V_{o(avg.)} = 0.955\ V_{L(m)} = 0.955 * 311 = 297$ V

b) $I_{o(avg.)} = V_{o(avg.)}/R = 297/50 = 5.94$ A

c) PIV $\geq V_{L(m)} = 311$ V

d) $I_{D(avg.)} = \dfrac{I_{o(avg.)}}{3} = 5.94/3 = 2.07$ A

e) $I_{D(m)} = I_m = V_m/R = 311/50 = 6.22$ A

$I_{o(RMS)} = 0.956 * I_m = 0.956 * 6.22 = 5.95$ A

(close to the average load current since the ripple is very small)

f) RMS diode current

$$I_{D(RMS)} = \frac{1}{\sqrt{3}}I_{o(avg.)} = \frac{5.94}{1.73} = 3.43 \text{ A}$$

g) Power to the load

$$P_L = I_{o(RMS)}^2\ R = 5.95^2 * 50 = 1770 \text{ W}$$

7.4 Twelve-Pulse Rectifier Circuits

To further reduce the ripple voltage in the DC output and increase the ripple frequency, we can increase the pulse number from six to twelve. A twelve-pulse rectifier can be constructed by connecting two six-pulse rectifiers in series, as shown in Figure 7.13. The three-phase AC sources supplying these two bridges are shifted by 30° with respect to one another. This can be achieved easily by phase shifting the AC sources using two three-phase transformers, one of which is Y-connected and the other Δ-connected on the secondary side. On the primary side, both transformers are connected in Y to the same three-phase source. With this connection, the secondary side phase voltages of the Δ transformer are shifted in phase by 30°, so all the secondary phase voltages of one transformer will be shifted in phase by the same 30° with respect to the corresponding phases in the other. There will also be a difference in the magnitude of the secondary phase voltage. The secondary voltage of the Δ transformer will be less by a factor of 3 than that of the Y transformer. However, this can be taken care of by having a different turn's ratio (a $\sqrt{3}$ for Y-Δ transformer). In this way, the secondary voltages can be identical in magnitude.

The resulting secondary line voltages v_{ab} and $v_{a'b'}$ are displaced by 30° as a result of the transformer connection. Because each of the six-pulse bridges operates independently, the output voltage v_o is the sum of v_{o1} and v_{o2}. Figure 7.14 shows the waveform of the output voltage. As shown, v_o is the twelve-pulse output, with a ripple frequency twelve times the source frequency.

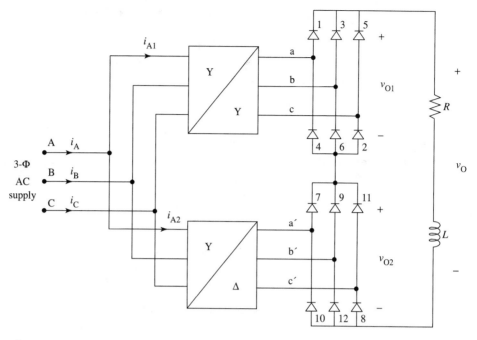

Figure 7.13
A twelve-pulse rectifier

Figure 7.14
Output voltage waveform

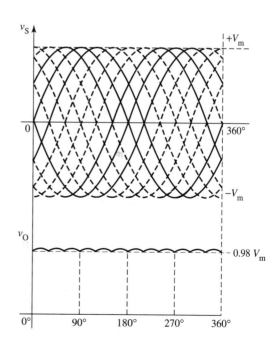

The average value of the output voltage is

$$V_{o(avg.)} = \frac{V_m * 6\sqrt{2}}{\pi(\sqrt{3}+1)} = 0.989\ V_m$$

7.31

The PIV rating for diodes is

$$\text{PIV rating} \geq 3\ V_s$$

7.32

Figure 7.15
Input current waveforms for
a twelve-pulse rectifier

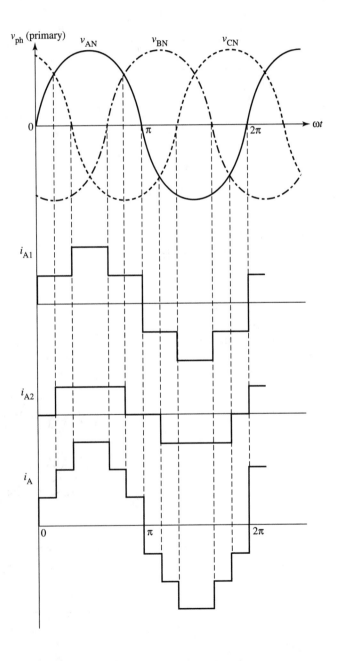

Figure 7.15 illustrates the combination of input current waveforms to give a resultant line current that is much closer to a sine wave than the six-pulse waveform.

7.5 Problems

7.1 What are the advantages of a six-pulse rectifier over a three-phase rectifier?

7.2 Draw the waveform for the output voltage if the diodes in Figure 7.1 are reversed.

7.3 A three-pulse rectifier delivers 20 A to a resistive load. If the load voltage is 120 V DC, find
a) the average diode current
b) the PIV rating of each diode
c) the RMS diode current

7.4 A six-pulse rectifier supplies 8.8 kW to a resistive load. If the load voltage is 220 V DC, find
a) the average diode current
b) the PIV rating of each diode
c) the RMS diode current

7.5 A three-pulse rectifier supplies a resistive load of 10 Ω from a 220 V source. Find
a) the average load voltage
b) the average load current
c) the maximum load current
d) the PIV rating of the diode
e) the maximum diode current
f) the average load power

7.6 Repeat problem 7.5 after adding a large inductance in series with the load resistance.

7.7 A three-pulse rectifier is connected to a 220 V source. If the rectifier supplies an average load current of 50 A, find
a) the DC load voltage
b) the diode average current
c) the maximum current in each diode
d) the RMS value of the line current

7.8 The six-pulse rectifier in Figure 7.6 is connected to a 220 V source. If the rectifier supplies an average load current of 50 A, find
a) the DC load voltage
b) the diode average current
c) the maximum current in each diode
d) the RMS value of the line current

7.9 A six-pulse rectifier supplies a resistive load of 20 Ω from a 220 V source. Find

a) the average load voltage

b) the average load current

c) the average diode current

d) the PIV rating of the diode

e) the average load power

7.10 Repeat problem 7.9 after adding a large inductance in series with the load resistance.

7.6 Equations

$$f_r = n f_s \tag{7.1}$$

$$V_{o(avg.)} = \frac{n}{\pi} V_m \sin\left(\frac{\pi}{n}\right) \tag{7.2}$$

$$V_{o(avg.)} = 0.827 \; V_m \tag{7.3}$$

$$V_{o(avg.)} = 0.477 \; V_{L(m)} \tag{7.4}$$

$$I_{o(avg.)} = \frac{n}{\pi} I_m \sin\left(\frac{\pi}{n}\right) \tag{7.5}$$

$$= 0.827 \; I_m \tag{7.6}$$

$$I_{D(avg.)} = I_{o(avg.)}/n = I_{o(avg.)}/3 \tag{7.7}$$

$$I_{o(m)} = 1.21 \; I_{o(avg.)} \tag{7.8}$$

$$I_{o(RMS)} = I_m \left[\frac{1}{2\pi}\left(\frac{\pi}{3} + \frac{1}{2}\sin\frac{2\pi}{3}\right)\right]^{1/2} \tag{7.9}$$

$$RF = \frac{\sqrt{2}}{n^2 - 1} \tag{7.10}$$

$$FF = \sqrt{n} \tag{7.11}$$

$$\text{PIV rating} \geq V_{L(m)} \tag{7.12}$$

$$I_{o(RMS)} = I_{o(avg.)} \tag{7.13}$$

$$I_{D(max)} = I_{o(max)} = I_{D(avg.)} \tag{7.14}$$

$$V_{o(avg.)} = 1.654 \; V_m \tag{7.15}$$

$$V_{o(avg.)} = 0.955 \; V_{L(m)} \tag{7.16}$$

$$I_{o(avg.)} = \frac{3}{\pi}\frac{V_m}{R} \tag{7.17}$$

$$I_{D(avg.)} = I_{o(avg.)}/3 \tag{7.18}$$

$$I_{D(RMS)} = \frac{1}{\sqrt{3}} I_{o(avg.)} \tag{7.19}$$

$$RF = \frac{\sqrt{2}}{[n^2 - 1]}$$ **7.20**

$$f_r = 6 f_s$$ **7.21**

diode conduction period = $2\pi/3$ **7.22**

$$PIV \geq V_{L(m)}$$ **7.23**

$$V_{o(avg.)} = \frac{3}{\pi} V_{L(m)}$$ **7.24**

$$I_{o(avg.)} = \frac{V_{o(avg.)}}{R}$$ **7.25**

$$I_{D(avg.)} = I_{o(avg.)}/3$$ **7.26**

$$I_{o(RMS)} = \frac{1}{\sqrt{3}} I_{o(avg.)}$$ **7.27**

$$I_{RMS} = 0.956 I_m$$ **7.28**

$$PIV \geq V_{L(m)}$$ **7.29**

peak-to-peak ripple = $0.189 V_s$ **7.30**

$$V_{o(avg.)} = \frac{V_m * 6\sqrt{2}}{\pi(\sqrt{3} + 1)} = 0.989 V_m$$ **7.31**

$$PIV \geq \sqrt{3} V_s$$ **7.32**

Three-Phase Controlled Rectifiers

8

Learning Objectives

After completing this chapter, the student should be able to

- discuss the advantages of three-phase controlled rectifiers as compared with single-phase controlled rectifiers
- describe with the help of waveforms the operation of a half-wave controlled rectifier with resistive and inductive loads
- describe with the help of waveforms the operation of a full-wave controlled

rectifier with resistive and inductive loads

- explain the operation of a full-wave half-controlled bridge rectifier
- explain the operation of a twelve-pulse controlled rectifier

8.1 Introduction

Single-phase controlled rectifiers have a high AC ripple content in their DC output voltage and are limited by the power capability of the single-phase source. When high power levels are required, three-phase controlled rectifiers are preferred because they provide an increased average DC output voltage and a reduced AC ripple component. When the diodes in the three-phase rectifier circuits of Chapter 7 are replaced by SCR, the circuits become fully controllable and the average output voltage can be varied by controlling the triggering inputs to the SCR gates in a suitable manner.

8.2 Half-Wave (Three-Pulse) Controlled Rectifiers

8.2.1 With a Resistive Load

Figure 8–1(a) shows a three-phase half-wave controlled rectifier with a resistive load. It is called a three-pulse circuit because the pulsation in DC voltage is three times the input frequency. Each SCR receives a firing pulse relative in time to its own phase voltage. The three gate pulses are displaced by 120° relative to each other, giving the same delay angle to each SCR.

If each SCR is triggered at the instant when the source makes its anode voltage positive with respect to its cathode (that is, 30° after the phase voltage crosses the zero axis), then the circuit behaves like a half-wave uncontrolled diode rectifier. However, if the firing of the SCRs is delayed from these crossing points, the output voltage waveform is altered.

The waveforms of the phase voltages v_{AN}, v_{BN}, and v_{CN} are shown in Figure 8.1(b). During the interval $\omega t = 30°$ to $150°$, the most positive voltage is v_{AN}. Therefore, SCR_1 is forward-biased in this interval and will conduct when triggered, while the remaining two SCRs are reverse-biased. SCR_1 will continue to conduct until $\omega t = 150°$, at which point voltage v_{BN} begins to become more positive than v_{AN}. SCR_2 now becomes forward-biased and will turn on if we apply a firing signal. When SCR_2 turns on, it automatically turns SCR_1 off by natural commutation. The same process is repeated at $\omega t = 270°$. Each SCR conducts for a period of 120° and blocks reverse voltage for 240°. When an SCR is on, it connects the input voltage terminal to the output terminal, so the output voltage is the same as the corresponding AC phase voltage. Therefore, the output voltage waveform consists of portions of the AC input voltage waveform. If the delay angle is zero, the output voltage consists of the peaks of the phase voltages and is at its maximum (see Figure 8.1(b)).

The firing angle α is measured from the intersection or crossover points of the corresponding phase voltages and not the zero crossing of the voltage waves. If each SCR turn-on is delayed by an angle α, the segments of the out-

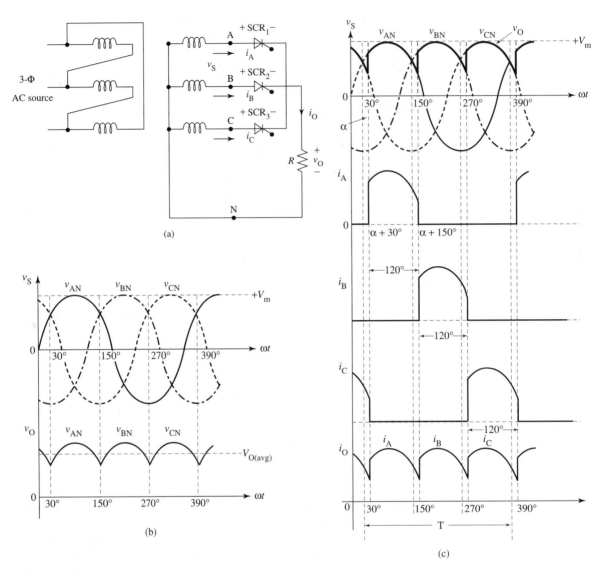

Figure 8.1
Half-wave controlled rectifier (a) circuit diagram (b) waveforms for voltage and current with $\alpha = 0°$ (c) waveforms with a small delay angle

put voltage waveform will also be delayed by the angle α, but the output will still have three pulses. The waveforms of the output voltage and current as a function of time for $\alpha < 30°$ are shown in Figure 8.1(c). As shown, the phase A SCR remains conducting until the phase B SCR is turned on. The output voltage and output current do not go to zero at any time; however, the average output voltage is decreased compared with the 0° case.

The average voltage is given by

$$V_{o(avg.)} = \frac{3\sqrt{3}}{2\pi} V_m \cos\alpha \qquad \text{for } 0° \leq \alpha \leq 30°$$

$$= 0.827 \, V_m \cos\alpha \qquad \qquad \textbf{8.1}$$

where V_m is the maximum value of the phase voltage.

Note that the effect of phase control is to add the $\cos\alpha$ term to the equation for a three-phase uncontrolled rectifier. From Equation 8.1, it is clear that the average value of the output voltage can be controlled by varying the firing angle α.

The average output current is

$$I_{o(avg.)} = \frac{V_{o(avg.)}}{R}$$

$$= \frac{3\sqrt{3}}{2\pi} \frac{V_m}{R} \cos\alpha \qquad \qquad \textbf{8.2}$$

The average SCR current is

$$I_{SCR(avg.)} = \frac{I_{o(avg.)}}{3} = \frac{\sqrt{3}}{2\pi} \frac{V_m}{R} \cos\alpha \qquad \qquad \textbf{8.3}$$

The RMS SCR current is

$$I_{SCR(RMS)} = \frac{I_{SCR(avg.)}}{\sqrt{3}} = \frac{1}{2\pi} \frac{V_m}{R} \cos\alpha \qquad \qquad \textbf{8.4}$$

The peak reverse voltage rating of the SCR is

$$\sqrt{3} \, V_m \text{ or } V_{L(m)} \qquad \qquad \textbf{8.5}$$

where $V_{L(m)}$ is the maximum value of the line voltage.

The conduction period for each SCR is one-third of a cycle or

$$\frac{2\pi}{3} \text{ or } 120° \qquad \qquad \textbf{8.6}$$

The ripple frequency is

$$f_r = 3 * \text{AC supply frequency} \qquad \qquad \textbf{8.7}$$

Example 8.1   A three-phase half-wave controlled rectifier connected to a three-phase, 208 V, 60 Hz AC source supplies power to a 10 Ω resistive load. If the delay angle is 20°, find

a) the maximum output current
b) the average output voltage
c) the average output current

d) the maximum SCR current
e) the average SCR current
f) the RMS SCR current
g) the maximum reverse voltage rating
h) the ripple frequency

Solution

$$V_{phase} = V_I/\sqrt{3} = 208/1.732 = 120 \text{ V}$$
$$V_m = \sqrt{2} * 120 = 170 \text{ V}$$

a) $I_m = V_m/R = 170/10 = 17$ A

b) $V_{o(avg.)} = 0.827 \ V_m \cos \alpha = (0.827)(170)(\cos 20°) = 132$ V

c) $I_{o(avg.)} = V_{o(avg.)}/R = 132/10 = 13.2$ A

Figure 8.2
Half-wave controlled
rectifier waveforms ($30° \leq \alpha$
$\leq 150°$)

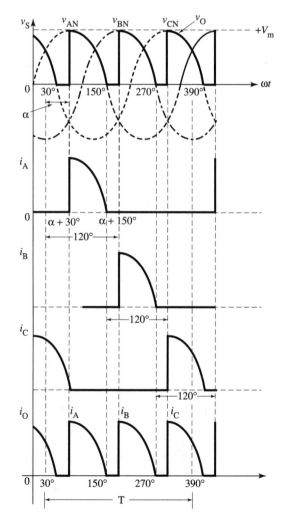

d) maximum SCR current = I_m = 17 A

e) SCR average current = $I_{SCR(avg.)}$ = $I_{o(avg.)}/3$ = 13.2/3 = 4.4 A

f) SCR RMS current = $I_{SCR(avg.)}/\sqrt{3}$ = 4.4/1.732 = 2.5 A

g) maximum reverse voltage = $V_{L(m)}$ = $\sqrt{2}$ (208) = 294 V

h) f_r = 3 * AC supply frequency = 180 Hz

When α becomes larger, 30° ≤ α ≤ 150°, the output current decreases to zero at some time and then tends to become negative. This is not possible with a resistive load, so the output current and voltage both remain zero until the next SCR is turned on. Figure 8.2 shows the voltage and current waveforms. The average output voltage is

$$V_{o(avg.)} = \frac{3\ V_m}{2\pi}\left(1 + \cos\left(\alpha + \frac{\pi}{6}\right)\right) \qquad \text{for } 30° \le \alpha \le 150° \qquad \textbf{8.8(a)}$$

$$= \frac{3\ V_m}{2\pi}(1 + 0.866 \cos \alpha - 0.5 \sin \alpha)$$

For values of $\alpha \ge 150°$, the average output voltage becomes zero:

$$V_{o(avg.)} = 0 \text{ V} \qquad \text{for } 150° \le \alpha \le 180° \qquad \textbf{8.8(b)}$$

Example 8.2   Repeat Example 8.1 for a delay angle of 100°.

Solution   With a delay angle of 100° (0.55 π), the SCR turns on after the supply voltage has reached its maximum value. Therefore, the maximum output voltage is less than V_m:

$$\text{maximum output voltage} = V_m \sin (30° + \alpha)$$
$$= 170 (\sin 130°)$$
$$= 130 \text{ V}$$

a) maximum output current = $\dfrac{\text{maximum output voltage}}{R}$

$$= 130/10$$
$$= 13 \text{ A}$$

b) From equation 8.8,

$$V_{o(avg.)} = \frac{3\ V_m}{2\ \pi}(1 + 0.866 \cos 100° - 0.5 \sin 100°)$$

$$= 0.17\ V_m$$
$$= 0.17 * 170$$
$$= 29 \text{ V}$$

c) $I_{o(avg.)}$ = $V_{o(avg.)}/R$ = 29/10 = 2.9 A

d) maximum SCR current = maximum output current = 13 A

e) SCR average current $= I_{SCR(avg.)} = I_{o(avg.)}/3 = 2.9/3 = 0.97$ A

f) SCR RMS current $= I_{SCR(avg.)}/\sqrt{3} = 0.97/1.732 = 0.56$ A

g) maximum reverse voltage $= \sqrt{2}\,(208) = 294$ V

h) ripple frequency $= f_r = 3 * 60$ Hz $= 180$ Hz

Figure 8.3
Waveforms with inductive (*RL*) load and continuous current

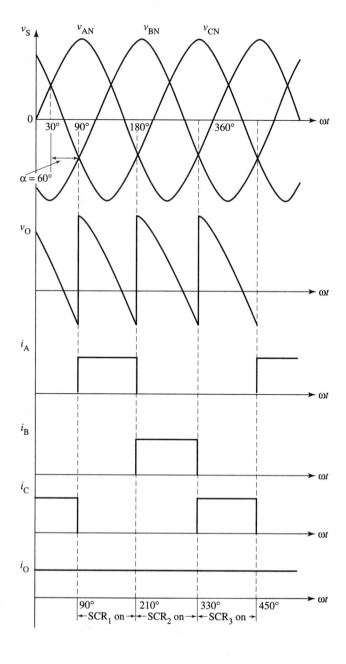

8.2.2 With an Inductive (*RL*) Load without a Freewheeling Diode

In the continuous constant-current mode where the inductive component of the load is sufficiently large, the output voltage waveform may be negative for some values of α. If the delay angle α is less than 30°, the output voltage is always positive and its average value is given by Equation 8.1. However, for a delay angle α greater than 30°, the output voltage becomes negative for a portion of each cycle. Figure 8.3 shows the waveforms for α = 60°.

The average output voltage is

$$V_{o(avg.)} = \frac{3\sqrt{3}}{2\pi} V_m \cos \alpha \qquad 0° \le \alpha \le 180°$$

$$= 0.827 \, V_m \cos \alpha \qquad \textbf{8.9}$$

This equation is the same as in the resistive case, but it is no longer limited to α < 30°. The earlier case was limited because the output voltage altered its shape for α > 30° with a resistive load. Now the output voltage continues along the same portion of the sinusoid for α > 30°.

The maximum average output voltage occurs at α = 0°:

$$V_{o(avg.)max} = 0.827 \, V_m \qquad \textbf{8.10}$$

The normalized average output voltage is

$$V_n = \frac{V_{o(avg.)}}{V_{o(avg.)max}} = \cos \alpha \qquad \textbf{8.11}$$

The average output voltage is zero for α = 90°. The control characteristic curve is shown in Figure 8.4.

Figure 8.4
Control characteristic of the DC output voltage

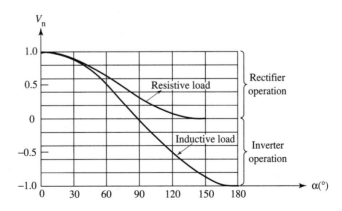

For a constant current, the RMS current in each SCR is

$$I_{SCR(RMS)} = I_{o(avg.)}/3 \qquad \textbf{8.12}$$

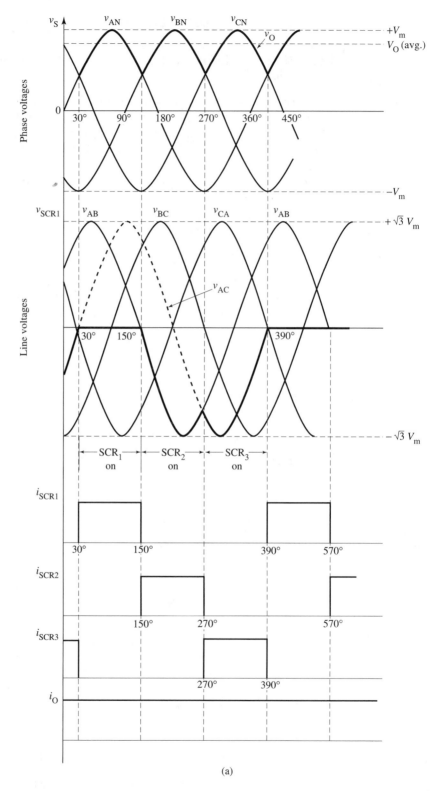

Figure 8.5
Voltage and current waveforms for Example 8.3 (a) $\alpha = 0°$

Example 8.3 For a three-pulse controlled rectifier with an *RL* load, plot the waveforms of the output voltage, the voltage across SCR$_1$, and the current through each SCR if the output current firing angle is

a) 0°
b) 45°
c) 90°
d) 135°
e) 180°

Figure 8.5
Voltage and current
waveform for Example 8.3
(b) α = 45°

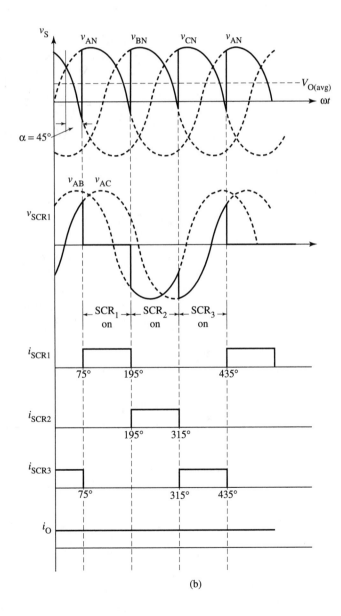

Solution   a) When $\alpha = 0°$, the AC input voltage with the highest instantaneous value is applied to the output terminal. The average output voltage is therefore at its positive maximum. Each SCR conducts current for a duration of 120°.

When SCR_1 is conducting, $v_{SCR1} = 0$ V. When SCR_2 is conducting, SCR_1 and SCR_3 are off, and v_{SCR1} is therefore v_{AB}. When SCR_3 is conducting, the voltage across SCR_1 is v_{AC}. The plot is shown in Figure 8.5(a).

b) When $\alpha = 45°$, the SCRs block forward voltage for 45°, and the average output voltage is reduced. Figure 8.5(b) shows the waveforms.

c) As the delay angle is increased to 90°, the SCRs block the forward and reverse voltage for equal periods and the average voltage is zero (see Figure 8.5(c)).

Figure 8.5
Voltage and current
waveform for Example 8.3
(c) $\alpha = 90°$

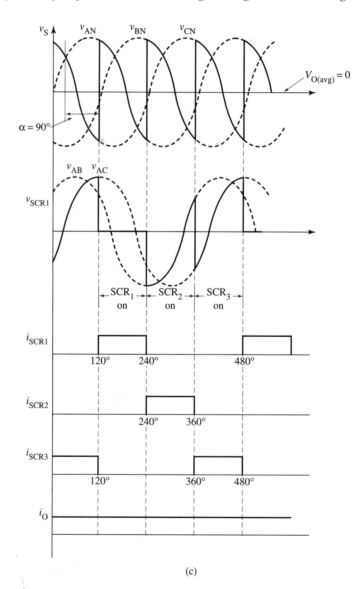

(c)

d) To operate with delay angles greater than 90°, we should have a counter-emf source as a load, for example, a DC motor under braking conditions. With $\alpha = 135°$, the SCRs block in the forward direction and the average output voltage becomes negative. This does not produce a negative current, since the SCRs can conduct only in the forward direction. The reversal of voltage polarity with the direction of current flow unchanged indicates a reversal of power flow. Under this condition, the circuit becomes an inverter, supplying power from the DC load side to the AC source side (see Figure 8.5(d)).

Figure 8.5
Voltage and current waveform for Example 8.3
(d) $\alpha = 135°$

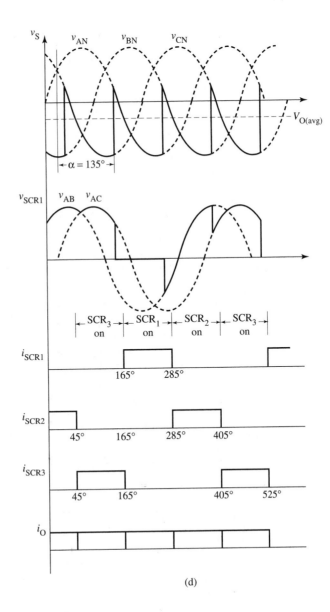

(d)

e) When α = 180°, the SCRs block forward voltage for the entire duration and the output voltage is at its negative maximum (Figure 8.5(e)).

From this example, it is clear that when 0° ≤ α ≤ 90°, the circuit operates in the rectifying mode, and for 90° ≤ α ≤ 180°, it operates in the inverting mode.

Figure 8.5
Voltage and current waveform for Example 8.3
(e) α = 180°

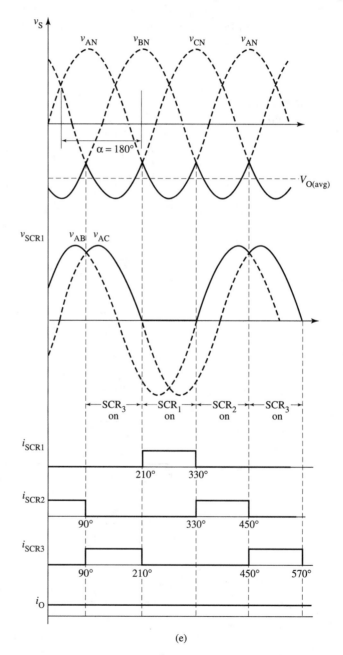

(e)

Example 8.4   A three-phase half-wave controlled rectifier is connected to a 220 V source. If the delay angle is 45° and the load resistance $R = 10\ \Omega$ find
a) the average output voltage
b) the average output current
c) the maximum output current
d) the maximum SCR current
e) the average SCR current
f) the SCR RMS current
g) the average power dissipation in the SCR, if the SCR has a forward voltage drop of 1.0V
h) the maximum reverse voltage rating

Solution   With a delay angle of 45°, the SCR turns on before the supply voltage reaches its maximum. Therefore, the maximum output voltage is equal to V_m, the maximum value of the phase voltage:

$$V_{Phase} = 220/\sqrt{3} = 127\ V$$
$$V_m = \sqrt{2}\ (127) = 180\ V$$

a) $V_{o(avg.)} = 0.827\ V_m \cos \alpha = (0.827)\ (180)\ (\cos 45°) = 105\ V$

b) average output current $= I_{o(avg.)} = V_{o(avg.)}/R = 105/10 = 10.5\ A$

c) maximum output current = average output current = 10.5 A

d) maximum SCR current = maximum output current = 10.5 A

e) SCR average current $= I_{o(avg.)}/3 = 10.5/3 = 3.5\ A$

f) SCR RMS current $= I_{SCR(RMS)} = I_{o(avg.)}/\sqrt{3} = 6.1\ A$

g) average power $= \dfrac{V_f * I_{o(avg.)}}{3} = \dfrac{1.0 * 10.5}{3} = 3.5\ W$

h) maximum reverse voltage $= \sqrt{2} * 220 = 311\ V$

8.2.3   With an Inductive (*RL*) Load and a Freewheeling Diode

A freewheeling diode (FWD) can be connected to the load in parallel to provide the load current with an alternative path that excludes the SCRs. Figure 8.6(a) shows a half-wave controlled rectifier with an FWD. In this circuit the output voltage cannot have a negative value. Thus inversion is not possible, and the average output voltage is the same as for the resistive case.

For small delay angles (0° ≤ α ≤ 30°), the instantaneous output voltage never becomes negative, so the FWD is always reverse-biased and has no effect. Equation 8.1 for average output voltage still applies. For delay angles greater than 30°, the output voltage tends to become negative during parts of the cycle, but the FWD prevents these negative excursions, periodically carrying the load

Figure 8.6
(a) Load with FWD

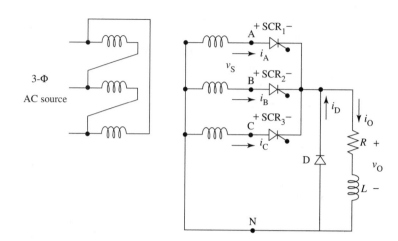

current that would otherwise have to flow through the rectifier. The average output voltage is given by Equation 8.7 for $30° < \alpha < 150°$. In this range of α, the FWD provides a path for the output current during three intervals in each cycle. Figure 8.6(b) shows the waveform for $\alpha = 75°$. In phase A, SCR_1 conducts from an angle of $105°$ ($75° + 30°$) to $180°$. The FWD conducts from $180°$ to the time when SCR_2 conducts at an angle of $225°$ ($75° + 150°$). As shown in the figure, the total conduction time for the SCR and one FWD period is still $120°$.

A delay angle greater than $150°$ would suggest a negative output voltage, which is not possible if an FWD is present.

8.3 Full-Wave (Six-Pulse) Bridge-Controlled Rectifier

The three-phase full-wave (or six-pulse) bridge rectifier is one of the most widely used high-power converters in power electronics. As shown in Figure 8.7, it is constructed with two three-phase, three-pulse rectifiers connected in series. SCRs 1, 3, and 5 are called a *positive group* since they are fired during the positive half-cycle of the voltage of the phases to which they are connected. Similarly, SCRs 2, 4, and 6 are fired during the negative cycles of the phase voltages and form the *negative group*. To provide a current path from the source side to the load side requires that two SCRs must be triggered simultaneously. Therefore, two pulses, $60°$ apart, are applied to each SCR in a cycle. When one element of the upper group and one element of the lower group of SCR, conduct the corresponding line voltage is applied directly to the load. For example, if SCR_3 and SCR_2 conduct simultaneously, then line voltage v_{BC} is applied across the load. The average voltage and current are controlled by the firing angle of the SCRs.

Figure 8.6
(b) voltage and current
waveforms for α = 75°

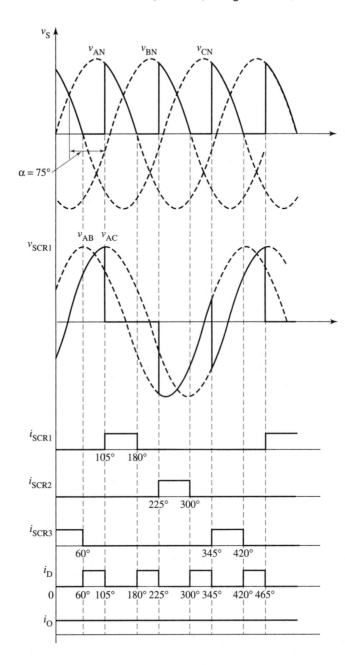

8.3.1 In Circuits with a Resistive Load (or with an Inductive Load and an FWD)

Consider the bridge circuit as two three-pulse groups in series, shifted 60° from each other. A simple method to analyze the waveforms is to obtain the output for each three-pulse group and then add them. If the SCRs are triggered at the

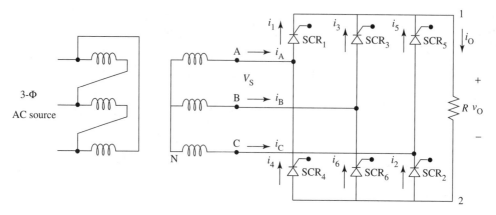

Figure 8.7
Six-pulse controlled bridge rectifier circuit

instant when the voltage tends to become positive in the forward direction, then the circuit will function in the same way as the diode bridge rectifier. The firing delay angle α of each SCR is measured from the crossover point of its respective phase voltage.

Consider the interval from 0° to 120°. In the positive group of SCRs connected to terminal 1, SCR$_1$ is connected to the most positive phase voltage, v_{AN}. With $\alpha = 0°$, SCR$_1$ will turn on and connect point 1 to A. In a similar manner, from 120° to 240°, SCR$_3$ turns on and connects point 1 to B. Over the interval 240° to 360°, point 1 is tied to C, and the cycle repeats. The voltage v_{1N} is therefore the positive peak of the phase voltages v_{AN}, v_{BN}, and v_{CN}.

In the negative group of SCRs connected to terminal 2, SCR$_4$ is connected to the most negative phase voltage during the interval from 180° to 300°. SCR$_4$ conducts and connects point 2 to A. Similarly, from 300° to 420° (or 60°), SCR$_6$

Table 8–1

Interval	Voltage at point 1	Voltage at point 2	Voltage v_{12}
0° to 60°	A	B	AB
60° to 120°	A	C	AC
120° to 180°	B	C	BC
180° to 240°	B	A	BA
240° to 300°	C	A	CA
300° to 360°	C	B	CB
360° to 420°	A	B	AB

turns on and connects point 2 to B. Over the interval from 60° to 180°, point 2 is tied to C. The voltage v_{2N} is therefore the negative peak of the phase voltages v_{AN}, v_{BN}, and v_{CN}. The output voltage v_o (= v_{12}) is simply $v_{1N} - v_{2N}$. Table 8–1 summarizes these results.

The output voltage waveform (see Figure 8.8) is smoother, consisting of six pulses of the input line voltages. The frequency of the output ripple voltage is six times the AC line frequency, and the magnitude is double that of a three-pulse rectifier. Each SCR conducts for 120° and blocks for 240° of each cycle. There are at least two SCRs always in conduction at the same time. With an ABC phase sequence, the firing order of the SCRs is SCRs 1 and 2, SCRs 2 and 3, SCRs 3 and 4, SCRs 4 and 5, and SCRs 5 and 6, and so forth, and the gate signals to the SCRs are 60° apart.

Figure 8.8
Output voltage waveforms of the bridge circuit with α = 0°

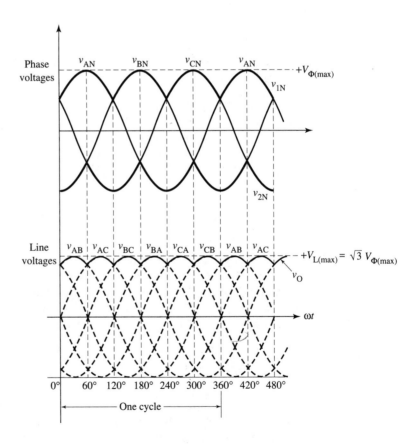

The voltage across the SCRs is easily determined from Table 8–1. For example, suppose the voltage across SCR$_1$ is needed. First note that v_{SCR1} is simply v_{A1}. Therefore, from 0° to 120°, $v_{A1} = 0$ since SCR$_1$ is conducting. From 120° to 240°, $v_{A1} = v_{AB}$ since 1 is tied to B, and from 240° to 360° $v_{A1} = v_{AC}$ since 1 is tied to C. The results are given in Table 8.2.

Table 8–2

Interval	V_{SCR1}	V_{SCR3}	V_{SCR5}	V_{SCR4}	V_{SCR6}	V_{SCR2}
0° to 60°	0	BA	CA	AB	0	CB
60° to 120°	0	BA	CA	AC	BC	0
120° to 180°	AB	0	CB	AC	BC	0
180° to 240°	AB	0	CB	0	BA	CA
240° to 300°	AC	BC	0	0	BA	CA
300° to 360°	AC	BC	0	AB	0	CB

Figure 8.9(a) shows the voltage across SCR$_1$. As shown, the maximum reverse voltage that can occur across an SCR is equal to the maximum instantaneous magnitude of the AC line voltage. It is important to note that the SCR should also be able to block forward voltages. The magnitude of this voltage depends on the firing angle. The larger the firing angle, the larger will be the forward voltage that the SCR must block.

The current in each SCR can also be determined from Table 8–1. The line currents are then obtained by applying KCL:

$$i_A = i_1 - i_4$$
$$i_B = i_3 - i_6$$
$$i_C = i_5 - i_2$$

The current waveforms are shown in Figure 8.9(b).

The SCR conduction now may be delayed by an angle α measured from the normal point of commutation. The average output voltage is then reduced. The circuit can operate in two different modes depending on the value of the delay angle. In the range $0° \leq \alpha \leq 60°$, the output voltage and current are continuous and the average output voltage is

$$V_{o(avg.)} = \frac{3\sqrt{3}}{\pi} V_m \cos \alpha \qquad \text{for } 0° \leq \alpha \leq 60° \qquad \textbf{8.13}$$

When the delay angle exceeds 60° (for $60° \leq \alpha \leq 120°$), the output voltage has a negative portion. If the inductive component of the load is sufficiently large, the current continues to flow. If the load is purely resistive or if an FWD is connected across an RL load, the negative portion of the output voltage is held to zero. The average value of the output voltage is then

$$V_{o(avg.)} = \frac{3\sqrt{3}}{\pi} V_m \left(1 + \cos\left(\alpha + \frac{\pi}{3}\right)\right) \qquad \text{for } 60° \leq \alpha \leq 120° \qquad \textbf{8.14}$$

Figure 8.9
Voltage and current waveforms for bridge circuit (a) voltage across SCR₁ (b) current waveforms

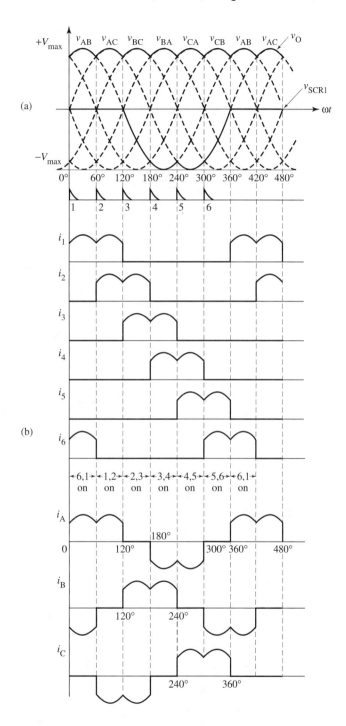

Note that when $\alpha = 60°$, Equations 8.13 and 8.14 give identical results. At $\alpha = 120°$, the average output voltage becomes zero.

$$V_{o(avg.)} = 0 \text{ V} \qquad \text{for } 120° \leq \alpha \leq 180° \qquad\qquad \textbf{8.15}$$

At $\alpha = 0°$, $V_{o(avg.)}$ is at its maximum, 1.65 V_m. The normalized average output voltage is

$$V_n = V_{o(avg.)}/V_{o(avg.)max} = 1 + \cos\left(\alpha + \frac{\pi}{3}\right)$$

The control characteristic curve is shown in Figure 8.10.

Figure 8.10
Control characteristic

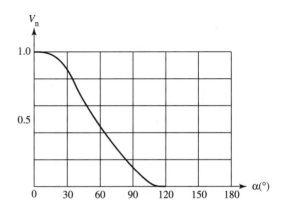

The average SCR current is

$$I_{SCR(avg.)} = I_{o(avg.)}/3 \qquad\qquad \textbf{8.16}$$

The RMS or effective value of the output current is

$$I_{o(RMS)} = \frac{\sqrt{3}\,V_m}{2\,R}\frac{\sqrt{2\pi + 3\sqrt{3}\cos 2\alpha}}{\pi} \qquad \text{for } 0° \leq \alpha \leq 120° \qquad \textbf{8.17}$$

$$I_{o(RMS)} = \frac{\sqrt{3}\,V_m}{2\,R}\frac{\sqrt{4\pi - 6\alpha - 3\sin(2\alpha - 60°)}}{\pi} \qquad \text{for } 60° \leq \alpha \leq 120° \quad \textbf{8.18}$$

The RMS value of the source line currents for both modes of operation is given by

$$I_{A(RMS)} = \sqrt{2/3} * I_{o(avg.)} \qquad\qquad \textbf{8.19}$$

The ripple frequency of the output is

$$f_r = \text{six times the source frequency} \qquad\qquad \textbf{8.20}$$

The ripple factor is

$$RF = \sqrt{\frac{I_{o(RMS)}^2}{I_{o(avg.)}^2} - 1} \qquad\qquad \textbf{8.21}$$

If the output current is an ideal constant DC waveform, $I_{o(RMS)} = I_{o(avg.)}$, and the ripple factor is zero.

The power dissipated in the load is

$$P_o = I^2_{o(RMS)} R \qquad\qquad\qquad 8.22$$

peak reverse voltage rating of the SCRs = $V_{L(m)}$ **8.23**

conduction period for each SCR = 120°   **8.24**

Example 8.5 For the six-pulse controlled-bridge rectifier with a delay angle of 0°, show the conduction path by drawing the equivalent circuit for each 60° interval.

Solution Figure 8.11 shows the waveform of the output voltage with α = 0°.

Figure 8.11
Waveforms of a bridge circuit with α = 0°

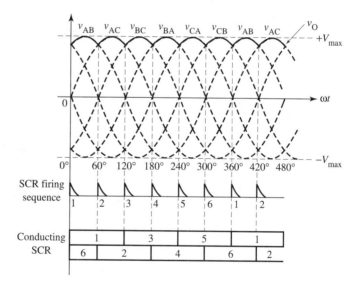

The operation of the circuit can be described by considering the following six periods:

Period 1: At 0°, SCR$_1$ is turned on, causing SCR$_5$ to turn off. From 0° to 60°, the largest line voltage is v_{AB}. Therefore, SCRs 1 and 6 will conduct at this time. v_{AB} appears across the load through SCRs 6 and 1.

Period 2: At 60°, SCR$_2$ is turned on, causing SCR$_6$ to turn off. From 60° to 120°, v_{AC} has the largest voltage; therefore, SCRs 1 and 2 will conduct. v_{AC} appears across the load through SCRs 1 and 2.

Period 3: At 120°, SCR$_3$ is turned on, causing SCR$_1$ to turn off. From 120° to 180°, v_{BC} appears across the load through SCRs 2 and 3.

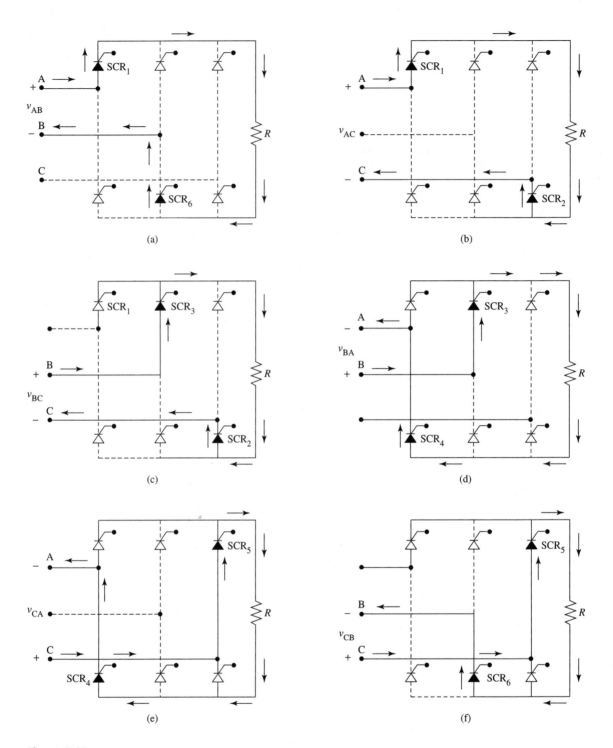

Figure 8.12
Changes in circuit configuration as the SCRs are switched sequentially (a) current path during period 1 (b) current path during period 2 (c) current path during period 3 (d) current path during period 4 (e) current path during period 5 (f) current path during period 6

Period 4: At 180°, SCR_4 is turned on, causing SCR_2 to turn off. From 180° to 240°, v_{BA} appears across the load through SCRs 3 and 4.

Period 5: At 240°, SCR_5 is turned on, causing SCR_3 to turn off. From 240° to 300°, v_{CA} appears across the load through SCRs 4 and 5.

Period 6: At 300°, SCR_6 is turned on, causing SCR_4 to turn off. From 300° to 360°, v_{CB} appears across the load through SCRs 5 and 6.

At 360°, the cycle is completed. SCR_1 is turned on, causing SCR_5 to turn off, and the sequence is repeated.

Figure 8.12 shows the output current paths in the circuit for each of these periods.

Example 8.6 Sketch the waveforms for output voltage, voltage across SCR_1, SCR currents i_1 and i_4, and line current i_A for the six-pulse controlled bridge rectifier supplying a purely resistive load with a delay angle of:
a) 30°
b) 60°
c) 90°
Assume ABC phase sequence.

Solution a) The output voltage waveform is shown in Figure 8.13(a). With $\alpha = 30°$, SCR_1 is forward-biased at $\omega t = 0°$. However, since the firing angle is 30°, it does not turn on until $\omega t = 0° + \alpha = 30°$, and then it continues to conduct for 60°. Before this instant, SCR_6 was turned on. Therefore, during the interval from 30° to 90°, SCRs 1 and 6 conduct the output current and the output terminals are connected to phase A and B (see Figure 8.14(a)). The output voltage is therefore equal to v_{AB}. At $\omega t = 60° + \alpha = 90°$, SCR_2 is fired and voltage v_{CB} immediately appears across SCR_6, which reverse-biases it and turns it off. The current from SCR_6 is transferred to SCR_2 (see Figure 8.14(b)) for a further 60° interval. The output terminals are connected to phase A through SCR_1 and phase C through SCR_2, making the output voltage equal to v_{AC}. At $\omega t = 150°$, SCR_1 is turned off by switching on SCR_3. Here, the line current i_A becomes zero, so the output current path is now provided by SCRs 2 and 3 for a further 60° interval. When $\omega t = 210°$, the most positive line voltage is v_{BA}, SCR_2 is reverse-biased, and conduction continues through the newly fired SCR_4 (see Figure 8.14(c)). The source current i_A now flows in the opposite direction. After a further 60°, at $\omega t = 270°$, the most positive voltage is v_{CA} and SCR_5 is turned on, causing SCR_3 to turn off. SCRs 4 and 5 then provide the path for the load current, which flows from phase C to phase A through the load (see Figure 8.14(d)). At $\omega t = 330°$, the line current becomes zero, so the output current path is now provided by SCRs 2 and 3 for a further 60° interval.

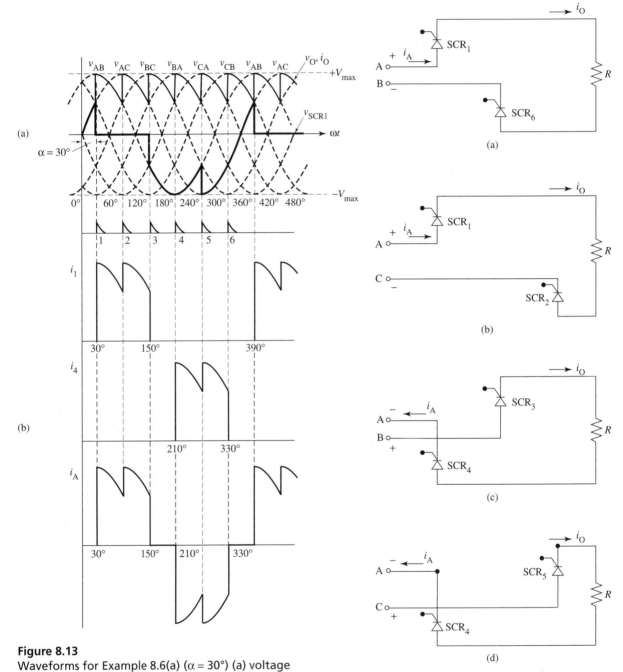

Figure 8.13
Waveforms for Example 8.6(a) ($\alpha = 30°$) (a) voltage
(b) current

Figure 8.14
Equivalent circuits of conduction (a) 30° to
90° (b) 90° to 150° (c) 210° to 270° (d) 270°
to 330°

Note that the triggering delay does not reduce the conduction period for each SCR, which is still 120° (one-third of a cycle), and each voltage segment has a duration of 60°. The current waveforms are shown in Figure 8.13(b). The line current i_A has a pulse width of 120°.

b) For $\alpha \geq 60°$, the current becomes discontinuous, as shown in Figure 8.15(a).

Figure 8.15
Voltage and current waveforms for Example 8.6 (parts (b) and (c)) (a) $\alpha = 60°$

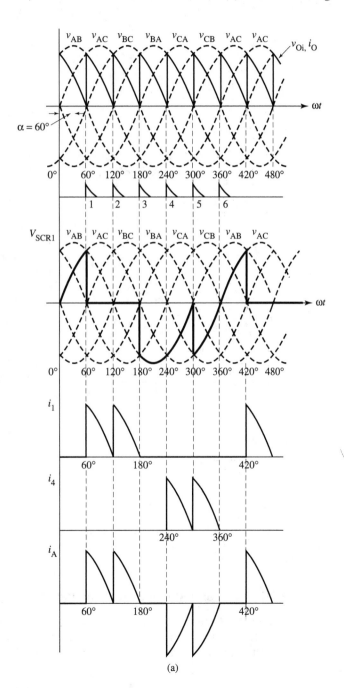

(a)

Figure 8.15
(b) $\alpha = 90°$

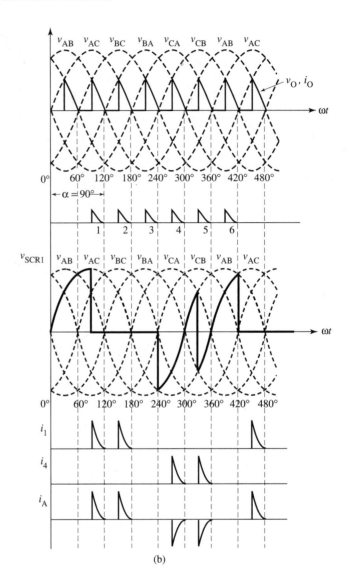

(b)

c) When the firing angle is increased to $\alpha = 90°$, the average output voltage decreases further, as shown in Figure 8.15(b).

8.3.2 With an Inductive (*RL*) Load without an FWD

The six-pulse bridge rectifier is most often used in applications where the load is highly inductive in nature. The effect of the load inductance is to smooth the output current to make it very close to a pure DC.

8.3.2.1 For $0° \leq \alpha \leq 60°$

For $0° \leq \alpha \leq 60°$, the output voltage is always positive, the output current is always present, and the output voltage is the same whether the load is inductive or a freewheeling diode is present. The graph of output voltage for $\alpha = 45°$ (see Figure 8.16) shows a six-pulse ripple.

Figure 8.16
Output voltage and voltage across SCR$_1$ for a firing angle $\alpha = 45°$

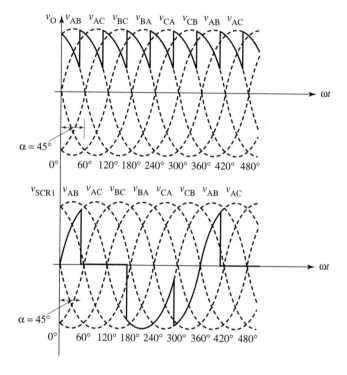

As before, each SCR begins conducting 45° later than it would if the rectifying devices were diodes. Each SCR conducts for a 60° interval, as in the similar situation without phase control. Therefore, the average output voltage is

$$V_{o(avg.)} = \frac{3\sqrt{3}}{\pi} V_m \cos \alpha$$

$$= \frac{3}{\pi} V_{L(m)} \cos \alpha$$

$$= 0.955 \, V_{L(m)} \cos \alpha \qquad \textbf{8.25}$$

Where $V_{L(m)}$ is the maximum value of the line voltage.

The output voltage therefore varies as a function of α, having a maximum value when $\alpha = 0°$. As we increase the delay angle, the output voltage decreases, becoming zero at 90°. The circuit acts as a rectifier when the delay angle is in the range $0° < \alpha < 90°$, the output voltage and current are positive, and power flow is from the AC source to the DC load. If the delay angle α is further increased, the DC output voltage will change signs and the circuit will operate as an inverter. The output voltage reaches its negative maximum at a firing angle of 180°. When the delay angle lies in the range $90° < \alpha < 180°$, the bridge circuit operates as an inverter, transferring power from the DC load to the AC source side. Therefore, a bridge rectifier with sufficient load inductance can operate as a rectifier or as an inverter with proper choice of firing angle.

The variation of the normalized average voltage with the firing angle is shown in Figure 8.17.

Figure 8.17
Average output voltage versus delay angle for continuous output current

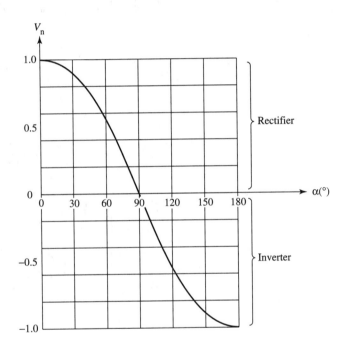

If the load has sufficient inductance, the output current essentially remains constant during the conduction period. The SCR currents have the same shape as the diode currents in Chapter 7, although they are shifted by an angle α to the right. Only the magnitude of the output current is changed due to the addition of cos α term for average voltage.

The average value of the output current is found by dividing the average output voltage by the load resistance:

$$I_{o(avg.)} = V_{o(avg.)}/R \qquad\qquad \textbf{8.26}$$

The average value of the SCR current is

$$I_{SCR(avg.)} = I_{o(avg.)}/3 \qquad\qquad \textbf{8.27}$$

The RMS value of the output current is

$$I_{o(RMS)} = I_{o(avg.)} \qquad\qquad \textbf{8.28}$$

The RMS value of the SCR current is

$$I_{SCR(RMS)} = I_{o(avg.)}/3 \qquad \textbf{8.29}$$

$$\text{peak reverse voltage rating} = V_{L(m)} \text{ or } \sqrt{3}\ V_m \qquad \textbf{8.30}$$

The RMS value of the output voltage is

$$V_{o(RMS)} = 2\ V_{L(m)} \left[\frac{1}{4} + \frac{3\ \sqrt{3}}{8\ \pi} \cos 2\alpha \right]^{1/2} \qquad \textbf{8.31}$$

Example 8.7 Determine the firing sequence in which the gate pulses must be applied for the six-pulse controlled bridge rectifier shown in Figure 8.7. Assume an ABC phase sequence.

Solution There are six SCRs, so at every 60° interval an SCR is fired in a particular sequence that keeps the source currents in balance. The output voltage at any instant is equal to one of the six line voltages v_{AB}, v_{AC}, v_{BC}, v_{BA}, v_{CA}, or v_{CB}. Thus two SCRs—one in the positive group (SCR$_1$, SCR$_3$, SCR$_5$) and one in the negative group (SCR$_4$, SCR$_6$, SCR$_2$)—must be triggered simultaneously. Let us assign a pair of SCRs for each phase of the source: SCRs 1 and 4 for phase A, SCRs 3 and 6 for phase B, and SCRs 5 and 2 for phase C. If we apply the gate pulse to SCR$_1$ at α, then SCR$_3$ must be fired 120° later at $\alpha + 120°$ and SCR$_5$ at $\alpha + 240°$.

Now SCR$_4$ must be fired 180° later than SCR$_1$, that is, at $\alpha + 180°$. SCR$_6$ is then fired 120° later than SCR$_4$, at $\alpha + 180° + 120° = \alpha + 300°$. Similarly, SCR$_2$ is fired at $\alpha + 180° + 240°$ or $\alpha + 60°$. Figure 8.18 shows the firing sequence as 1, 2, 3, 4, 5, 6, 1, ...

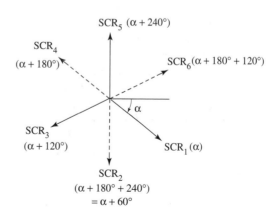

Figure 8.18
Firing sequence

This scheme has one drawback. Since two SCRs, one in the negative group and the other in the positive group, must conduct for the output current to flow, the firing pulses will not be sufficient for starting the bridge. In our scheme, only one SCR receives the trigger pulse at a time. Therefore, when the first gate signal is received by SCR_1, for example, it cannot turn on the bridge, since the other SCR (SCR_2) has not received its gate signal. After $60°$, when SCR_2 does get the gate signal, its own signal is removed. One method to solve this problem is to provide pulses of long duration, so that a pulse applied to an SCR gate does not end before the next pulse arrives on the gate of the SCR in the other group. The second method is to provide double pulses to each SCR—the idea is to simultaneously provide a pulse to the conducting partner in the opposite group whenever a pulse is given to any SCR. This method is shown in Figure 8.19 for the negative and positive group, respectively. Every SCR is provided with two successive pulses, the second one timed $60°$ after the first.

Figure 8.19
Double-pulse firing scheme

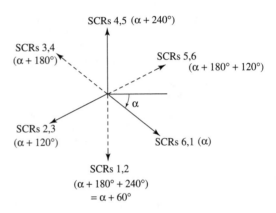

Example 8.8 A six-pulse controlled bridge rectifier is connected to a three-phase 220 V AC supply. If the load resistance is 20 Ω with a large inductive component and the delay angle is $30°$, find
a) the average output voltage
b) the average output current
c) the RMS output current
d) the average SCR current
e) the RMS SCR current
f) the voltage rating of the SCRs
g) the average output power

Solution a) $V_{L(m)} = \sqrt{2} * 220 = 311$ V

Using Equation 8.25

$$V_{o(avg.)} = 0.955 \; V_{L(m)} \cos \alpha$$
$$= 0.955 * 311 * \cos 30$$
$$= 257 \text{ V}$$

b) $I_{o(avg.)} = V_{o(avg.)}/R = 257/20 = 12.9$ A

c) $I_{o(RMS)} = I_{o(avg.)} = 12.9$ A

d) The average SCR current is

$$I_{o(avg.)}/3 = 12.9/3 = 4.3 \text{ A}$$

e) The RMS SCR current is

$$I_{o(avg.)}/\sqrt{3} = 12.9/1.732 = 7.5 \text{ A}$$

f) The maximum voltage across the SCRs is

peak reverse voltage $= V_{L(m)} = 311$ V

g) The average output power is

$$(I_{oRMS})^2 R = 12.9^2 * 20 = 3328 \text{ W}$$

Example 8.9 For a six-pulse controlled rectifier with an RL load, plot the waveform for the output voltage if the firing angle is
a) 15°
b) 60°

Figure 8.20
Voltage and current waveforms in the bridge rectifier (a) $\alpha = 15°$
(b) $\alpha = 60°$

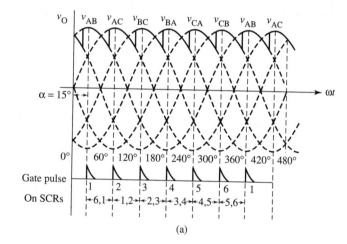

(a)

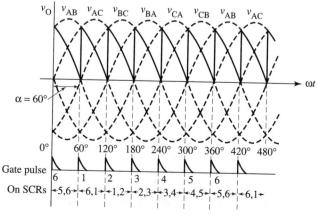

(b)

Solution Since the delay angle is ≤ 60°, the output voltage waveform is the same for either a purely resistive load, a purely inductive load, or any combination of *RL* loads.

Example 8.10 For a six-pulse controlled rectifier with an *RL* load, plot the waveforms for the output current and the line currents i_A, i_B, and i_C if the firing angle is:
a) 0°
b) 30°
c) 60°

Solution The SCR currents i_1, i_2, i_3, i_4, i_5, and i_6 flow for 120° with peak values of $I_{o(avg.)}$. The AC line currents can be easily determined by applying KCL at the appropriate nodes. Therefore, line currents i_A, i_B, and i_C are

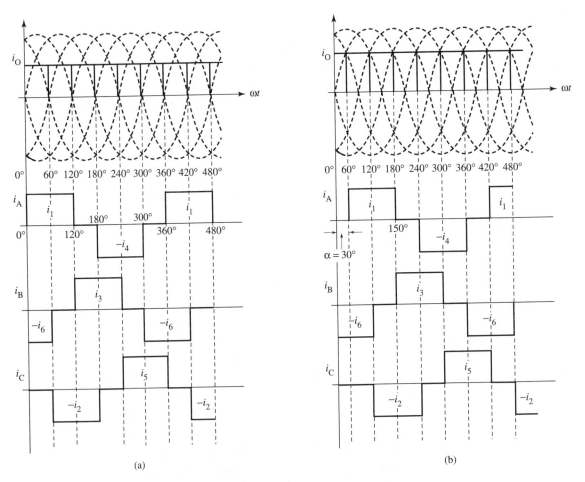

(a)   (b)

Figure 8.21
Voltage and current waveforms in the bridge rectifier (a) α = 0° (b) α = 30°

Figure 8.21
(c) $\alpha = 60°$

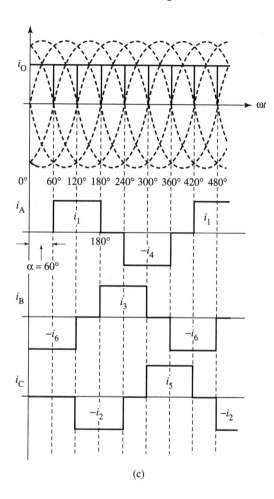

(c)

$$i_A = i_1 - i_4$$
$$i_B = i_3 - i_6$$
$$i_C = i_5 - i_2$$

These line currents also have peak values of $I_{o(avg.)}$, but they flow in positive and negative pulses of 120°.

The output current and line currents are plotted in Figure 8.21. The output current is continuous for all values of α. With a smooth output current, there is no ripple component at all and the current ripple factor is ideally zero.

Example 8.11 A six-pulse controlled bridge rectifier with an RL load is fed from a three-phase 220 V AC source. If the load resistance is 10 Ω, calculate the average output voltage and power dissipation if the firing angle is

a) 30°

b) 60°

Solution The phase voltage is

$$V_{\text{phase}} = V_L/\sqrt{3} = 220/1.732 = 127 \text{ V}$$

The maximum value of the phase voltage is

$$V_m = \sqrt{2}\,(127) = 180 \text{ V}$$

a) The average output voltage is

$$V_{o(\text{avg.})} = \frac{3\sqrt{3}}{\pi} V_m \cos \alpha = 258 \text{ V}$$

The average power dissipation is

$$P_{(\text{avg.})} = I^2_{o(\text{avg.})}\,R = \frac{V^2_{o(\text{avg.})}}{R} = \frac{258^2}{10} = 6656 \text{ W}$$

b) $V_{o(\text{avg.})} = 149 \text{ V}$

   $P_{(\text{avg.})} = 2216 \text{ W}$

8.3.2.2 For 60° ≤ α ≤ 120°

When α is increased beyond 60°, the output voltage becomes negative for portions of the cycle. If there is no FWD present, the average output voltage can still be obtained by using Equation 8.14, and the other equations for current are also valid. If a freewheeling diode is present, the average voltage equation changes form to conform to the period for which the output voltage is zero. The graph in Figure 8.22 shows the output voltage waveform for $\alpha = 90°$.

Figure 8.22
Output voltage waveforms
for α = 90°

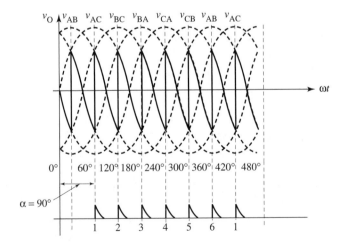

The average output voltage is

$$V_{o(\text{avg.})} = \frac{3\,V_{L(m)}}{\pi}\left[1 + \cos\left(\alpha + \frac{\pi}{3}\right)\right] \qquad \text{discontinuous current} \qquad \textbf{8.32}$$

The normalized average output voltage is

$$V_n = V_{o(avg.)}/V_{o(avg.)max} = \left[1 + \cos\left(\alpha + \frac{\pi}{3}\right)\right]$$

The control characteristic curve is shown in Figure 8.23.

Figure 8.23
Control characteristic

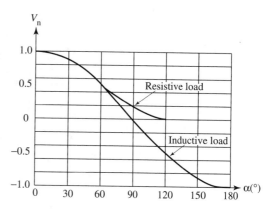

For the case of an inductive load without an FWD, the SCR average current is given by

$$I_{SCR(avg.)} = I_{o(avg.)}/3 \qquad \qquad \textbf{8.33}$$

where $I_{o(avg.)}$ is the average output current (assuming negligible ripple current in the output). If α is greater than 60° and an FWD is present, the FWD allows a path for the output current for the angle $\alpha - 60°$ twice in each 120° period. The SCR therefore conducts for the period $120° - 2(\alpha - 60°)$ or for the period $240° - 2\alpha$. The average SCR current is given by

$$I_{SCR(avg.)} = \frac{(240° - 2\alpha)}{360°} I_{o(avg.)} \qquad \qquad \textbf{8.34}$$

With a delay angle α less than 60° or when no FWD is present, the line current includes both a positive and a negative pulse equal to the output current for 120° periods. These pulses are separated by two 60° periods of zero value. The RMS value of the line current is given by

$$I_{RMS} = 0.816\, I_{o(avg.)} \qquad \qquad \textbf{8.35}$$

For a delay angle α greater than 60° with an FWD present, the line current includes two positive pulses, both of $120° - \alpha$ duration and two similar negative pulses. The remaining periods are zero. The RMS value of the line current is given by

$$I_{RMS} = \sqrt{\frac{(120° - \alpha)}{90}(I_{o(avg.)})^2} \qquad \qquad \textbf{8.36}$$

Example 8.12 For a six-pulse controlled rectifier, plot the waveforms for the output voltage, the voltage across SCR_1, the current through SCRs 1 and 4, and line current i_A if the firing angle is

a) 75°
b) 120°
c) 135°
d) 150°

Figure 8.24
Voltage and current
waveforms (a) $\alpha = 75°$

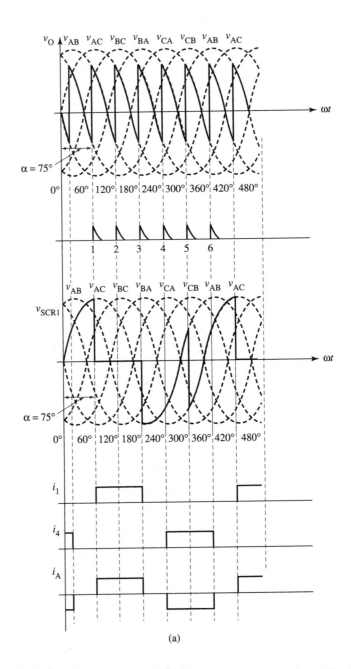

(a)

Figure 8.24
(b) $\alpha = 120°$

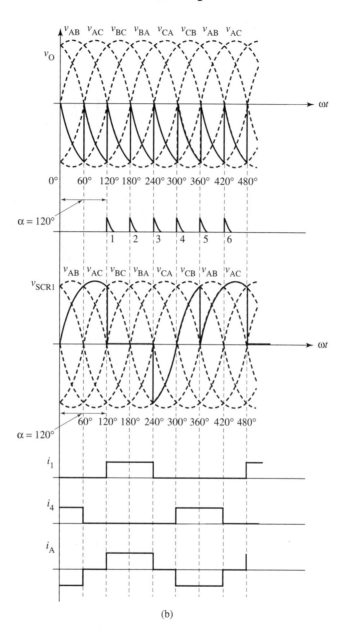

(b)

Solution An increase in α beyond 60° results in the output voltage becoming negative, with a negative maximum at $\alpha = 180°$. The circuit operates in the inversion mode if there is a source of negative voltage at the load. Figure 8.24 shows the waveforms for the various firing angles.

From this example, we can conclude that when $0° \le \alpha \le 90°$, the circuit operates in the rectifying mode. For $90° \le \alpha \le 180°$, the circuit operates in the inverting mode if a source of negative voltage is present at the load.

Figure 8.24
(c) α = 35°

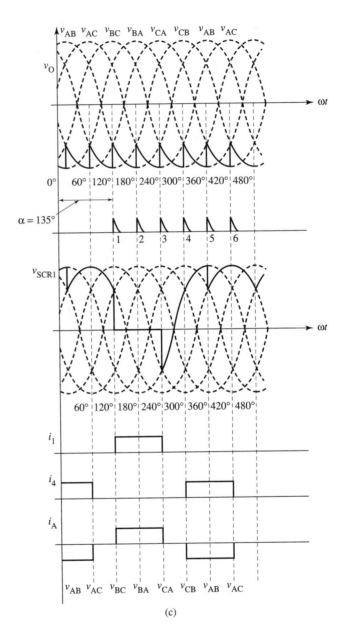

(c)

8.4 Full-Wave Half-Controlled Bridge Rectifiers with FWD

Fully controlled converters are extensively used in adjustable-speed DC drives with medium to large motors. These circuits operate with both positive and negative DC output voltages. In the rectifying mode, they convert AC power to DC power, and in the inversion mode they do the reverse. In many applications the

Figure 8.24
(d) $\alpha = 150°$

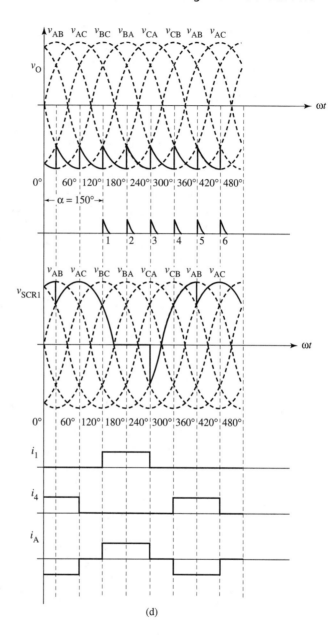

(d)

circuit is required to operate only in the rectifying mode, that is, to convert AC power to DC power. Converters that operate in this manner are called half-controlled or semiconverters.

Figure 8.25(a) shows a half-controlled converter circuit. The circuit includes a freewheeling diode to help sustain continuous output current. A cost advantage is obtained by using diodes instead of SCRs in the lower half of the bridge; however, the output ripple frequency is only three times the line frequency. The output voltage includes a contribution from the controlled upper half-bridge plus a contribution from the uncontrolled lower half-bridge. The SCRs turn on when fired and are turned off either by firing another SCR or by the action of the FWD. With a delay angle $\alpha \leq 60°$, the output voltage waveform contains six pulses per cycle. The output current is always continuous, the voltage never becomes negative, and the FWD plays no part. The output voltage waveform is equal to v_{AB} or v_{AC} whenever SCR$_1$ is on. The firing sequence and the voltage and current waveforms, assuming a large inductive load, are shown in Figure 8.25(b). Each SCR and diode pair conducts for a full 120°. The ripple frequency is only three times the AC source frequency—this is a drawback of a semiconverter as compared with a fully-controlled bridge, which has a ripple frequency six times the source frequency.

For delay greater than 60°, the output voltage tends to become negative during part of the cycle. The freewheeling diode acts to make the output voltage zero for this period and turns off the conducting SCR and diode pair. Continuous output current can still flow through the FWD path. Figure 8.25(c) shows the voltage and current waveforms. The firing sequence of the SCRs and the period of conduction of the diodes is also shown.

Assume that the output current is continuous and ripple-free. At $30° + \alpha$, SCR$_1$ turns on and SCR$_1$ and D$_2$ conduct the output current, making the output voltage equal to v_{AC}. When the output voltage tends to be negative, the freewheeling diode is forward-biased and starts conducting. The output current will now freewheel through the FWD, making the output voltage zero. When SCR$_3$ is turned on, the output current conducts through SCR$_3$ and D$_4$, making the output voltage equal to v_{BA}. The process is repeated every 120° whenever an SCR is turned on.

The average value of the output voltage is

$$V_{o(avg.)} = \frac{3}{2\pi} V_{L(m)} (1 + \cos \alpha) \qquad \text{for } 0° \leq \alpha \leq 180° \qquad \textbf{8.37}$$

At $\alpha = 0°$, $V_{o(avg.)}$ is at a maximum

$$V_{o(avg.)max} = 0.955 \ V_{L(m)} \qquad \textbf{8.38}$$

Figure 8.25
(a) Half-controlled bridge circuit

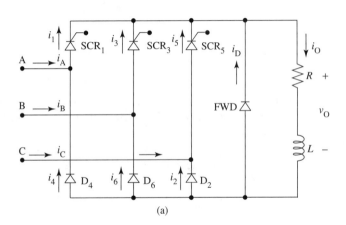

(a)

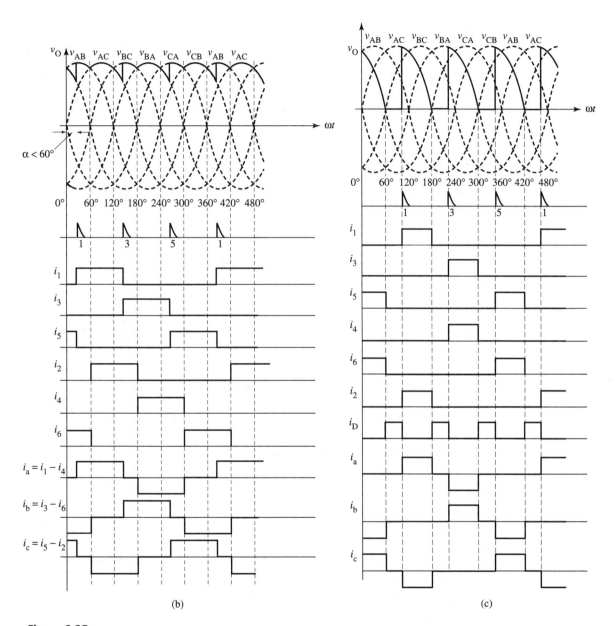

(b)   (c)

Figure 8.25
(b) voltage and current waveforms for a small firing angle (c) waveforms for large firing angle

The normalized average output voltage is

$$V_n = V_{o(avg.)}/V_{o(avg.)max} = (1 + \cos \alpha)/2 \qquad \textbf{8.39}$$

The variation of the normalized average output voltage with α is shown in Figure 8.26. Note that the output voltage cannot be negative, therefore the semi-converter cannot operate in the inverting mode.

Figure 8.26
Average output voltage versus delay angles for continuous output current

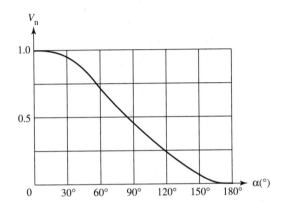

The RMS value of the output voltage is

$$V_{o(RMS)} = V_{L(m)} \left[\frac{3}{4\pi} \left(\pi - \alpha + \frac{1}{2} \sin 2\alpha \right) \right]^{1/2} \qquad \text{for } 0° \leq \alpha \leq 180° \qquad \textbf{8.40}$$

The average value of the SCR current is

$$I_{SCR(avg.)} = I_{o(avg.)}/3 \qquad \textbf{8.41}$$

The RMS value of the SCR and diode current is

$$I_{SCR(RMS)} = I_{o(avg.)}/\sqrt{3} \qquad \textbf{8.42}$$

Example 8.13 A six-pulse half-controlled bridge rectifier is connected to a three-phase 220 V AC source. Calculate the firing angle if the terminal voltage of the rectifier is 240 V. What is the maximum value of the DC output voltage?

Solution

$$V_{o(avg.)} = 240 \text{ V}$$
$$V_{L(m)} = \sqrt{2} \, (220) = 311 \text{ V}$$

$$V_{o(avg.)} = \frac{3}{2\pi} V_{L(m)} (1 + \cos \alpha)$$

$$1 + \cos \alpha = \frac{V_{o(avg.)} \, 2\pi}{3 \, V_{L(m)}} = 1.62$$

$$\cos \alpha = 0.62$$

$$\alpha = 52°$$

Maximum DC voltage is obtained with $\alpha = 0°$

$$V_{o(max)} = \frac{3}{2\pi} (311)(1 + 1) = 297 \text{ V}$$

Example 8.14 Sketch the waveforms for output voltage and the input line current i_A for a half-controlled rectifier with a delay angle of

a) 0°

b) 60°

c) 120°

Also find the ripple frequency in each case if the source frequency is 60 Hz.

Solution a) With $\alpha = 0°$, the output is at a maximum as it is for a full-controlled bridge. Each SCR conducts for 120°, and the input line current consists of alternating components of rectangular pulses of 120° durations. Figure 8.27(a) shows the waveforms.

The ripple frequency is

$$6 f = 6 * 60 = 360 \text{ Hz}$$

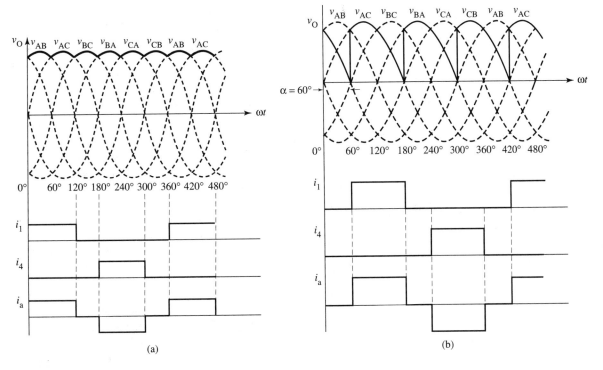

(a)   (b)

Figure 8.27
Voltage waveforms for Example 8.14 (a) $\alpha = 0°$ (b) $\alpha = 60°$

Figure 8.27
(c) $\alpha = 120°$

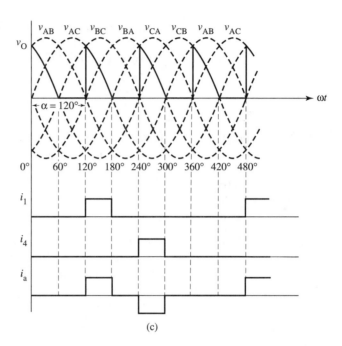

(c)

b) With $\alpha = 60°$, the average output voltage is the difference in voltage between the positive and negative group. The input line current is delayed by 60° and then flows for a duration of 120°. Figure 8.27(b) show the waveforms.

The ripple frequency is

$$3\,f = 3 * 60 = 180 \text{ Hz}$$

c) Figure 8.27(c) shows the waveforms for $\alpha = 120°$.

The ripple frequency is

$$3\,f = 3 * 60 = 180 \text{ Hz}$$

8.5 Twelve-Pulse Bridge Converters

A six-pulse bridge rectifier reduces the output ripple considerably as compared with that of a three-pulse rectifier. In applications such as DC power transmission, a further reduction in output ripple is achieved with a twelve-pulse converter. A twelve-pulse converter can be obtained by connecting two six-pulse bridges in series or parallel. Figure 8.28(a) and (b) shows these two connections using transformers with a 30° phase shift between the two voltages of the Y- and Δ-connected secondary windings. Figure 8.28(c) shows the output voltage waveform of the series-connected twelve-pulse bridge. The resultant output voltage is the sum of the individual converter output voltages. Since each converter gives a six-pulse output, the resultant voltage is a twelve-

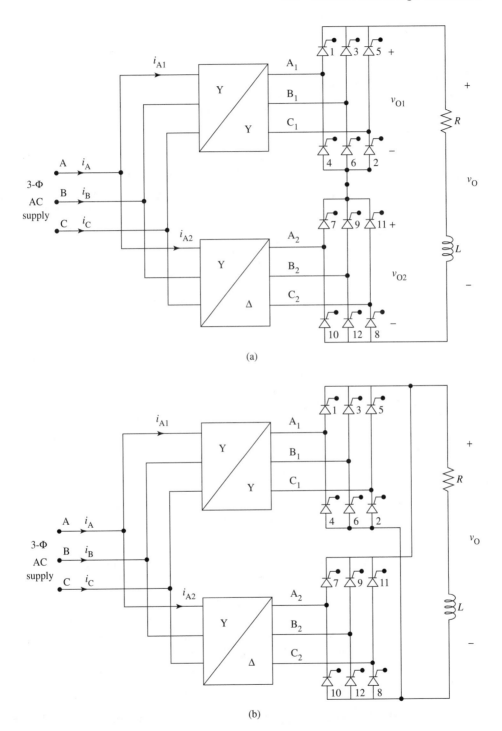

(a)

(b)

Figure 8.28
Twelve-pulse connection using two six-pulse bridges (a) series (b) parallel

Figure 8.28
(c) output voltage
waveforms for the series-
connected bridge

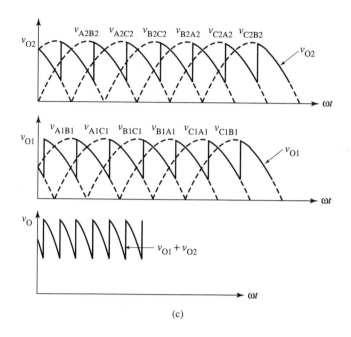

(c)

pulse output. Due to the series connection, the same output current flows through the two converters.

The equations for output voltage for a resistive load are:

$$V_{o(avg.)} = \sqrt{2}\, V_{L(m)} \frac{12}{\pi} \sin \alpha \qquad \text{for } 0° \leq \alpha \leq 15° \qquad \textbf{8.43}$$

$$V_{o(avg.)} = \sqrt{2}\, V_{L(m)} \frac{6}{\pi} [(1 - \tfrac{1}{2}\sqrt{3}) \sin \alpha + \tfrac{1}{2} \cos \alpha]$$
$$\text{for } 15° \leq \alpha \leq 90° \qquad \textbf{8.44}$$

$$V_{o(avg.)} = \sqrt{2}\, V_{L(m)} \frac{6}{\pi} (1 + \cos (\alpha + 60°)) \qquad \text{for } 90° \leq \alpha \leq 120° \qquad \textbf{8.45}$$

■——————————————————————————————————————

8.6 Problems

8.1 A three-pulse rectifier is supplied from a 208 V source. If the load current is 10 A and the firing angle is 0°, find
   a) the average output voltage
   b) the maximum output current
   c) the peak reverse voltage rating of the SCRs
   d) the power dissipation in each SCR, if the forward voltage drop is 1.1 V

8.2 Repeat problem 8.1 for a firing angle of 30°.

8.3 For a three-pulse controlled rectifier with a resistive load, plot the waveforms for the output voltage, the output current, and the voltage across the SCRs if the firing angle is
a) 30°
b) 90°
c) 120°

8.4 For a three-pulse controlled rectifier with an inductive load, plot the waveforms for the output voltage, the output current, and the voltage across the SCRs if the firing angle is 100°.

8.5 For a three-pulse controlled rectifier with an *RL* load, plot the waveforms for the output voltage, the output current, and the voltage across the SCRs if the firing angle is 100°.

8.6 For a three-pulse controlled rectifier with an *RL* load, plot the waveforms for the output voltage if the firing angle is 175°.

8.7 A six-pulse bridge rectifier is connected to a three-phase 220 V AC source. Calculate the DC output voltage if the firing angle is 30°.

8.8 For a six-pulse bridge rectifier with a resistive load, plot the waveforms for the output voltage, the output current, and the voltage across SCRs 1 and 3 if the firing angle is:
a) 0°
b) 60°
c) 90°
d) 120°

8.9 For a six-pulse bridge rectifier with a resistive load, plot the waveforms for the line currents i_A, i_B, and i_C and the current through each SCR if the firing angle is
a) 0°
b) 90°

8.10 A six-pulse bridge circuit is connected to a three-phase 208 V AC source. Find
a) the average output voltage if the delay angle is 75°
b) the delay angle required to produce 75 V DC when the circuit is operating in the rectifier mode
c) the delay angle required to produce 75 V DC when the circuit is operating in the inverter mode

8.11 For a half-controlled six-pulse bridge rectifier with a freewheeling diode across the resistive load, plot the waveforms for the output voltage and output current if the firing angle is
a) 0°
b) 60°
c) 90°
d) 120°

8.12 For a half-controlled six-pulse bridge rectifier with an *RL* load, plot the waveforms for the output voltage, output current, and line currents i_A, i_B, and i_C if the firing angle is 120°.

8.13 For a half-controlled six-pulse bridge rectifier with a freewheeling diode across an *RL* load, plot the waveforms for SCR currents with a delay angle of 120°.

8.14 Repeat problem 8.11 without a freewheeling diode.

8.15 A three-pulse rectifier is shown in Figure 8.29. With a delay angle of 90° for SCRs 1 and 2, plot the waveform for the output voltage.

Figure 8.29
See problem 8.15

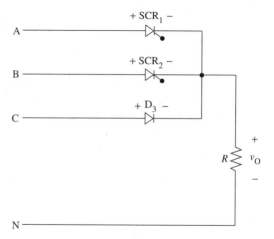

8.16 A six-pulse rectifier is shown in Figure 8.30. Plot the waveform for the output voltage if the delay angle is 30°.

Figure 8.30
See problem 8.16

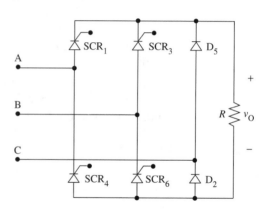

8.17 A six-pulse bridge circuit is connected to a 3.8 KV source. The load draws a current of 100 A. If the delay angle is 60°, find
a) the average output voltage
b) the active power input
c) the reactive power absorbed by the circuit

8.18 A six-pulse bridge circuit is fed from a 220 V source. The load draws a current of 100 A. If the delay angle is 60°, find
a) the average output voltage
b) the active power input
c) the reactive power absorbed by the circuit

8.19 A six-pulse bridge converter connected to a three-phase 120 V AC source is operating in the inverting mode. The load consists of a 100 V DC source having an internal resistance of 0.1 Ω. If the battery delivers a constant current of 70 A, find
a) the delay angle
b) the active power supplied to the source
c) the reactive power absorbed by the converter
d) the peak reverse voltage rating of the SCR
e) the average current through the SCR

8.20 A 5 HP separately-excited DC shunt motor rated at 120 V, 1800 rpm is fed from a fully controlled six-pulse bridge converter. The converter is supplied from a 208-V AC source. If the rated armature current is 20 A and the armature resistance is 0.1 Ω find
a) the delay angle for the motor to operate at the rated load
b) the delay angle if the starting current is limited to 25 A
c) the reactive power absorbed by the converter when operating at rated load
d) the SCR voltage rating

8.21 For a six-pulse controlled bridge rectifier with a resistive load, find the ripple factor when α = 60°.

8.7 Equations

$$V_{o(avg.)} = \frac{3\sqrt{3}}{2\pi} V_m \cos\alpha \qquad \text{for } 0° \leq \alpha \leq 30° \tag{8.1}$$

$$I_{o(avg.)} = \frac{3\sqrt{3}}{2\pi} \frac{V_m}{R} \cos\alpha \tag{8.2}$$

$$I_{SCR(avg.)} = \frac{I_{o(avg.)}}{3} = \frac{\sqrt{3}}{2\pi} \frac{V_m}{R} \cos\alpha \tag{8.3}$$

$$I_{SCR(RMS)} = \frac{I_{SCR(avg.)}}{\sqrt{3}} = \frac{1}{2\pi} \frac{V_m}{R} \cos\alpha \tag{8.4}$$

peak reverse voltage rating of SCR = $\sqrt{3}\ V_m$ **8.5**

conduction period for each SCR = one-third of a cycle = $\dfrac{2\ \pi}{3}$ **8.6**

$f_r = 3 *$ AC supply frequency **8.7**

$V_{o(avg.)} = \dfrac{3\ V_m}{2\pi}\left(1 + \cos\left(\alpha + \dfrac{\pi}{6}\right)\right)$ for $30° \le \alpha \le 150°$ **8.8(a)**

$V_{o(avg.)} = 0$ V for $150° \le \alpha \le 180°$ **8.8(b)**

$V_{o(avg.)} = \dfrac{3\ \sqrt{3}}{2\ \pi}\ V_m \cos\alpha$ for $0° \le \alpha \le 180°$ **8.9**

$V_{o(avg.)max} = 0.827\ V_m$ **8.10**

$V_n = \dfrac{V_{o(avg.)}}{V_{o(avg.)max}} = \cos\alpha$ **8.11**

$I_{SCR(RMS)} = I_{o(avg.)}/3$ **8.12**

$V_{o(avg.)} = \dfrac{3\ \sqrt{3}}{\pi}\ V_m \cos\alpha$ for $0° \le \alpha \le 60°$ **8.13**

$V_{o(avg.)} = \dfrac{3\ \sqrt{3}}{\pi}\ V_m\left(1 + \cos\left(\alpha + \dfrac{\pi}{3}\right)\right)$ for $60° \le \alpha \le 120°$ **8.14**

$V_{o(avg.)} = 0$ V for $120° \le \alpha \le 180°$ **8.15**

$I_{SCR(avg.)} = I_{o(avg.)}/3$ **8.16**

$I_{o(RMS)} = \dfrac{\sqrt{3}}{2\ R}V_m\ \dfrac{\sqrt{2\pi + 3\ \sqrt{3}\cos 2\alpha}}{\pi}$ for $0° \le \alpha \le 120°$ **8.17**

$I_{o(RMS)} = \dfrac{\sqrt{3}}{2\ R}V_m\ \dfrac{\sqrt{4\pi - 6\alpha - 3\sin(2\alpha - 60°)}}{\pi}$

for $60° \le \alpha \le 120°$ **8.18**

$i_{A(RMS)} = \sqrt{2/3} * I_{o(avg.)}$ **8.19**

$f_r =$ six times the source frequency **8.20**

$RF = \sqrt{\dfrac{I^2_{o(RMS)}}{I^2_{o(avg.)}} - 1}$ **8.21**

$P_o = I^2_{o(RMS)}\ R$ **8.22**

PRV rating of SCRs = $V_{L(m)}$ **8.23**

conduction period for each SCR = $120°$ **8.24**

$V_{o(avg.)} = \dfrac{3\ \sqrt{3}}{\pi}\ V_m \cos\alpha$ **8.25**

$I_{o(avg.)} = V_{o(avg.)}/R$ **8.26**

$I_{SCR(avg.)} = I_{o(avg.)}/3$ **8.27**

$I_{o(RMS)} = I_{o(avg.)}$ **8.28**

$I_{SCR(RMS)} = I_{o(avg.)}/\sqrt{3}$ **8.29**

PRV rating = $\sqrt{3}\ V_m$ **8.30**

$$V_{\text{o(RMS)}} = 2 \ V_{\text{L(m)}} \left[\frac{1}{4} + \frac{3 \sqrt{3}}{8 \ \pi} \cos 2\alpha \right]^{1/2} \tag{8.31}$$

$$V_{\text{o(avg.)}} = \frac{3 \ V_{\text{L(m)}}}{\pi} \left[1 + \cos \left(\alpha + \frac{\pi}{3} \right) \right] \qquad \text{discontinuous current} \tag{8.32}$$

$$I_{\text{SCR(avg.)}} = I_{\text{o(avg.)}}/3 \tag{8.33}$$

$$I_{\text{SCR(avg.)}} = \frac{(240 - 2 \ \alpha)}{360} I_{\text{o(avg.)}} \tag{8.34}$$

$$i_{\text{RMS}} = 0.816 \ I_{\text{o(avg.)}} \tag{8.35}$$

$$I_{\text{RMS}} = \sqrt{\frac{(120 - \alpha)}{90} (I_{\text{o(avg.)}})^2} \tag{8.36}$$

$$V_{\text{o(avg.)}} = \frac{3 \ V_{\text{L(m)}}}{2\pi}(1 + \cos \alpha) \qquad \text{for } 0° \leq \alpha \leq 180° \tag{8.37}$$

$$V_{\text{o(avg.)max}} = 0.955 \ V_{\text{L(m)}} \tag{8.38}$$

$$V_{\text{n}} = V_{\text{o(avg.)}} / V_{\text{o(avg.)max}} = (1 + \cos \alpha)/2 \tag{8.39}$$

$$V_{\text{o(RMS)}} = V_{\text{L(m)}} \left[\frac{3}{4\pi} \left(\pi - \alpha + \frac{1}{2} \sin 2\alpha \right) \right]^{1/2} \qquad \text{for } 0° \leq \alpha \leq 180° \tag{8.40}$$

$$I_{\text{SCR(avg.)}} = I_{\text{o(avg.)}}/3 \tag{8.41}$$

$$I_{\text{SCR(RMS)}} = I_{\text{o(avg.)}}/\sqrt{3} \tag{8.42}$$

$$V_{\text{o(avg.)}} = \sqrt{2} \ V_{\text{L(m)}} \frac{(12)}{\pi} \sin \alpha \qquad \text{for } 0° \leq \alpha \leq 15° \tag{8.43}$$

$$V_{\text{o(avg.)}} = \sqrt{2} \ V_{\text{L(m)}} \frac{(6)}{\pi} ((1 - \frac{1}{2} \sqrt{3}) \sin \alpha + \frac{1}{2} \cos \alpha)$$
$$\text{for } 15° \leq \alpha \leq 90° \tag{8.44}$$

$$V_{\text{o(avg.)}} = \sqrt{2} \ V_{\text{L(m)}} \frac{(6)}{\pi} (1 + \cos (\alpha + 60)) \qquad \text{for } 90° \leq \alpha \leq 120° \tag{8.45}$$

DC Choppers

Chapter Outline

Learning Objectives

After completing this chapter, the student should be able to

- explain what the term *chopper* means
- discuss the principle of basic DC choppers
- describe the basic operation of a step-down (buck) chopper
- describe the basic operation of a step-up (boost) chopper
- explain the operating principle of a buck-boost chopper

9.1   Introduction

A DC-to-DC converter or *chopper,* as it is commonly called, is used to obtain a variable DC voltage from a constant-voltage DC source. The average value of the output voltage is varied by changing the proportion of the time during which the output is connected to the input. This conversion can be achieved with a combination of an inductor or/and capacitor and a solid-state device operated in a high-frequency switching mode. In high-voltage and high-current applications, the switching devices used in chopper circuits are thyristors. When power transistors (BJTs, or MOSFETS) or GTO thyristors are used, they can be turned off easily by controlling the base or gate current. Thyristors used in DC circuits must be turned off using forced commutation since they lack the facility of natural commutation that is available in AC circuits.

The switching technique used in DC choppers is called *pulse-width modulation* (PWM). There are two fundamental kinds of chopper circuits: the step-down or buck chopper and the step-up or boost chopper. The buck chopper produces an output voltage that is less than or equal to the input voltage; the boost chopper provides an output voltage that is greater than or equal to the input voltage.

Choppers are used in many industrial applications where a constant DC source is available. Typical applications include DC motor control for electric traction, switching power supplies, inverters for uninterruptible power supplies (UPS), and battery-operated equipment. In this chapter, we will focus mainly on the basic circuits and operating principles of DC choppers.

9.2   The Principles of Basic DC Choppers

The fundamental principle of a basic chopper is illustrated in Figure 9.1. A switch is connected in series with the DC voltage source (V_i) and the load. The switch S can be a power transistor, an SCR, or a GTO thyristor. It is assumed throughout this chapter that the switching devices are ideal. Ideal switches have the following characteristics:

1. They have zero resistance (zero voltage drop) when on.
2. They have infinite resistance (zero leakage current) when off.
3. They can switch from either state in zero time.

Ideally, the power loss in the chopper is zero, so the output power is equal to the input power:

$$V_o I_o = V_i I_i$$

9.1

Figure 9.1
Basic DC chopper

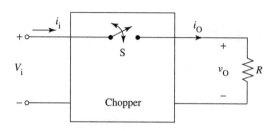

where

V_o = average output voltage

V_i = input voltage

I_o = average output current

I_i = average input current

We assume that the output voltage is adjustable in a certain range from zero to the input level. Let us operate the switch so that it is on (closed) for a time T_{ON} and off (open) for a time T_{OFF} in each cycle with a fixed period T. The resulting output voltage waveform (shown in Figure 9.2) is a train of rectangular pulses of duration T_{ON}.

Figure 9.2
Waveform of v_0 for Figure 9.1

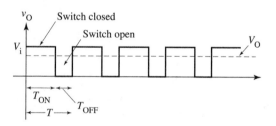

Figure 9.2 shows that the instantaneous voltage across the load is either zero (S off) or V_i (S on). The average (DC) output voltage over a cycle is given by:

$$V_o = \frac{T_{ON}}{T_{ON} + T_{OFF}} V_i$$

$$V_o = \frac{T_{ON}}{T} V_i$$

9.2

where T is the period ($T_{ON} + T_{OFF}$). The chopper switching frequency is $f = 1/T$.

If we use the idea of duty cycle (d), which is the ratio of the pulse width T_{ON} to the period of the waveform,

$$d = \frac{T_{ON}}{T}$$

9.3

then

$$V_o = d \, V_i \qquad\qquad \textbf{9.4}$$

From Equation 9.4, it is obvious that the output voltage varies linearly with the duty cycle. Figure 9.3 shows the output voltage as d varies from zero to one. It is therefore possible to control the output voltage in the range zero to V_i.

Figure 9.3
Output voltage v_0 as a function of duty cycle

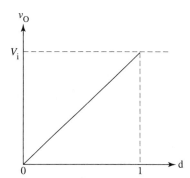

If the switch S is a transistor, the base current will control the on and off period of the transistor switch. If the switch is a GTO thyristor, a positive gate pulse will turn it on and a negative gate pulse will turn it off. If the switch is an SCR, a commutation circuit is required to turn it off.

The load current waveform is similar to Figure 9.2, and its average value is given by

$$I_o = \frac{V_o}{R} = \frac{d \, V_i}{R} \qquad\qquad \textbf{9.5}$$

The effective (RMS) value of the output voltage is

$$V_{o(RMS)} = \sqrt{\frac{V_i^2 \, T_{ON}}{T}}$$

$$= V_i \sqrt{\frac{T_{ON}}{T}} \qquad\qquad \textbf{9.6}$$

$$= V_i \sqrt{d}$$

The average output voltage can be varied in one of the following ways:

1. **Pulse-width modulation** (PWM). In this method, the pulse width T_{ON} is varied while the overall switching period T is kept constant. Figure 9.4 shows how the output waveforms vary as the duty cycle is increased.

2. **Pulse-frequency modulation** (PFM). In this method, T_{ON} is kept constant while the period (frequency) is varied. As shown in Figure 9.5, the output voltage reduces as frequency is decreased, being high at higher frequencies.

Figure 9.4
Output voltage waveforms with fixed switching frequency

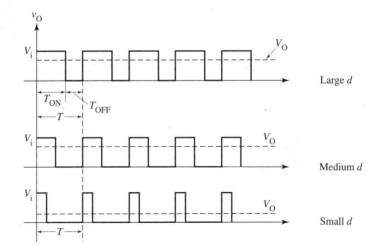

Figure 9.5
Output voltage waveforms with variable switching frequency

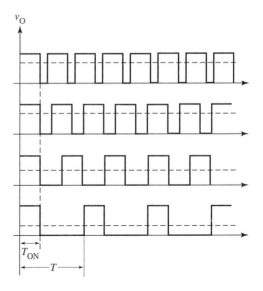

For either PWM or PFM control, the output voltage is zero when switch S is open, and it is equal to the input voltage when the switch is closed for a time longer than the normal switching cycle.

In the PFM method, it is necessary to reduce the chopper switching frequency to obtain a lower output voltage. This may result in discontinuity at low frequencies. Moreover, a reduction in frequency increases the output current ripple, thereby increasing losses and heating in the load. On the other hand, the losses in the components become very high at higher frequencies. The PWM method has the advantage of low ripple, which means smaller filter components.

9.3   Step-Down (Buck) Choppers

The DC chopper circuit of Figure 9.1 is not very practical. It is suitable only for supplying resistive loads where a smooth output current is not required. A much more practical arrangement (shown in Figure 9.6(a) includes an inductor L and a diode D, which are added to eliminate current' pulsations. This circuit provides a smooth DC current to practical loads like a DC motor.

When the switch S is closed, the diode D is off, since it is reverse-biased. It will stay off as long as S remains on. The equivalent circuit configuration is shown in Figure 9.6(b). The input current builds up exponentially and flows through the inductor L and the load. The output voltage is equal to V_i. The switch S is kept on for a time T_{ON} and then turned off.

When the switch is opened, the current through the inductor starts decaying to zero (it cannot change instantaneously). This causes an induced voltage with opposite polarity across the inductor. The inductor voltage forward-biases the diode, and the current flowing through the inductor now freewheels through the diode D and the load. The purpose of the diode therefore is to provide a path for the load current when S is off. Therefore, turning off S

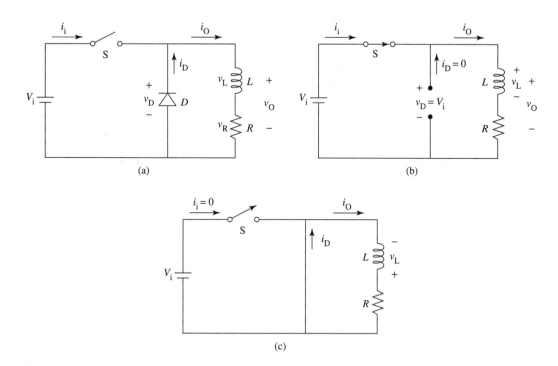

(a)

(b)

(c)

Figure 9.6
(a) Basic step-down chopper circuit (b) equivalent circuit for the on state
(c) equivalent circuit for the off state

automatically turns on D. The new circuit configuration is shown in Figure 9.6(c). The voltage across the load is zero, and the current decays toward zero as long as S remains off, that is, for a period T_{OFF}. The energy stored in L is delivered to the load. This circuit arrangement permits the use of a simple filter inductance L to provide a satisfactory smooth DC load current for many applications. When the switching frequency is high, a relatively small inductance is sufficient to reduce the ripple to an acceptable degree.

9.3.1   Continuous Current Mode

Figure 9.7(a) shows the waveform of the voltage across the load, which is also the voltage appearing across the FWD(D). This voltage is equal to the input voltage V_i when the switch is on and diode D is reverse-biased. When the switch is open, the output voltage is held at zero by the action of the FWD, which provides a path for the load current. As the average voltage across inductor L is negligible when it has no resistive component, the output voltage must be the average voltage across the diode. Therefore, Equation 9.2 also applies here.

Figure 9.7(b) shows the diode current. It is the same as the load current (Figure 9.7(d)) during T_{OFF}. During T_{ON}, the output current i_o is the same as the input current i_i. When the switch is open (T_{OFF}), the load current falls from its maximum value I_{max} to a final value I_{min}. During this interval, current flows through the inductor, the load, and the FWD. When the current has fallen to a value I_{min}, the switch closes. The current in the diode immediately stops flowing, and the current supplied by the source is now I_{min}. The current then starts increasing, and when it reaches the value I_{max}, after a time T_{ON}, the switch reopens. The FWD again provides a path for the load current, and the cycle repeats. The load current therefore oscillates between I_{max} and I_{min}. The ripple included in the output current reduces as the chopper switching frequency increases. The average value of the inductor current is given by:

$$I_L = \frac{I_{max} + I_{min}}{2} \tag{9.7}$$

Now,

$$I_L = I_o = \frac{V_o}{R}$$

Therefore

$$\frac{I_{max} + I_{min}}{2} = \frac{V_o}{R}$$

$$I_{max} + I_{min} = 2\frac{V_o}{R} \tag{9.8}$$

The voltage across the inductor is

$$V_L = V_o = L \frac{di_o}{dt}$$

$$\frac{di_o}{dt} = \frac{V_o}{L}$$

$$\Delta i_o = \frac{V_o}{L} \Delta t$$

With the switch open (T_{OFF}),

$$\Delta i_o = I_{max} - I_{min} = \frac{V_o}{L} T_{OFF} \qquad\qquad \textbf{9.9}$$

Adding Equations 9.8 and 9.9,

$$2 I_{max} = 2 \frac{V_o}{R} + \frac{V_o}{L} T_{OFF}$$

$$I_{max} = \frac{V_o}{R} + \frac{V_o}{2 L} T_{OFF} \qquad\qquad \textbf{9.10}$$

Similarly, I_{min} is given by

$$I_{min} = \frac{V_o}{R} - \frac{V_o}{2 L} T_{OFF} \qquad\qquad \textbf{9.11}$$

The peak-to-peak ripple current is

$$I_{p-p} = I_{max} - I_{min}$$

$$= T_{OFF} \frac{V_o}{L} \qquad\qquad \textbf{9.12}$$

The average diode current is

$$I_D = \frac{I_{OFF} I_o}{T} \qquad\qquad \textbf{9.13}$$

Although the load current of a chopper is basically constant, the input current still consists of a train of sharp pulses. Figure 9.7(c) shows the waveform for the source current. A capacitor filter is often used in parallel with the input power source to smooth the input current.

The current and voltage waveforms shown in Figure 9.7 assume a switching frequency and a load such that $T \ll \tau$. Here τ is the *circuit time constant,* which depends on the ratio L/R. Note that the output current variation is linear and the ripple is quite small due to the large time constant. The output current i_o is always present, so this mode of operation is called the *continuous current mode* of operation. The output current is continuous since the inductor absorbs energy during T_{ON} and discharges during T_{OFF}. As a result, the average voltage across L becomes zero in steady state, and the voltage across the load resistor must be equal to $V_{o(avg.)}$.

Figure 9.7
Voltage and current
waveforms for $T_{ON} \approx T_{OFF}$
and $T \ll \tau$ (a) output
voltage (b) FWD current
(c) source current (d) load
current (e) voltage across
the inductor

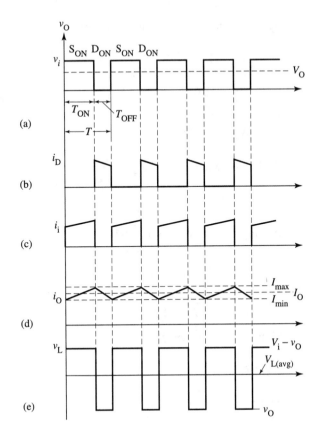

The average values of output voltage, output power, and input power are given by

$$V_o = V_i d \qquad\qquad \textbf{9.14}$$
$$P_o = V_o \, I_o$$
$$P_i = V_i \, I_i$$

Since we are considering ideal elements (with no losses), the DC power drawn from the source must equal the DC power absorbed by the load.

$$P_o = P_i$$
$$V_o I_o = V_i I_i$$
$$I_o = V_i * \frac{I_i}{V_o}$$
$$\quad = V_i * \frac{I_i}{V_i d}$$
$$I_o = \frac{I_i}{d} \qquad\qquad \textbf{9.15}$$

Equations 9.14 and 9.15 are very similar to the basic transformer equations, where the duty cycle d corresponds to the turns ratio a of the transformer. Therefore, a DC chopper achieves in DC circuits what a transformer does in AC circuits. By changing the duty cycle in a step-down chopper, we can obtain output voltages that are less than or equal to the input voltage. Furthermore, the output current is stepped up when we step down the voltage.

9.3.2 Discontinuous Current Mode

For low values of d, especially with low inductance, the load current decreases and may fall to zero during the part of each cycle when the switch is off. The current again builds up from zero when the switch turns on in the next cycle. The load current is said to be *discontinuous*.

Figure 9.8 shows the current and voltage waveforms when T_{ON} is approximately equal to τ ($\tau = L/R$). The voltage waveform v_o is the same as in Figure 9.2; however, the output current i_0 cannot jump to V_i/R due to the inductive nature of the load L—instead it rises exponentially to V_i/R. Similarly, when the transistor is off, the same current, flowing through the freewheeling diode, decays to zero.

Figure 9.8
Voltage and current waveforms with $T_{ON} < T_{OFF}$ and $T_{ON} \approx \tau$

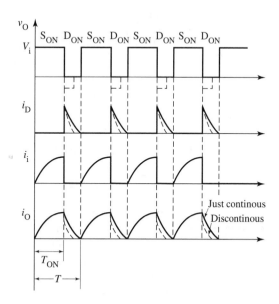

As can be seen in Figure 9.8, the load current flows in pulses and the chopper operates in the discontinuous mode. Such an operation is usually undesirable. This mode of operation can be avoided by a proper choice of chopping frequency or by choosing a suitable value of inductance. The minimum value of inductance required to ensure continuous current is obtained by setting $I_{min} = 0$ in Equation 9.11:

$$I_{min} = 0 = \frac{V_o}{R} - \frac{V_o}{2L} T_{OFF}$$

$$\frac{V_o}{R} = T_{OFF} \frac{V_o}{2L}$$

$$2L = T_{OFF} R$$

or

$$L = \frac{T_{OFF}}{2} * R \qquad\qquad\qquad 9.16$$

Therefore, having a high inductance in the load is desirable to avoid the discontinuous mode. Moreover, the peak-to-peak ripple of the load current decreases as the value of the inductance increases. When the load inductance ideally becomes infinite, we get a current that is ripple-free (pure DC).

Example 9.1   In Figure 9.6, the switching frequency is 25 Hz and $T_{ON} = 3$ ms. If the average value of the output current is 40 A, determine the average source current.

Solution   Rearranging Equation 9.15,

$$I_i = d\, I_o$$

$$I_i = \frac{T_{ON}}{T_{ON} + T_{OFF}} I_o$$

$$= \frac{T_{ON}}{T} I_o$$

$$= T_{ON}\, f_c I_o$$

$$= 0.003 * 25 * 40$$

$$= 3A$$

Example 9.2   In Figure 9.6, the input voltage $V_i = 100$ V, the load resistance $R = 10\ \Omega$, and $L = 100$ mH. The switching frequency is $f = 1$ KHZ and the on time is 0.5 ms. If the average source current is 1 A, determine
a) the average load voltage
b) the output current
c) the output power
d) the minimum value of L required

Solution   a) The duty cycle is

$$d = \frac{T_{ON}}{T} = \frac{0.5}{1} = 0.5 \text{ or } 50\%$$

The average voltage is

$$V_o = d\,V_i = (0.5)(100) = 50 \text{ V}$$

b) The output current is

$$I_o = \frac{I_i}{d} = \frac{1}{0.5} = 2A$$

c) The output power is

$$P_o = V_o I_o = 100 \text{ W}$$

d) $T_{OFF} = T - T_{ON} = 0.5$ ms

$$L = \frac{T_{OFF} R}{2} = \frac{0.5\ (10^{-3})\ 10}{2} = 2.5 \text{ mH}$$

Example 9.3   A DC buck chopper operates at a frequency of 1 KHz from a 100 V DC source supplying a 10 Ω resistive load. The inductive component of the load is 50 mH. If the average output voltage is 50 V, find
a) the duty cycle
b) T_{ON}
c) the RMS value of the load voltage
d) the average value of the load current
e) I_{max} and I_{min}
f) the input power
g) the peak-to-peak ripple current
h) the peak-to-peak ripple current if the frequency is increased to 5 kHz
i) the peak-to-peak ripple current if the inductance is increased to 250 mH

Solution   a) $V_o = d\ V_i$

The duty cycle is

$$d = \frac{V_o}{V_i} = \frac{50}{100} = 0.5 \text{ or } 50\%$$

b) The period T is

$$T = 1/f = 1/1000 = 1 \text{ ms}$$
$$d = T_{ON}/T$$
$$T_{ON} = T\,d = (1 \text{ ms})\ (0.5) = 0.5 \text{ ms}$$

c) $V_{o(RMS)} = V_i \sqrt{d} = 100 \sqrt{0.5} = 70.7 \text{ V}$

d) $I_o = \dfrac{V_o}{R} = \dfrac{50}{10} = 5 \text{ A}$

e) $I_{max} = \dfrac{V_o}{R} + \dfrac{V_o}{2\,L}\,T_{OFF}$

$= I_o + \dfrac{V_o}{2\,L}\,T_{OFF}$

$= 5 + \dfrac{50 * 0.5\ (10^{-3})}{2 * 50\ (10^{-3})} = 5 + 0.25 = 5.25\ A$

$I_{min} = 5 - 0.25 = 4.75\ A$

f) $I_{i(avg.)} = \dfrac{(I_{max} + I_{min})}{2}\,d = \dfrac{(5.25 + 4.75)}{2}\,0.5 = 2.5\ A$

$P_i = V_i I_{i(avg.)} = 100 * 2.5 = 250\ W$

g) The peak-to-peak ripple current is

$I_{p-p} = I_{max} - I_{min} = 5.25 - 4.75 = 0.5\ A$

h) $T = 1/5(10^3) = 200\ \mu s$

$T_{ON} = T_{OFF} = 100\ \mu s$

$I_{max} = 5 + \dfrac{100(10^{-6}) * 50}{2 * 50(10^{-6})} = 5 + 0.5 = 5.05\ A$

$I_{min} = 5 - 0.05 = 4.95\ A$

$I_{p-p} = 5.05 - 4.95 = 0.1$

Notice the large reduction in ripple current as the chopping frequency is increased. In fact, the reduction is in the same proportion as the increase in frequency.

i) $I_{max} = 5 + [0.5\ (10^{-3}) * 50]/[2 * 250(10^{-3})] = 5 + 0.05 = 5.05\ A$

$I_{min} = 5 - 0.05 = 4.95\ A$

I_{p-p} is the same as in part (h).

This example shows that we can reduce the ripple current by either increasing the chopping frequency or increasing the value of the inductor. In practice, a higher switching frequency is favored since it reduces the size of the filter inductance for the same ripple magnitude. However, higher switching frequencies would increase the power losses in the switches. With present-day devices, typical switching frequencies used in DC chopper applications range from 20 to 50 kHz.

Example 9.4   The chopper shown in Figure 9.9 operates at a frequency of 100 HZ with an on time of 4 ms. The average value of the load current is 20 A, with a peak-to-peak ripple of 4 A. Calculate the average value of the source and diode currents and plot the two waveforms.

Solution   The period is

$$T = \frac{1}{f} = \frac{1}{100} = 10 \text{ ms}$$

$$T_{ON} = 4 \text{ ms}$$

$$T_{OFF} = T - T_{ON} = (10 - 4)\text{ms} = 6 \text{ ms}$$

Figure 9.9
See Example 9.4

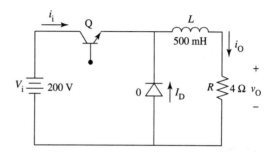

$$\tau = L/R = 0.5/4 = 0.125 \text{ s}$$

$\tau \gg T$, therefore the load current is linear.

Now,

$$I_o = 20 \text{ A}$$

and

peak-to-peak ripple = 4 A

$$I_{max} = 20 + 2 = 22 \text{ A}$$

$$I_{min} = 20 - 2 = 18 \text{ A}$$

The load current waveform (Figure 9.10) can now be plotted.

Figure 9.10
Load current waveform (not to scale)

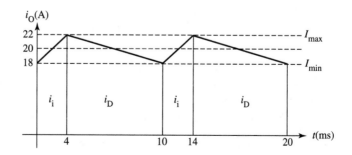

The output power is

$$P_o = I_o^2 R = 20^2 * 4 = 1600 \text{ W}$$

The input power is

$$P_i = P_o = 1600 \text{ W}$$

$$I_i = \frac{1600}{200} = 8 \text{ A}$$

Applying KCL at node 1,

$$I_i + I_D = I_o$$
$$I_D = I_o - I_i$$
$$= 20 - 8$$
$$= 12 \text{ A}$$

Plots of i_i and i_d are shown in Figure 9.11.

Figure 9.11
Source and diode current
waveforms (not to scale)

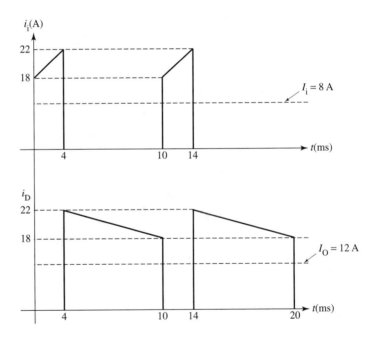

Example 9.5   A step-down chopper operates at a frequency of 4 KHz from a 25 V DC source supplying a 5 Ω resistive load. If the output voltage is 15 V and the current is discontinuous, find
a) the duty cycle
b) the minimum value of L required
c) the power from the source
d) the power to the load
e) I_{max} and I_{min}

Solution a) $V_o = d V_i$

$d = 15/25 = 0.6$

b) $L = T_{OFF} * R/2$

$d = T_{ON}/T = 0.6$

$T_{ON} = T * 0.6 = 0.6/f = 0.6/4(10^3) = 150 \ \mu s$

$T = T_{ON} + T_{OFF} = 250 \ \mu s$

$T_{OFF} = 100 \ \mu s$

$L = 100 \ (10^{-6}) * 5/2 = 0.25 \ mH$

c) The average source current is

$I_i = I_o \ d$

Here,

$I_o = V_o/R = 15/5 = 3 \ A$

Therefore,

$I_i = 3 * 0.6 = 1.8 \ A$

The power from the source is

$P_i = V_i * I_i = 25 * 1.8 = 45 \ W$

d) The power supplied to the load is

$V_o^2/R = 15^2/5 = 45 \ W$

e) $I_{max} = I_o + T_{OFF} \dfrac{V_o}{2 \ L} = 3 + \dfrac{100 \ (10^{-6}) * 15}{2 * 0.25 \ (10^{-3})} = 3 + 3 = 6 \ A$

$I_{min} = I_o - T_{OFF} \dfrac{V_o}{2 \ L} = 3 - 3 = 0 \ A$ since the *current is discontinuous*.

Example 9.6 A step-down chopper operates at a fixed frequency of 100 Hz from a 220 V DC source supplying a load with 1 Ω resistance and 10 mH inductance. If the output voltage is 60 V, find

a) T_{ON}

b) T_{OFF}

c) the average output current
and draw the waveforms of i_o and i_D.

Solution a) $V_o = d V_i$

$d = 60/220 = 0.27$

$T_{ON}/T = 0.27$

$T_{ON} = 0.27/f = 0.27/100 = 2.7 \ ms$

b) $T_{OFF} = T - T_{ON} = 10(10^{-3}) - 2.7(10^{-3}) = 7.3 \ ms$

c) $I_{o(avg.)} = 60/1 = 60$ A

$$I_{max} = 60 + \frac{7.3 \ (10^{-3}) * 60}{2 * 10 \ (10^{-3})} = 60 + 21.9 = 81.9 \ \text{A}$$

$I_{min} = 60 - 21.9 = 38.1$ A

Plots of i_o and i_D are shown in Figure 9.12.

Figure 9.12
Current waveforms (not to scale)

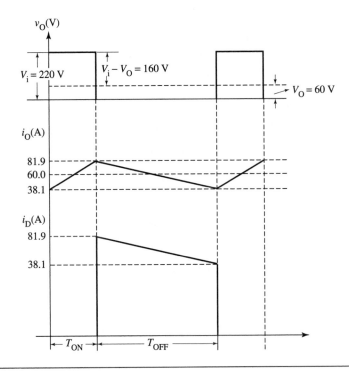

9.4 Step-Up (Boost) Choppers

In the step-up circuit, the output voltage can be varied from the source voltage up to several times the source voltage. The basic circuit of the step-up chopper is shown in Figure 9.13. Inductor L is used to provide a smooth input current.

Figure 9.13
Basic step-up chopper circuit

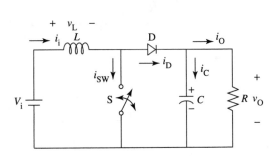

Some ripple component is actually included in the input current, but it is regarded as negligibly small when the switching action is repeated at a high frequency. The solid-state switch, which is operated in the PWM fashion, can be a transistor or an SCR.

When the switch S is turned on (see Figure 9.14(a)), the inductor is connected to the supply. The voltage across the inductor v_L jumps instantaneously to the source voltage V_i, but the current through the inductor i_i increases linearly and stores energy in the magnetic field. When the switch opens (see Figure 9.14(b)), the current collapses and the energy stored in the inductor is transferred to the capacitor through the diode D. The induced voltage V_L across the inductor reverses, and the inductor voltage adds to the source voltage to increase the output voltage. The current that was flowing through S now flows through L, D, and C to the load. Therefore, the energy stored in the inductor is released to the load. When S is again closed, D becomes reverse-biased, the capacitor energy supplies the load voltage, and the cycle repeats.

Figure 9.14
Step-up chopper equivalent circuit (a) on state (b) off state

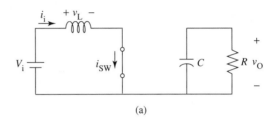

(a)

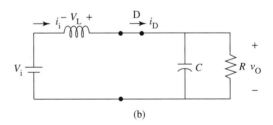

(b)

The voltage across the load (and the capacitor) is:

$$v_o = V_i + v_L$$

v_o will always be higher than V_i because the polarity of v_L is always the same as that of V_i.

The voltage and current waveforms for $d = 0.5$ are shown in Figure 9.15. The diode current i_D behaves as follows:

$i_D = 0$ when the switch is closed (on)

$i_D = i_i$ when the switch is open (off)

Thus, the diode current will pulsate as illustrated in Figure 9.15.

Figure 9.15
Voltage and current
waveforms ($d = 0.5$) for
step-up chopper.

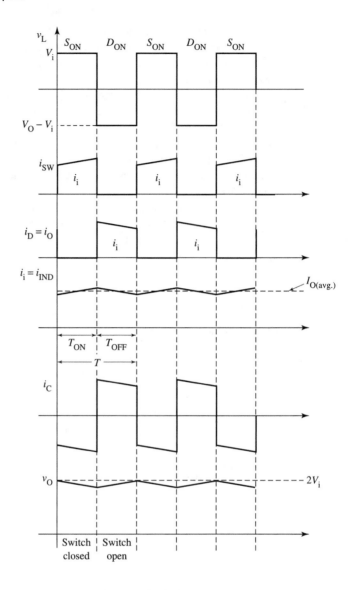

If the inductance L is quite large, the source current i_i is ripple-free and considered constant. The energy stored in the inductor during the interval (T_{ON}) when the chopper is on is

$$W_{ON} = V_i I_i T_{ON}$$

Similarly, assuming that the capacitance is large enough to neglect the voltage ripple, the output voltage v_o can be considered constant. During the time (T_{OFF}) that the chopper is off, the energy transferred by the inductor to the load is:

$$W_{OFF} = (V_o - V_i) * I_i * T_{OFF} \qquad\qquad 9.17$$

Since we are neglecting losses, the energy transferred by inductance during T_{ON} must equal the energy gained by it during the period T_{ON}.

$$W_{ON} = W_{OFF}$$

$$V_i I_i T_{ON} = (V_o - V_i) I_i T_{OFF}$$

or $\quad V_o = V_i = V_i \dfrac{T_{ON}}{T_{OFF}}$

$$V_o = V_i + V_i \dfrac{T_{ON}}{T_{OFF}}$$

$$= V_i \left(1 + \dfrac{T_{ON}}{T_{OFF}} \right)$$

$$= V_i \left(\dfrac{T_{OFF} + T_{ON}}{T_{OFF}} \right)$$

$$= V_i \dfrac{(T)}{T_{OFF}}$$

Since $\quad T = T_{OFF} + T_{ON}$

$$T_{OFF} = T - T_{ON}$$

$$V_o = V_i \left(\dfrac{T}{T - T_{ON}} \right)$$

$$= V_i \left(\dfrac{1}{1 - \dfrac{T_{ON}}{T}} \right)$$

$$V_o = V_i \left(\dfrac{1}{1 - d} \right) \qquad\qquad\qquad \textbf{9.18}$$

If the switch is open ($d = 0$), output voltage is equal to input voltage. As d increases, the output voltage becomes larger than the input voltage. Therefore, the output or load voltage is always higher than the input voltage if the switch S is operated at an appropriately high frequency.

The average value of source current can be obtained from:

$$P_i = P_o \qquad \text{neglecting losses}$$

$$V_i I_i = \dfrac{V_o^2}{R} \qquad\qquad\qquad\qquad\qquad \textbf{9.19}$$

$$I_i = \dfrac{V_o^2}{V_i} * \dfrac{1}{R}$$

Capacitor C in Figure 9.13 reduces the ripple in the output voltage and smoothes the current supplied to the load. If the capacitor is large enough, the output current will have a negligible ripple component and will equal the time average of the diode current. Therefore, the average value of the current I_o from Figure 9.15 is given by

$$I_o = I_i \frac{T_{OFF}}{T}$$

$$= I_i(1 - d)$$

9.20

Equations 9.18 and 9.20 clearly show that the circuit functions as a step-up DC transformer. As d (or T_{ON}) approaches zero, the output voltage V_o equals V_i, the source voltage. On the other hand, when d approaches unity (or $T_{ON} = T$), the output voltage approaches infinity. In practice, however, the maximum voltage cannot be infinite due to the power loss associated with the nonideal components. Because of these practical limitations, the output voltage is limited to around five times the input voltage.

The chopping frequency is limited by the SCR and the forced commutation circuit. A higher chopper frequency can be achieved by using power transistors or gate-turnoff thyristor (GTOs).

Let us determine the equations for the maximum and minimum input current.

The input power is

$$P_i = V_i * I_i$$

The output power is

$$P_o = V_o^2/R$$

Neglecting losses, the output power must be the same as the power supplied by the source:

$$V_i * I_i = \frac{V_o^2}{R} = \frac{V_i^2}{(1 - d)^2 R}$$

$$I_i = \frac{V_i}{(1 - d)^2 R}$$

9.21

Now from Equation 9.7,

$$I_L = \frac{I_{max} + I_{min}}{2}$$

Now, $I_L = I_i$
Therefore,

$$\frac{I_{max} + I_{min}}{2} = I_i$$

$$I_{max} + I_{min} = 2 I_i$$

9.22

The voltage across the inductor is

$$v_L = V_i = L \frac{di_i}{dt}$$

$$\frac{di_i}{dt} = \frac{V_i}{L}$$

With the switch closed (T_{ON}),

$$\Delta I_i = \frac{V_i}{L} T_{ON}$$

or

$$I_{max} - I_{min} = \frac{V_i}{L} T_{ON} \qquad \textbf{9.23}$$

Adding Equations 9.22 and 9.23,

$$2\, I_{max} = 2\, I_i + \frac{V_i T_{ON}}{L}$$

$$I_{max} = I_i + \frac{V_i}{2\,L} T_{ON}$$

$$= \frac{V_i}{(1-d)^2 R} + \frac{V_i}{2\,L} T_{ON} \qquad \textbf{9.24}$$

$$I_{max} = V_i \left[\frac{1}{R(1-d)^2} + \frac{T_{ON}}{2\,L} \right]$$

Similarly, I_{min} is given by

$$I_{min} = V_i \left[\frac{1}{R(1-d)^2} - \frac{T_{ON}}{2\,L} \right] \qquad \textbf{9.25}$$

The peak-to-peak ripple in the input current I_i is given by

$$I_{p-p} = I_{max} - I_{min} = \frac{V_i\, T_{ON}}{L} \qquad \textbf{9.26}$$

For continuous current conditions, the minimum value of inductance required is obtained by setting Equation 9.25 equal to zero.

$$I_{min} = 0 = V_i \left[\frac{1}{R(1-d)^2} - \frac{T_{ON}}{2\,L} \right]$$

$$\frac{1}{R(1-d)^2} = \frac{T_{ON}}{2\,L}$$

Solving for L,

$$L = \frac{R\, T_{ON}}{2} (1-d)^2 \qquad \textbf{9.27}$$

Example 9.7   The step-up chopper shown in Figure 9.13 is supplied from a 110 V DC source. The voltage required by the load is 440 V. If the switch is turned on for 0.25 ms, find the chopper frequency.

Solution
$$V_o = V_i \frac{1}{1-d}$$

$$440 = 110 \ (1/1 - d)$$

$$1 - d = 110/440 = 0.25$$

$$d = 1 - 0.25 = 0.75$$

$$d = T_{ON}/T = T_{ON} * f$$

or

$$f = d/T_{ON} = 0.75/0.25 \ (10^{-3}) = 3 \text{ kHz}$$

Example 9.8   The step-up chopper shown in Figure 9.13 operates at a frequency of 1 kHz. The source voltage is 100 V DC and the load resistance is 2 Ω. If the average value of the load current is 100 A, find
a) the power dissipated by the load resistor
b) the duty cycle
c) the average value of source current
d) the average value of the switch current

Solution   a) The average power dissipated by the load is

$$P_o = I_o^2 R = 100^2(2) = 20 \text{ kW}$$

b) From Equation 9.18,

$$V_o = V_i \left(\frac{1}{1-d} \right)$$

The average output voltage can be determined from

$$P_o = V_o^2/R$$
$$V_o = \sqrt{P_o R} = \sqrt{20 \ (10^{+3}) * 2} = 200 \text{ V}$$

Now, $1 - d = V_i/V_o = 100/200 = 0.5$

$$d = 0.5$$

c) The power input is

$$P_i = V_i * I_i$$

The average source current is

$$I_i = P_i/V_i$$

If we neglect losses,

$$P_i = P_o = 20 \text{ kW}$$

Therefore,

$$I_i = 20 \ (10^3)/100 = 200 \text{ A}$$

d) The average diode current is the same as average load current, since the average capacitor current is zero.

$$I_D = I_o = 100 \text{ A}$$

The current through the switch can be found by applying KCL:

$$I_i = I_{SW} + I_D$$

or

$$I_{SW} = I_i - I_D = 200 - 100 = 100 \text{ A}$$

Example 9.9 In Figure 9.13, $V_i = 50$ V, $V_o = 75$ V, $L = 2$ mH, $R = 2 \ \Omega$, and $T_{ON} = 1$ ms.
a) Determine the duty cycle.
b) Determine the switching frequency.
c) Determine $I_{i(ON)}$ and $I_{o(ON)}$.
d) Plot the waveform for i_{SW} and i_D.

Solution a) From Equation 9.18,

$$V_o = V_i \left(\frac{1}{1-d} \right)$$

$$75 = 50 \left(\frac{1}{1-d} \right)$$

$$\frac{75}{50} = \frac{1}{1-d}$$

$$1.5 = \frac{1}{1-d}$$

$$1.5 - 1.5 \ d = 1$$

$$-1.5 \ d = -.5$$

$$d = 1/3$$

$$= 0.33$$

b) $$d = T_{ON}/T$$

$$0.33 = \frac{1 * 10^{-3}}{T}$$

$$T = 3 \text{ ms}$$

$$f_c = \frac{1}{T}$$

$$= \frac{1}{3 * 10^{-3}}$$

$$= 333.3 \text{ HZ}$$

c) From Equation 9.19,

$$I_i = \frac{V_o^2}{V_i} * \frac{1}{R}$$

$$= \frac{75^2}{50} * \frac{1}{2}$$

$$= 56.25 \text{ A}$$

and from Equation 9.20,

$$I_o = I_i (1 - d)$$

$$= 56.25 (1 - 0.33)$$

$$= 37.7 \text{ A}$$

d) Plots of i_{SW} and i_D are shown in Figure 9.16.

$$v_L = V_i = L \, di_i/dt$$

$$\frac{di_i}{dt} = \frac{V_i}{L}$$

$$\frac{\Delta I_i}{\Delta t} = \frac{V_i}{L}$$

$$\Delta I_i = \frac{V_i}{L} T_{ON}$$

$V_i \, T_{ON}$ is the volt-sec on the inductor.

Figure 9.16
Current waveforms for
Example 9.9 (not to scale)

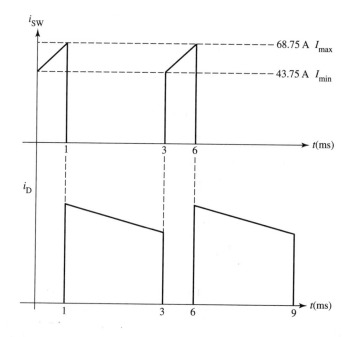

Therefore, $\Delta I_i = 50 \ (10^{-3})/2 \ (10^{-3}) = 25$ A

$\qquad\qquad I_{max} = 56.25 + 25/2 = 68.75$ A

$\qquad\qquad I_{min} = 56.25 - 25/2 = 43.75$ A

Example 9.10   The step-up chopper shown in Figure 9.13 supplies power to a 20 Ω resistive load at 120 V. The source voltage is 40 V DC, and the load inductance is 0.2 mH. If the chopper frequency is 4 kHz, find

a) the duty cycle
b) the average value of the source current
c) T_{ON}
d) I_{max}
e) I_{min}
f) the average value of the diode current

Solution   a) From Equation 9.18,

$$V_o = V_i \left(\frac{1}{1 - d} \right)$$

$$1 - d = 40/120 = 0.33$$

$$d = 0.67$$

b) $I_i = \dfrac{V_i}{(1 - d)^2 \ R} = 40/(1 - 0.67)^2 \ 20 = 18.4$ A

c) $T = 1/f_c = 250$ μs

$\qquad d = T_{ON}/T = 0.67$

$\qquad T_{ON} = 0.67 * 250 \ (10^{-6}) = 167.5$ μs

d) $I_{max} = V_i \left[\dfrac{1}{R \ (1 - d)^2} + \dfrac{T_{ON}}{2 \ L} \right]$

$\qquad\quad = 40 \left[\dfrac{1}{20 \ (1 - 0.67)^2} + \dfrac{167.5 \ (10^{-6})}{2 * 0.2(10^{-3})} \right]$

$\qquad\quad = 40 \ [0.46 + 0.42] = 35.2$ A

e) $I_{min} = V_i \left[\dfrac{1}{R \ (1 - d)^2} - \dfrac{T_{ON}}{2 \ L} \right]$

$\qquad\quad = 40 \ [0.46 - 0.42] = 1.6$ A

f) The average diode current is the same as the average load current, since the average capacitor current is zero.
Therefore,

$$I_D = I_o$$

$$\quad = V_o/R$$

$$\quad = 120/20$$

$$\quad = 6 \text{ A}$$

Example 9.11   The step-up chopper shown in Figure 9.13 supplies power to a load having 1.5 Ω resistance and 0.8 mH inductance. The source voltage is 50 V DC and the load voltage is 75 V. If the on time is 1.5 ms, find
a) the chopper switching frequency
b) the average value of the source current
c) I_{max}
d) I_{min}
Then draw the waveforms of voltage across the inductor v_L and current through the inductor i_i.

Solution   a) From Equation 9.18,

$$V_o = V_i \left(\frac{1}{1-d} \right)$$

$$1 - d = 50/75$$
$$= 0.67$$
$$d = 0.33$$
$$T_{ON}/T = 0.33$$
$$T = T_{ON}/0.33$$
$$= 1.5(10^{-3})/0.33$$
$$= 4.5 \text{ ms}$$

The chopper switching frequency is

$$f = 1/4.5(10^{-3}) = 222 \text{ HZ}$$

b) $$I_i = \frac{V_i}{(1-d)^2 R} = 50/(1-0.33)^2 \, 1.5 = 74.25 \text{ A}$$

Figure 9.17
Plots of v_L and i_i (not to scale)

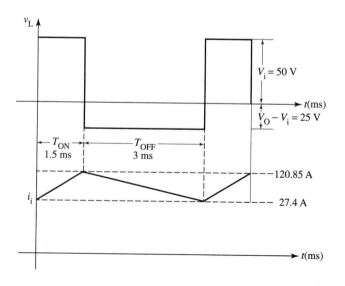

c) $\quad I_{max} = V_i \left[\dfrac{1}{R\,(1-d)^2} + \dfrac{T_{ON}}{2\,L} \right]$

$\qquad = 50 \left[\dfrac{1}{1.5(1-0.33)^2} + \dfrac{1.5\,(10^{-3})}{2*0.8(10^{-3})} \right]$

$\qquad = 50\,(1.48 + 0.937) = 120.85 \text{ A}$

d) $\quad I_{min} = V_i \left[\dfrac{1}{R\,(1-d)^2} - \dfrac{T_{ON}}{2\,L} \right]$

$\qquad = 50\,(1.48 - 0.937)$

$\qquad = 27.4 \text{ A}$

e) Plots of v_L and i_i are shown in Figure 9.17.

9.5 Buck-Boost Choppers

A buck-boost DC-to-DC chopper circuit combines the concepts of the step-up and step-down choppers. The output voltage can be either higher than, equal to, or lower than the input voltage. A reversal of the output voltage polarity may also occur. The circuit configuration is shown in Figure 9.18(a). The switch can

Figure 9.18
(a) Buck-boost DC chopper
(b) equivalent circuit with
switch on (c) equivalent
circuit with switch off

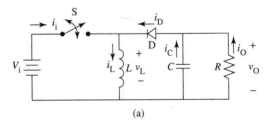

(a)

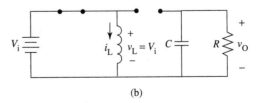

(b)

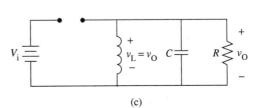

(c)

be any type of controlled switching device such as a power transistor, a GTO thyristor, or an IGBT.

When S is on, the diode D is reverse-biased and i_D is zero. The circuit can be simplified as shown in Figure 9.18(b). The voltage across the inductor is equal to the input voltage, and the current through the inductor i_L increases linearly with time. When S is off, the source is disconnected. The current through the inductor cannot change instantly, so it forward-biases the diode and provides a path for the load current. The output voltage becomes equal to the inductor voltage. The circuit can be simplified as shown in Figure 9.18(c). The voltage and current waveforms are shown in Figure 9.19.

With the switch *on* (T_{ON}),

$$W_{ON} = V_i I_i T_{ON}$$

Figure 9.19
Voltage and current waveforms for buck-boost chopper

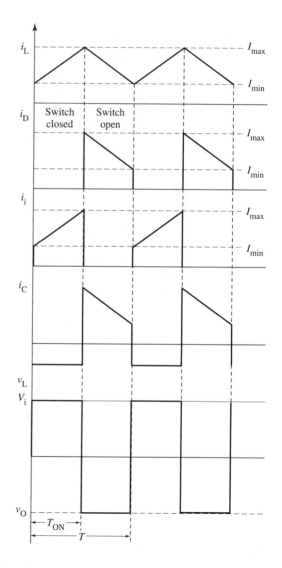

With the switch off (T_{OFF})

$$W_{OFF} = V_o \, I_i \, T_{OFF}$$

Ignoring losses,

$$W_{ON} = W_{OFF}$$
$$V_i I_i T_{ON} = V_o I_i T_{OFF}$$

or $\quad V_o = V_i \dfrac{T_{ON}}{T_{OFF}}$

Now, $d = \dfrac{T_{ON}}{T}$

$$T_{ON} = d \, T$$

and $\quad T = T_{ON} + T_{OFF}$

$$T_{OFF} = T - T_{ON}$$
$$= T\left(1 - \dfrac{T_{ON}}{T}\right)$$
$$= T(1 - d)$$

Substituting in V_o,

$$V_o = V_i \dfrac{d \, T}{(1 - d) \, T}$$

9.28

$$V_o = \dfrac{d}{1 - d} \, V_i$$

The output voltage can be controlled by changing the duty cycle d. Depending on the value of d, the output voltage can be higher than, equal to, or lower than the input voltage. When $d > 0.5$, the output voltage is greater than the input voltage and the circuit operates in the step-up mode. If $d < 0.5$, the output voltage is less than the input voltage and the circuit acts like a step-down chopper. The buck-boost chopper can transfer from operating in the step-down mode to operating in the step-up mode very smoothly and quickly by changing only the control signals for switch S.

Now, from Equation 9.7,

$$I_L = \dfrac{I_{max} + I_{min}}{2}$$

$$I_i = I_L \, d = \left(\dfrac{I_{max} + I_{min}}{2}\right) d$$

The average power input is

$$P_i = V_i * I_i$$
$$= \left(\dfrac{I_{max} + I_{min}}{2}\right) d \, V_i$$

The output power is

$$P_o = \frac{V_o^2}{R}$$

If we neglect power losses, the power input must equal the power output:

$$\left(\frac{I_{max} + I_{min}}{2}\right) d\, V_i = \frac{V_o^2}{R}$$

$$I_{max} + I_{min} = \frac{2\, V_o^2}{R\, d\, V_i}$$

Substituting V_o from Equation 9.28,

$$I_{max} + I_{min} = \frac{2\, d^2\, V_i^2}{R\, d\, (1-d)^2\, V_i} = \frac{2\, d\, V_i}{R\, (1-d)^2} \qquad \textbf{9.29}$$

With the switch closed (T_{ON}),

$$\Delta I_L = \frac{V_i}{L} T_{ON}$$

or $$I_{max} - I_{min} = \frac{V_i}{L} T_{ON} = \frac{V_i}{L} d\, T \qquad \textbf{9.30}$$

Adding Equations 9.29 and 9.30,

$$2\, I_{max} = \frac{2\, d\, V_i}{R\, (1-d)^2} + \frac{V_i}{L} d\, T$$

$$I_{max} = V_i \left[\frac{1}{R\,(1-d)^2} + \frac{T}{2\, L}\right] d \qquad \textbf{9.31}$$

Similarly, I_{min} is given by

$$I_{min} = V_i \left[\frac{1}{R\,(1-d)^2} - \frac{T}{2\, L}\right] d \qquad \textbf{9.32}$$

The peak-to-peak ripple in the input current I_i is given by

$$I_{p-p} = I_{max} - I_{min} = \frac{V_i T\, d}{L} \qquad \textbf{9.33}$$

For continuous current conditions, the minimum value of the inductance required is obtained by setting Equation 9.32 equal to zero:

$$I_{min} = 0 = V_i \left[\frac{1}{R\,(1-d)^2} - \frac{T}{2\, L}\right] d$$

$$\frac{1}{R\,(1-d)^2} = \frac{T\, d}{2\, L}$$

Solving for L,

$$L = \frac{R\, T\, d\, (1-d)^2}{2} \qquad \textbf{9.34}$$

Example 9.12 The buck-boost chopper shown in Figure 9.18 supplies power to a load having 1.5 Ω resistance and 0.8 mH inductance. The source voltage is 50 V DC and the load voltage is 75 V. If the on time is 1.5 ms, find

a) the chopper switching frequency
b) I_{max}
c) I_{min}
d) the average value of the input current
e) the average value of the diode current
f) the peak-to-peak ripple in the input current
g) the minimum inductance required for continuous current operation

Solution a) From Equation 9.28,

$$V_o = V_i \left(\frac{d}{1-d} \right)$$

$$75 = 50 \left(\frac{d}{1-d} \right)$$

$$d = 0.6$$

$$\frac{T_{ON}}{T} = 0.06$$

$$T = \frac{1.5\,(10^{-3})}{0.6} = 2.5 \text{ ms}$$

$$f = 1/T = 400 \text{ Hz}$$

b) from Equation 9.31,

$$I_{max} = 50 \left[\frac{1}{1.5\,(1-0.6)^2} + \frac{2.5\,(10^{-3})}{2 * 0.8\,(10^{-3})} \right] 0.6$$

$$= 50\,(4.2 + 1.6)\,0.6$$

$$= 144 \text{ A}$$

c) $I_{min} = 50\,(4.2 - 1.6)\,0.6 = 78$ A

d) $I_i = \left(\frac{I_{max} + I_{min}}{2} \right) d = \frac{(144 + 78)}{2}\,0.6 = 66$ A

e) The average value of the diode current is

$$I_D = I_o = \frac{V_o}{R} = \frac{75}{1.5} = 50 \text{ A}$$

f) $I_{p-p} = I_{max} - I_{min} = 144 - 78 = 66$ A

g) $L = \dfrac{R\,T\,d\,(1-d)^2}{2} = \dfrac{1.5 * 2.5\,(10^{-3})\,0.6\,(1-0.6)^2}{2} = 180 \text{ μH}$

9.6   Problems

9.1   What is meant by the term *chopper*? Give some typical applications.

9.2   Explain the basic operating principle of a step-down chopper.

9.3   Explain the basic operating principle of a step-up chopper.

9.4   For a step-down chopper circuit, the following information is available: $V_i = 120$ V, $L = 0.8$ mH, $R = 0.1$ Ω, $T = 2$ ms, and $T_{ON} = 1$ ms. If the chopper operates in the continuous current mode, find
a) the average output voltage
b) the average output current
c) the maximum value of the output current
d) the minimum value of the output current
e) the minimum value of the current if the current is discontinuous

9.5   A buck chopper circuit is operating with a frequency of 1 kHz. The source voltage is 110 V, and the load voltage is 75 V. Calculate T_{ON} and T_{OFF}.

9.6   A buck chopper circuit is connected to 100 V DC source and has a load voltage of 50 V. If the average output current is 10 A, find
a) the duty cycle
b) the average and RMS values of the input current
c) the average and RMS values of the current through the diode
d) the peak-to-peak ripple

9.7   The battery charger circuit shown in Figure 9.20 operates at a frequency of 2 KHZ. Find.
a) the duty cycle and the on time
b) the average value of i_i, i_D, and i_L
c) the power delivered to the 12 V battery

Figure 9.20
See Problem 9.7

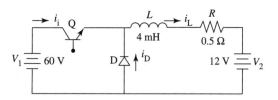

9.8   A step-down chopper operates at a frequency of 1 kHz with a duty cycle of 0.40. If trhe load resistance is 4 Ω and the inductance is 0.8 mH, find
a) the average value of the output current
b) I_{max}
c) I_{min}
d) the peak-to-peak ripple

9.9 Referring to Figure 9.6, calculate V_o, I_i, I_D, and P_o if $V_i = 500$ V, $T = 15$ ms, $T_{OFF} = 10$ ms, and $R = 1$ Ω.

9.10 A step-down chopper circuit operating at 8 kHz supplies 100 W at 15 V to a load resistor. If the source voltage is 25 V, find the value of L required for continuous current.

9.11 Referring to Figure 9.7, show that the average or DC value of
a) the load current is

$$I_o = \frac{I_{max} + I_{min}}{2}$$

b) the source current is

$$I_i = \frac{T_{ON}}{T} \frac{(I_{max} + I_{min})}{2}$$

9.12 In Figure 9.6, $V_i = 100$ V, $L = 10$ mH, $R = 2$ Ω, $T = 10$ ms, and $T_{ON} = 5$ ms. Plot the load current exactly and show that it is discontinuous.

9.13 Repeat problem 9.12 with $T_{ON} = 0.1$ ms and show that the current is continuous.

9.14 For a step-up chopper circuit, the following information is available: $V_i = 20$ V, $L = 10$ μH, $R = 0.1$ Ω, $T = 10$ μs, and $d = 0.4$. If the chopper operates in the continuous mode, find
a) the average output voltage
b) the maximum value of the output current
c) the minimum value of the output current

9.15 A boost chopper circuit is operating with a frequency of 10 kHz. The source voltage is 50 V and the load voltage is 70 V. Calculate T_{ON} and T_{OFF}.

9.16 A boost chopper circuit is connected to a 30 V DC source, and the load voltage is 50 V. If switching frequency is 50 kHz, find
a) the duty cycle
b) the minimum value of L to ensure a continuous inductor current

9.17 A buck-boost chopper supplied from a 20 V source is operating at a frequency of 50 kHz with a duty cycle of 0.40. If the load resistance is 20 Ω, find
a) the output voltage
b) the minimum value of L for a continuous inductor current

9.18 Referring to Figure 9.18, calculate V_o, I_{max}, I_{min}, and the value of L for continuous inductor current if $V_i = 50$ V, $T = 15$ μs, and $T_{OFF} = 10$ μs.

9.19 A buck-boost chopper circuit operating at 8 kHz supplies 100 W at 25 V to a load resistor. If the source voltage is 15 V, find the value of L required for continuous current.

9.7 Equations

$$V_o I_o = V_i I_i \tag{9.1}$$

$$V_o = \frac{T_{ON}}{T_{ON} + T_{OFF}} \, V_i \tag{9.2}$$

$$d = \frac{T_{ON}}{T} \tag{9.3}$$

$$V_o = d \, V_i \tag{9.4}$$

$$I_o = \frac{V}{R} = \frac{d \, V_i}{R} \tag{9.5}$$

$$V_{o(RMS)} = V_i \sqrt{d} \tag{9.6}$$

$$I_L = \frac{I_{max} + I_{min}}{2} \tag{9.7}$$

$$I_{max} + I_{min} = 2 \, \frac{V}{R} \tag{9.8}$$

$$I_{max} - I_{min} = T_{OFF} \, \frac{V_o}{L} \tag{9.9}$$

$$I_{max} = \frac{V_o}{R} + T_{OFF} \, \frac{V_o}{2 \, L} \tag{9.10}$$

$$I_{min} = \frac{V_o}{R} - T_{OFF} \, \frac{V_o}{2 \, L} \tag{9.11}$$

$$I_{p-p} = \frac{T_{OFF} \, V_o}{L} \tag{9.12}$$

$$I_D = \frac{T_{OFF} \, I_o}{T} \tag{9.13}$$

$$V_o = V_i d \tag{9.14}$$

$$I_o = \frac{I_i}{d} \tag{9.15}$$

$$L = \frac{T_{OFF}}{2} * R \tag{9.16}$$

$$W_{OFF} = (V_o - V_i) \, I_i \, T_{OFF} \tag{9.17}$$

$$V_o = V_i \left(\frac{1}{1 - d} \right) \tag{9.18}$$

$$I_i = \frac{V_o^2}{V_i} * \frac{1}{R} \tag{9.19}$$

$$I_o = I_i \, (1 - d) \tag{9.20}$$

$$I_i = \frac{V_i}{(1 - d)^2 \, R} \tag{9.21}$$

$$I_{max} + I_{min} = 2\, I_i \qquad\qquad \textbf{9.22}$$

$$I_{max} - I_{min} = \frac{V_i}{L} T_{ON} \qquad\qquad \textbf{9.23}$$

$$I_{max} = V_i \left(\frac{1}{R\,(1-d)^2} + \frac{T_{ON}}{2\,L} \right) \qquad\qquad \textbf{9.24}$$

$$I_{min} = V_i \left(\frac{1}{R\,(1-d)^2} - \frac{T_{ON}}{2\,L} \right) \qquad\qquad \textbf{9.25}$$

$$I_{p-p} = I_{max} - I_{min} = \frac{V_i T_{ON}}{L} \qquad\qquad \textbf{9.26}$$

$$L = \frac{R\, T_{ON}}{2} (1-d)^2 \qquad\qquad \textbf{9.27}$$

$$V_o = \frac{d}{1-d} V_i \qquad\qquad \textbf{9.28}$$

$$I_{max} + I_{min} = \frac{2\, d\, V_i}{R\,(1-d)^2} \qquad\qquad \textbf{9.29}$$

$$I_{max} - I_{min} = \frac{V_i}{L}\, T_{ON} = \frac{V_i}{L} d\, T \qquad\qquad \textbf{9.30}$$

$$I_{max} = V_i \left(\frac{1}{R\,(1-d)^2} + \frac{T}{2\,L} \right) d \qquad\qquad \textbf{9.31}$$

$$I_{min} = V_i \left(\frac{1}{R\,(1-d)^2} - \frac{T}{2\,L} \right) d \qquad\qquad \textbf{9.32}$$

$$I_{p-p} = I_{max} - I_{min} = \frac{V_i T\, d}{L} \qquad\qquad \textbf{9.33}$$

$$L = \frac{R\, T\, d\, (1-d)^2}{2} \qquad\qquad \textbf{9.34}$$

Inverters

10

Chapter Outline

Learning Objectives

After completing this chapter, the student should be able to

- explain what an inverter is
- list some applications of inverters
- describe the basic operation of a voltage source inverter

- explain the operation of half-bridge and full-bridge voltage source inverters
- describe inverter voltage control techniques
- describe the principle of pulse width modulation
- describe the principle of sinusoidal pulse width modulation

- describe the basic principle of three-phase inverters
- describe the basic operation of single-phase and three-phase current source inverters

10.1   Introduction

Inverters are static circuits (that is, they have no moving parts) that convert DC power into AC power at a desired output voltage or current and frequency. The output voltage of an inverter has a periodic waveform that is not sinusoidal but can be made to closely approximate this desired waveform. There are many types of inverters, and they are classified according to number of phases, use of power semiconductor devices, commutation principles, and output waveforms. We will first look at the single-phase inverter. Secondly, we will discuss voltage source inverters (VSI) and current source inverters (CSI), and finally we will cover the principles of three-phase inverters. Inverters are used in many industrial applications, including speed control of induction and synchronous motors, induction heating, aircraft power supplies, uninterruptible power supplies (UPS), and high-voltage DC transmission.

10.2   The Basic Inverter

The basic circuit for generating a single-phase alternating voltage from a DC power supply is shown in Figure 10.1. The circuit is also known as an H(half)-bridge inverter because it uses two semiconductor switches. Switches S_1 and S_2 connect the DC source to the load (and disconnect) alternately, thus producing an AC rectangular voltage waveform.

Figure 10.1
Basic circuit for half-bridge inverter

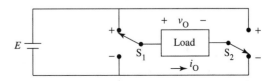

Table 10.1

State	S_1	S_2	Output voltage
1	+	−	$+E$
2	−	−	0
3	−	+	$-E$
4	+	+	0

Since each switch has positive and negative terminals, the combination of the two switches provides the four states shown in Table 10.1.

When states 1 and 3 are repeated alternately, a *square-wave* voltage is generated across the load, as shown in Figure 10.2(a). If states 2 and 4, which makes the load voltage zero, are used, the *step-wave* or *quasi-square-wave* waveform of Figure 10.2(b) is obtained.

Figure 10.2
Switching sequence in the H-bridge inverter (a) square-wave output (b) step-wave output

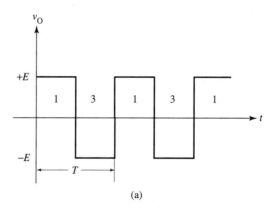

(a)

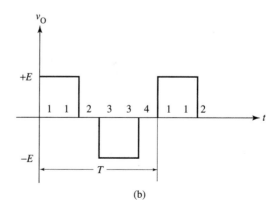

(b)

The frequency of the alternating voltage is determined by the switching rate. If the switching period is T seconds, the frequency f is:

$$f = 1/T \quad \text{(Hz)} \qquad \qquad \textbf{10.1}$$

The rectangular AC output voltage of the inverter is adequate for some applications. However, the sine-wave output voltage is the ideal waveform for many applications. Two methods can be used to make the output closer to a sinusoid. One is to use a filter circuit on the output side of the inverter. This filter must be capable of handling the large power output of the inverter, so it must be large and will therefore add to the cost and weight of the inverter. Moreover, the efficiency will be reduced due to the additional power losses in the filter.

The second method, pulse-width modulation (PWM), uses a switching scheme within the inverter to modify the shape of the output voltage waveform. PWM is discussed in Section 10.5.

10.3 Voltage Source Inverters (VSI)

A voltage source inverter (VSI) is the most commonly used type of inverter. In a VSI the input DC voltage source is essentially constant and independent of the load current drawn. The input DC voltage may be from an independent source such as a battery or may be the output of a controlled rectifier. A large capacitor is placed across the DC input line to the inverter. The capacitor ensures that any switching events within the inverter do not significantly change the DC input voltage. The capacitor charges and discharges as necessary to provide a stable output. The inverter converts the input DC voltage into a square-wave AC output source.

10.3.1 Half-Bridge VSI

The half-bridge inverter, which is used for low-power applications, is the basic building block of inverter circuits. Figure 10.3(a) shows a single-phase half-bridge VSI configuration that uses two switches (S_1 and S_2) and two DC power supplies. The switching device can be a power transistor (a BJT or a MOSFET), a GTO thyristor, or an SCR with its commutation circuit. Diodes D_1 and D_2 are freewheeling diodes.

Figure 10.3(b) shows the output voltage waveform with a resistive load. The switches are turned on and off alternately, with one switch on while the other is off. During the period 0 to $T/2$, switch S_1 is closed, which makes $v_o = +E$. At $T/2$, S_1 is opened and S_2 is closed. During the period $T/2$ to T, the output voltage $v_o = -E$. Therefore, the output voltage has an alternating rectangular waveform of frequency $f = 1/T$. By controlling T, we can control the frequency of the inverter output voltages. However, care must be taken not to turn both switches on, as they will short circuit the DC source.

Figure 10.3
Half-bridge voltage source
inverter (a) circuit diagram
(b) output waveforms with
an *R* load (c) waveform with
an *RL* load

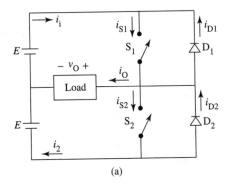

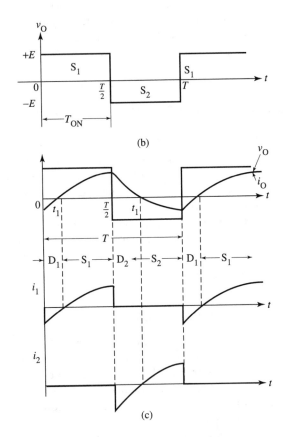

If switches S_1 and S_2 are each closed for an interval T_{ON}, the half-wave average output voltage is given by

$$V_{o(avg.)} = E \frac{T_{ON}}{T/2} = 2 E \frac{T_{ON}}{T} = 2 E d \qquad \textbf{10.2}$$

where *d*, the duty cycle, is

$$d = T_{ON}/T \qquad \textbf{10.3}$$

The RMS value of the output voltage is given by

$$V_{o(RMS)} = \sqrt{2d}\, E \qquad\qquad\qquad \textbf{10.4}$$

The output current waveform depends on the nature of the load. If the load is resistive, the current waveform is similar to the voltage waveform and the half-wave average output current is given by

$$I_{o(avg.)} = V_{o(avg.)}/R \qquad\qquad\qquad \textbf{10.5}$$

and

$$\text{average current in the switch} = I_{o(avg.)}/2 \qquad\qquad \textbf{10.6}$$

The average power absorbed by the load is

$$P_L = \frac{V^2_{o(RMS)}}{R} = 2\,d\,\frac{E^2}{R} \qquad\qquad\qquad \textbf{10.7}$$

If the load is inductive, the output current cannot reverse at the same instant that the output voltage changes its polarity, so the freewheeling diode provides a path for the load current to flow in the same direction. The voltage and current waveforms with an RL load are shown in Figure 10.3(c). The output current i_o lags the output voltage. During the interval 0 to $T/2$, the output voltage v_o is positive. Therefore, either S_1 or D_1 is conducting during this interval. However, during the interval 0 to t_1, the output current i_o is negative, indicating that D_1 must be conducting during this interval. Now, during the interval t_1 to $T/2$, i_o is positive, so S_1 must be conducting. At $T/2$, S_1 is turned off and the current transfers to D_2. At t_2, D_2 turns off and S_2 is turned on to take over conduction from t_2 to T. The cycle then repeats. Figure 10.3(c) also shows the conducting devices during the various intervals. Note that the freewheeling diodes conduct only when voltage and current are of opposite polarities.

The DC source currents i_1 and i_2 can be obtained from the output current waveform as shown in Figure 10.3(c). During the positive half-period of the output voltage, the current is supplied by the upper section of the DC source, and during the negative half-period it is supplied by the lower section.

Example 10.1   The single-phase half-bridge inverter shown in Figure 10.4(a) produces a step-wave output across a resistive load. $E = 100$ V, $d = 50\%$, and load resistance $R = 1\ \Omega$.

a) Plot the waveforms for the output voltage (v_o), the voltages across the SCRs (v_{SCR1} and v_{SCR2}), and the source currents (i_1 and i_2).
b) Find the maximum forward voltage that the switch must withstand.
c) Find the average load current.
d) Find the average switch current.
e) Find the power delivered to the load.

Solution   a) The waveforms are shown in Figure 10.4(b). If we turn on SCR$_1$ at $t = 0$ for an interval T_{ON}, the load current flows from the source to the load. At time $t = T/2$, while SCR$_1$ remains off, SCR$_2$ is turned on to conduct load current in the reverse direction. SCR$_2$ remains on for an interval T_{ON} before it is

Figure 10.4
(a) Circuit diagram for
Example 10.1 (b) waveforms
of voltage and current

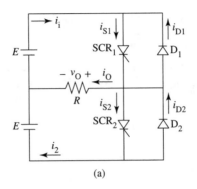

(a)

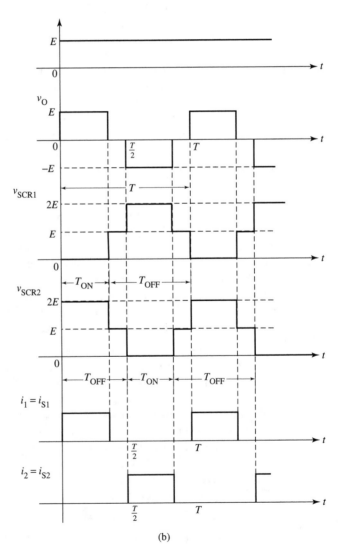

(b)

turned off. SCR_1 and SCR_2 remain off for the rest of the period T. The cycle is then repeated. The output voltage has an alternating rectangular waveform of amplitude E. Since the load is resistive, the output current is also rectangular in shape with a magnitude of E/R.

b) The voltage across the open switch is

$$2 * \text{DC supply voltage} = 2\,(100) = 200 \text{ V}$$

c) The average load voltage over a half-cycle is

$$V_{o(\text{avg.})} = 2\,E\,d = 2\,(100)(0.5) = 100 \text{ V}$$

The average load current is

$$V_{o(\text{avg.})}/R = 100/1 = 100 \text{ A}$$

d) average current in the switch = average load current/2 = $100/2 = 50$ A

e) The RMS value of the output voltage is

$$V_{o(\text{RMS})} = \sqrt{2d}\,E = \sqrt{2 * 0.5}\,(100) = 100 \text{ V}$$

The power delivered to the load is

$$P_L = V_{o(\text{RMS})}^2/R = 100^2/1 = 10 \text{ KW}$$

10.3.2 Full-Bridge VSI

10.3.2.1 With a Resistive Load

A full-bridge VSI can be constructed by combining two half-bridge VSIs. Figure 10.5 shows the basic circuit for a single-phase full-bridge voltage source inverter. Four switches and four freewheeling diodes are required. The amplitude of the output voltage and therefore the output power are twice that of the half bridge. The switches are turned on and off in diagonal pairs, so either switches S_1 and S_4 or S_2 and S_3 are turned on for a half-cycle ($T/2$). Therefore, the DC source is connected to the load alternately in opposite directions. The output frequency is controlled by the rate at which the switches open and close. If the pairs of switches are turned on at equal intervals, the output voltage waveform will be

Figure 10.5
Full-bridge voltage source inverter

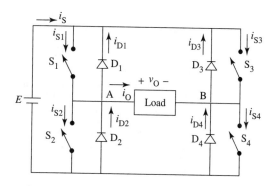

a square wave with a peak amplitude of E, as shown in Figure 10.6(a). The switching sequence is given in Table 10.2.

Figure 10.6
Switching sequence and
output voltage waveform
for the bridge inverter
(a) squarewave output
(b) step-wave output

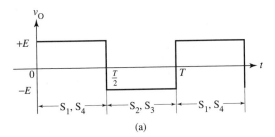

(a)

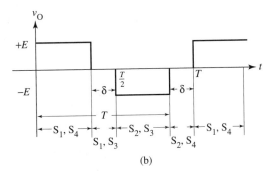

(b)

Comparing the waveforms of Figure 10.6(a) and 10.3(b) shows that the output voltage waveforms of the half-bridge and full bridge inverter are identical. Therefore, the same equations apply.

When the switching state is changed while going from one state to the other, both pairs of switches must be in the off state for a short time to avoid the possibility of short-circuiting the DC source in the transient state in which the two switches can be simultaneously closing. Therefore, switching from the on state to the off must be done as quickly as possible, while the switching from off to on must be carried out with an appropriate delay and take a definite time.

Table 10.2   Switching Sequence for a Squarewave Output

State	S_1	S_2	S_3	S_4	Output Voltage
1	On	Off	Off	On	$+E$
2	Off	On	On	Off	$-E$
3	On	Off	Off	On	E
4	Off	On	On	Off	$-E$

Table 10.3   Switching Sequence for a Step-Wave Output

State	S_1	S_2	S_3	S_4	Output Voltage
1	On	Off	Off	On	$+E$
2	On	Off	Off	On	$+E$
3	On	Off	On	Off	0
4	Off	On	On	Off	$-E$
5	Off	On	On	Off	$-E$
6	Off	On	Off	On	0
7	On	Off	Off	On	$+E$
8	On	Off	Off	On	$+E$

We can control the AC voltage by using a third switch state during which the output voltage is zero. The output waveform is the step wave shown in Figure 10.6(b). In the third switch state, switches S_1 and S_3 or S_2 and S_4 close for a time δ, during which $v_o = 0$. The switching sequence is given in Table 10.3.

The average value of the output voltage is given by

$$V_{o(avg.)} = E \frac{(T/2) - \delta}{T/2} = E\left(1 - \frac{\delta}{T/2}\right) = E\left(1 - \frac{2\delta}{T}\right) \qquad \textbf{10.8}$$

The RMS value of the output voltage is given by

$$V_{o(R.M.S.)} = E \sqrt{1 - \frac{2\,\delta}{T}}$$

Therefore, the magnitude of the output voltage may be controlled somewhat by delaying the turn-on of the appropriate pair of switches after the conducting pair has been turned off.

10.3.2.2   With an Inductive (RL) Load

Figure 10.7(a) shows a voltage source bridge inverter that uses SCRs as switches and supplies an *RL* load. The output voltage is a rectangular waveform shown with a 50% duty cycle. The output current waveform has an exponential form. When the output voltage is positive, the output current rises exponentially. During the next cycle, when the output voltage is negative, the output current decays exponentially.

The function of the freewheeling diodes is to provide a return path for the load current when the switches are off. For example, just after SCR$_2$ and SCR$_3$ turn off at $t = 0$, diodes D$_1$ and D$_4$ turn on. The load current starts at a negative value and rises exponentially at a rate given by the load time constant ($\tau = L/R$). The DC source current during this period is inverted and actually flow to

Figure 10.7
Full bridge inverter with an
RL load (a) circuit diagram
(b) waveforms with an *RL*
load

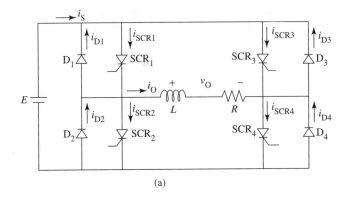

(a)

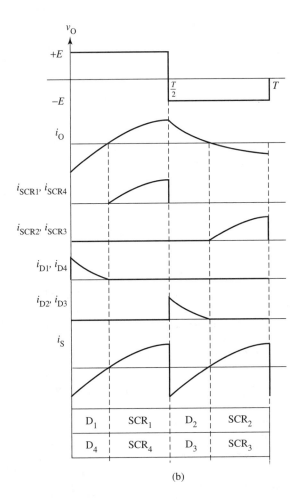

(b)

the DC source. When the output current reaches zero, D_1 and D_4 turn off and SCR_1 and SCR_4 turn on. The output voltage and output current are both positive, producing positive power. The current continues to grow and reaches its maximum value at $t = T/2$, when SCR_1 and SCR_4 are turned off. The output voltage reverses but the output current continues to flow in the same direction. The output current can flow only through diodes D_2 and D_3, which connect the DC source to the load, giving a reverse voltage. The energy stored in the inductor returns to the DC source, and the output current now falls from its maximum value and decays to zero. Once the load current stops, SCR_2 and SCR_3 can conduct to supply power to the load. The current reaches its negative maximum value at $t = T$, and the cycle repeats. Figure 10.7(b) shows the voltage and current waveforms. Also shown on the waveforms are the devices that conduct during the various intervals. Note from the source current waveform that the source current is positive when the switches conduct current and power is delivered by the source. The source current is negative when the diodes conduct current and power is absorbed by the source.

Example 10.2   A single-phase full-bridge inverter (Figure 10.7(a)) produces a step-wave output, as shown in Figure 10.8(a), across a resistive load. $E = 200$ V, $d = 50\%$, and load resistance $R = 2$ Ω.

a) Plot the waveforms for the output current (i_o), the voltages across the SCRs (V_{SCR1}, V_{SCR2}, V_{SCR3}, and V_{SCR4}), the SCR currents (i_{SCR1}, i_{SCR2}, i_{SCR3}, and i_{SCR4}) and the source current (i_s).

b) Find the maximum forward voltage that the switch must withstand.

c) Find the average load current.

d) Find the average switch current.

e) Find the power delivered to the load.

f) Find the average source current.

Solution   a) If each diagonal pair of SCRs is turned on alternately for an interval T_{ON} over a period T, the voltage and current waveforms are as shown in Figure 10.8(b). If we close (turn on) SCR_1 and SCR_4, at $t = 0$ for an interval T_{ON}, the load current flows from the source to the load. At time $t = T/2$, while SCR_1 and SCR_4 remain off, SCR_2 and SCR_3 are closed to conduct load current in the reverse direction. SCR_2 and SCR_3 remain on for an interval T_{ON} before they are turned off for the rest of the period T. The cycle is then repeated. The output voltage is an alternating rectangular waveform with an amplitude $E = 200$ V. Since the load is resistive, the output current is also rectangular in shape with a magnitude of $E/R = 100$ A.

b) When either pair of switches is on, both the switches in the other nonconducting pair must block the full supply voltage E. When all the switches are off, each switch must block half the supply voltage.

   The maximum forward voltage that the switches must withstand is $E = 200$ V.

Figure 10.8
Voltage and current
waveforms for Example 10.2

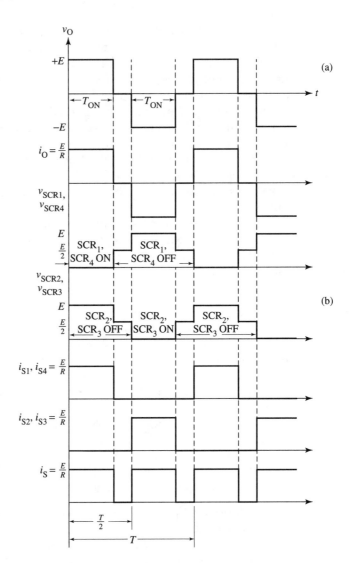

c) The average output voltage over a half-cycle is

$$V_{o(avg.)} = 2\,E\,d = 2\,(200)(0.5) = 200 \text{ V}$$

The average load current is

$$V_{o(avg.)}/R = 200/2 = 100 \text{ A}$$

d) average current in the switch = average load current/2 = 100/2 = 50 A

e) The RMS value of the output voltage is

$$V_{o(RMS)} = \sqrt{2d}\,E = \sqrt{2 * 0.5}\,200 = 200 \text{ V}$$

The power delivered to the load is

$$P_L = V_{o(RMS)}^2/R = 200^2/2 = 20 \text{ KW}$$

Figure 10.9
Single-phase bridge inverter
with an *RL* load (a) step-
wave output (b) current and
voltage waveforms

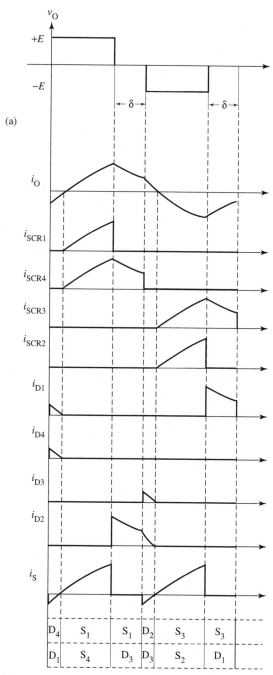

f) The power supplied by the source is the same as the power delivered to the load:

$$P_S = P_L = E \, I_S$$

$$I_S = \frac{20 \, (10^3)}{200} = 200 \text{ A}$$

Example 10.3 A single-phase bridge inverter (Figure 10.7(a)) produces a step-wave output across an *RL* load. Plot the waveforms for output voltage, output current, switch current, diode currents, and source current.

Solution A step-wave output is shown in Figure 10.9(a). Here, the operation of the second leg, in which SCR_2 and SCR_3 are present, is delayed by $\delta°$ from the operation of the first leg, where SCR_1 and SCR_4 are present. By varying the delay angle δ, the output pulse width can be controlled, thereby controlling the output voltage.

At the instant on the output voltage waveform when SCR_2 is turned on, the output current transfers to diode D_2, but because switch SCR_4 is still on, the output current follows a path through diode D_2 and SCR_4, effectively shorting the load and giving zero output voltage.

When SCR_3 is turned on, the only path for the output current is through diode D_3; this connects the DC source to the output with opposite polarity. SCR_3 and SCR_2 turn on immediately after the output current becomes zero. The SCR currents in the first arm (i_{SCR1} and i_{SCR2}) and the second arm (i_{SCR3} and i_{SCR4}) are not identical, as can be seen in Figure 10.9(b). The same is true for the diode currents.

10.4 Inverter Voltage Control Techniques

Most inverter applications require some means of controlling AC output voltage. Various methods can be used to control the output voltage of an inverter, but they can be classified into the following three broad categories:

1. control of the DC input voltage supplied to the inverter

2. control of the AC output voltage of the inverter

3. control of the voltage within the inverter

10.4.1 Control of the DC Input Voltage Supplied to the Inverter

For a given switching pattern, the inverter output voltage is directly proportional to its input voltage. Therefore, varying the DC input voltage supplied to the inverter is the simplest means of controlling the output voltage. If the power source is DC, then using a chopper is the main method of obtaining a variable DC voltage. However, in those cases where the DC voltage is obtained from AC voltage, controlling the AC voltage is easier. This can be accomplished by using controlled rectifiers or uncontrolled rectifiers and obtaining a variable DC output voltage with a chopper.

10.4.2   Control of the AC Output Voltage of the Inverter

In this method, introducing an AC regulator between the inverter and the load controls the AC voltage of the inverter and thus controls the inverter output voltage.

10.4.3   Control of the Voltage within the Inverter

Pulse-width modulation (PWM) is the most common method for controlling the voltage within the inverter. In this method, the output voltage is a pulse-width modulated wave and the voltage is controlled by varying the duration of the output voltage pulses. This method is presented in detail in Section 10.5.

Figure 10.10
PWM output voltage
waveform

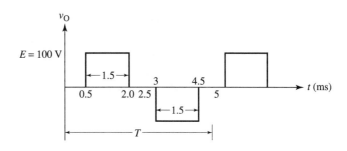

Example 10.4   The output voltage waveform of a PWM inverter feeding a resistive load is shown in Figure 10.10. Determine the RMS value of the output voltage. Also find the new duty cycle required to reduce the output voltage by 75%.

Solution   The duty cycle of the pulse is

$$d = T_{ON}/T = 1.5 \ (10^{-3})/5 \ (10^{-3}) = 0.3$$

The mean square value of the output voltage is

$$V_o^2 = E^2 * \frac{T_{ON}}{T/2} = 2 \ E^2 * \frac{T_{ON}}{T} = 2 \ E^2 * d$$

The RMS value of the output voltage is

$$V_{o(RMS)} = E \sqrt{2d} = 100 \ \sqrt{0.6} = 77.5 \ V$$

The duty cycle required to cut the output voltage to 0.75 (77.5) = 58.1 V is

$$\sqrt{2d} = \frac{V_{o(RMS)}}{E}$$

or

$$d = \frac{V^2_{o(RMS)}}{2\,E^2} = \frac{58.1^2}{2(100^2)} = 0.17$$

Figure 10.11
Single pulse-width
modulation output
waveforms (a) without
modulation (b) modulated
waveform

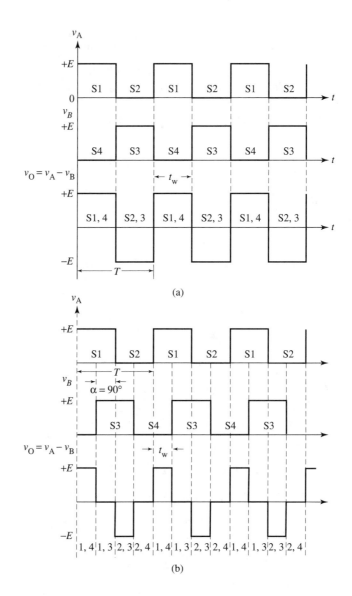

10.5   Pulse-Width Modulation (PWM)

The three most commonly used methods for pulse-width modulation fall into the following groups:

1. Single pulse-width modulation

2. Multiple pulse-width modulation

3. Sinusoidal pulse-width modulation

10.5.1   Single Pulse-Width Modulation

In this method of voltage control, the output voltage waveform consists of a single pulse in each half-cycle of the required output voltage. For a given frequency ($f = 1/T$), the pulse width t_w can be varied to control the AC output voltage. The output voltage waveform of a single-phase bridge inverter (see Figure 10.5) without modulation is shown in Figure 10.11(a). Here switches S_1 and S_4 are on for one half-cycle and S_2 and S_3 are on for the other half-cycle to give maximum output voltage.

Voltage control is achieved by varying the phase of S_3 and S_4 with respect to S_1 and S_2. Figure 10.11(b) shows the output voltage waveform when the conduction interval of S_3 and S_4 is advanced by an angle $\delta = 90°$. The output voltage is obtained by adding the two square-wave voltages, which are shifted in phase with respect to each other. The output voltage consists of alternating pulses with a pulse width of ($180° - \delta$) = $90°$.

The output voltage can be smoothly adjusted from its maximum (0° delay) to zero (180° delay) by either phase-advancing or -delaying the turn-on of one pair of switches with respect to the other.

10.5.2   Multiple Pulse-Width Modulation

Instead of reducing the pulse width to control the output voltage, the output of the inverter can be switched on and off rapidly several times during each half-cycle to produce a train of constant magnitude pulses.

Figure 10.12 shows the idea of multiple pulse-width modulation. The output voltage waveform consists of m pulses for each half-cycle of the required output voltage. If f is the output frequency of the inverter, the frequency of the pulses (f_p) is given by

$$f_p = 2 f m$$

Therefore, the number of pulses per cycle is

$$2m = f_p/f$$

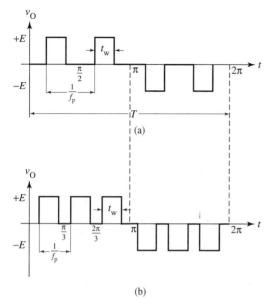

Figure 10.12
Multiple pulse-width modulation waveforms
(a) $m = 2$ (b) $m = 3$

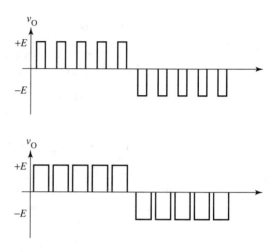

Figure 10.13
Variable duty cycle with fixed $m = 5$

Figure 10.12(a) shows the output voltage waveform for $m = 2$. The pulse width t_w should be less than $\pi/2$. In Figure 10.12(b), for $m = 3$, it is clear that $t_w < \pi/3$. In general, the pulse width $t_w \leq \pi/m$.

An alternative approach for controlling the magnitude of the output voltage is to keep m constant and vary the pulse width t_w (see Figure 10.13).

10.5.3 Sinusoidal Pulse-Width Modulation (SPWM)

In sinusoidal pulse-width modulation, the output voltage is controlled by varying the on-off periods so that the on periods (pulse width) are longest at the peak of the wave. Figure 10.14 shows a general SPWM pattern. The switching times are determined as shown in Figure 10.15(a). $v_R(t)$ is a *reference* modulating sinusoidal wave of amplitude V_m and frequency f_m, which is equal to the desired output frequency of the inverter. A high-frequency triangular *carrier*

Figure 10.14
Sinusoidal pulse-width
modulation pattern

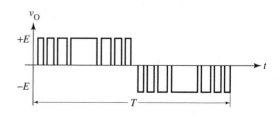

wave $v_c(t)$ with an amplitude V_c and frequency f_c is compared with the reference sine wave. The switching points are determined by the intersection of the $v_c(t)$ and $v_R(t)$ waves. The pulse width t_w is determined by the time during which $v_c(t) < v_R(t)$ in the positive half-cycle of $v_R(t)$ and $v_c(t) > v_R(t)$ in the negative half-cycle of $v_R(t)$.

Two control parameters that regulate the output voltage are the chopping ratio and the modulation index. The frequency ratio f_c/f_m is known as the chopping carrier ratio N. It determines the number of pulses in each half-cycle of the inverter output voltage. The ratio V_m/V_c is called the modulation index M ($0 \leq M \leq 1$). It determines the width of the pulses and therefore the RMS value of the inverter output voltage. M is usually adjusted by varying the amplitude of the reference wave while keeping the carrier wave amplitude fixed. The inverter output frequency is varied by varying the reference wave frequency. Figure 10.15(a) is drawn for $N = 6$ and $M = 1$. Maximum output voltage occurs with $M = 1$; when $M = 0.5$, the output is halved (Figure 10.15(b)).

Rather than using a triangular carrier wave with an alternating offset, as shown in Figure 10.15, a triangular carrier without an offset can be used. In this

Figure 10.15
Sinusoidal pulse width modulation waveforms
(a) $M = 1$ (b) $M = 0.5$

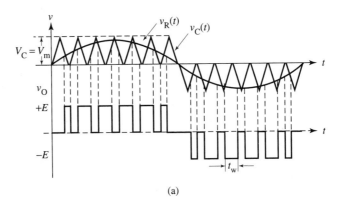

(a)

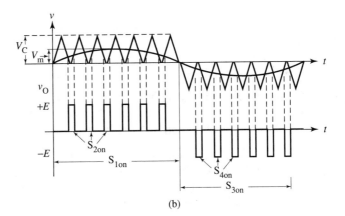

(b)

case, a triangular carrier wave of frequency f_c and a reference modulating sine wave of frequency f_m (the same as the required output frequency of the inverter) are used to regulate the output voltage. Varying the amplitude of the reference sine wave varies the pulse width and controls the effective magnitude of the output waveform.

The carrier wave and the sinusoidal reference wave are shown in Figure 10.16(a). The output voltage waveform v_o is drawn in Figure 10.16(b). Note that the number of output pulses in one full cycle is six. There are also six carrier voltage waves during this period. Therefore, the pulse repetition frequency of the inverter is the same as the carrier frequency.

Figure 10.16
Sinusoidal pulse width modulation waveforms (a) reference signals (b) output voltage waveform

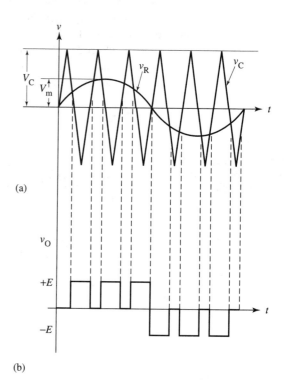

10.6 Pulse-Width Modulated (PWM) Inverters

10.6.1 Single-Phase Full-Bridge Inverters

In a pulse-width modulated inverter, the output voltage waveform has a constant amplitude whose polarity reverses periodically to provide the output fundamental frequency. The source voltage is switched at regular intervals to produce a variable output voltage. The output voltage of the inverter is controlled by varying the pulse width of each cycle of the output voltage. Figure 10.17(a) shows a single-phase bridge inverter using BJTs as switches. Switches Q_3 and Q_4 of the right leg of the inverter are turned on after an angle α with respect to the turning-on of switches Q_1 and Q_2 of the left leg of the inverter. The switching sequence is shown in Table 10.4. The resulting output voltage v_o, shown in Figure 10.17(b), has a pulse width t_w of α. By changing the shift angle α, the inverter output voltage can be changed.

Inverters using transistors or GTO thyristors can operate at much higher switching frequencies than inverters using conventional SCRs and therefore are used in pulse-width modulated inverters. In a pulse-width modulated inverter operating at a high switching frequency, the switching losses will be relatively high, reducing efficiency and creating heat-removal problems.

Figure 10.17
Basic pulse-width
modulated waveform
(a) circuit diagram (b) PWM
output waveform

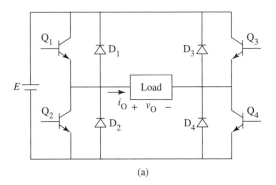

(a)

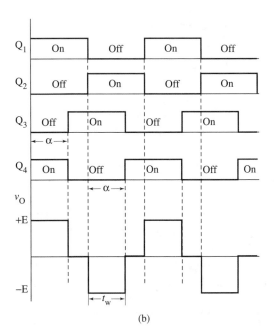

(b)

Table 10.4

Q$_1$	Q$_2$	Q$_3$	Q$_4$	V$_o$
On	Off	Off	On	$+E$
On	Off	On	Off	0
Off	On	On	Off	$-E$
Off	On	Off	On	0
On	Off	Off	On	$+E$
On	Off	On	Off	0

10.6.2 Single-Phase Half-Bridge Inverters

With a half-bridge inverter, an output voltage of zero is not possible—the output voltage can be only positive or negative. Therefore, the output voltage is allowed to reverse instead of being zero. Figure 10.18 shows a PWM waveform for a half-bridge inverter. We can control the output voltage by controlling the width 2δ.

Figure 10.18
Pulse-width modulation in a half-bridge inverter

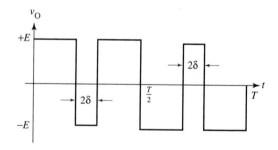

Sinusoidal pulse-width modulation (Figure 10.19) is commonly used with half-bridge inverters. A rectified sinusoidal reference signal is compared with a triangular carrier wave. For the time during which the reference signal is higher than the carrier wave, the switches are operated to produce positive-going pulses; otherwise, negative-going pulses are produced:

When $v_R > v_c$, S$_1$ is on and $v_o = +E$
When $v_R < v_c$, S$_2$ is on and $v_o = -E$

Switch conduction is also shown in Figure 10.19.

Figure 10.19
Sinusoidal pulse-width modulation in a half-bridge inverter

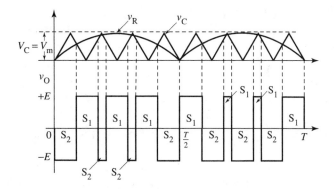

Example 10.5 A single-phase full-bridge inverter uses PWM for voltage control. Plot the output voltage waveform if the carrier frequency is a sawtooth voltage waveform synchronized to the fundamental and the reference wave is a DC voltage level.

Figure 10.20
PWM waveforms for full-bridge inverter

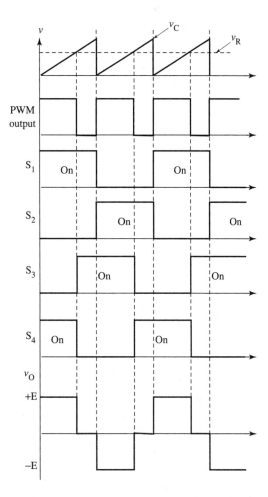

Solution   The PWM output is obtained by comparing the DC level with the sawtooth voltage waveforms. The magnitude of the DC voltage determines the pulse width of the output voltage. The output pulses shown in Figure 10.20 have equal width.

10.7   Other Basic Types of Single-Phase Inverters

The H-bridge inverter, which uses four switching devices, is the most common type of inverter. Other types of inverters are shown in Figure 10.21.

Figure 10.21
Single-phase inverter
(a) center-tap inverter
(b) center-tap inverter in series (c) with a large capacitor

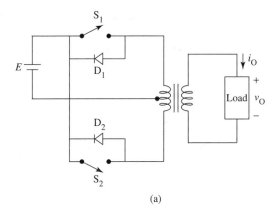

(a)

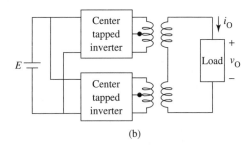

(b)

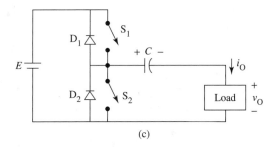

(c)

10.7.1   Center-Tap Inverters

Figure 10.21(a) shows a transformer-coupled inverter. This type of inverter features electrical isolation between the load and the power supply. The load voltage can be adjusted through the turn ratio of the transformer. Because the SCRs are alternately turned on and off, the source voltage is connected to each half of the transformer primary winding in turn, producing an alternating voltage in the secondary winding. By combining two such inverters in series as shown in Figure 10.21(b) and varying the relative firing instants, a step-wave output can be produced.

The circuit in Figure 10.21(c) employs a large capacitor C. The charged capacitor C works like a DC battery, having a voltage of $E/2$ when the time constant ($\tau = RC$ for a resistive load) is much higher than the reciprocal of the switching frequency. Therefore, when S_1 is on and S_2 is off, as shown in Figure 10.22, the voltage applied to the load is $E - E/2 = E/2$. When the switching states are reversed, the load voltage is the capacitor voltage itself ($-E/2$).

Figure 10.22
Principle of the single-phase inverter of Figure 10.21(c)
(a) S_1 is on and S_2 is off
(b) S_1 is off and S_2 on

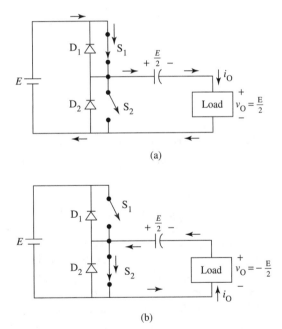

10.8   The Basic Principle of the Three-Phase Bridge VSI Inverter

A three-phase inverter circuit changes DC input voltage to a three-phase variable-frequency variable-voltage output. The input DC voltage can be from a DC source or a rectified AC voltage. A three-phase bridge inverter can be constructed by combining three single-phase half-bridge inverters. The basic circuit is shown in Figure 10.23. It consists of six power switches with six associated freewheeling diodes. The switches are opened and closed periodically in the proper sequence to produce the desired output waveform. The rate of switching determines the output frequency of the inverter. Many sequences for operating these three switches are possible, but there are two fundamental modes that complete one cycle with six switchings: one is known as the 120° conduction type and the other, the 180° conduction type.

Figure 10.23
Three-phase bridge inverter circuit diagram

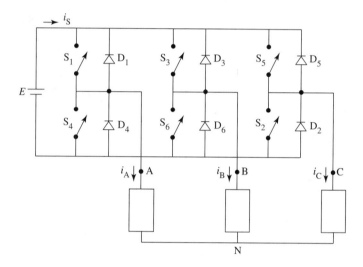

10.8.1   The 120° Conduction Type

The basic three-phase inverter bridge in Figure 10.23 can be controlled so that each switch conducts for a period of 120°. In this situation, only two switches are conducting at any time, one from the positive group (S_1, S_3, and S_5) and the other from the negative group (S_2, S_4, and S_6), The two on switches connect two of the load terminals to the DC supply terminals, while the third terminal remains floating. There are six intervals in one cycle of the AC voltage waveform. The switches are turned on at 60° intervals of the output voltage waveform in the appropriate sequence to obtain voltages v_{AB}, v_{BC}, and v_{CA}. The rate of switching determines the output frequency.

To eliminate the possibility of shorting out the DC source, we must make sure that two switches in the same leg are not on simultaneously. Therefore, an interval of 60° elapses between the end of conduction in switch S_1 and the beginning of the conduction in switch S_4, which is in the same leg as S_1. The same is true for switches S_3 and S_6 and switches S_5 and S_2.

The phase voltages across the load, v_{AN}, v_{BN}, and v_{CN}, can be determined for various 60° durations with a Y-connected resistive load. These voltages can be obtained by considering the equivalent circuits of the various inverter-load combinations for the six intervals as shown in Figure 10.24. The results are summarized in Table 10.5. The switching sequence is S_1 and S_2, S_2 and S_3, S_3 and S_4, S_4 and S_5, S_5 and S_6, S_6 and S_1, ...

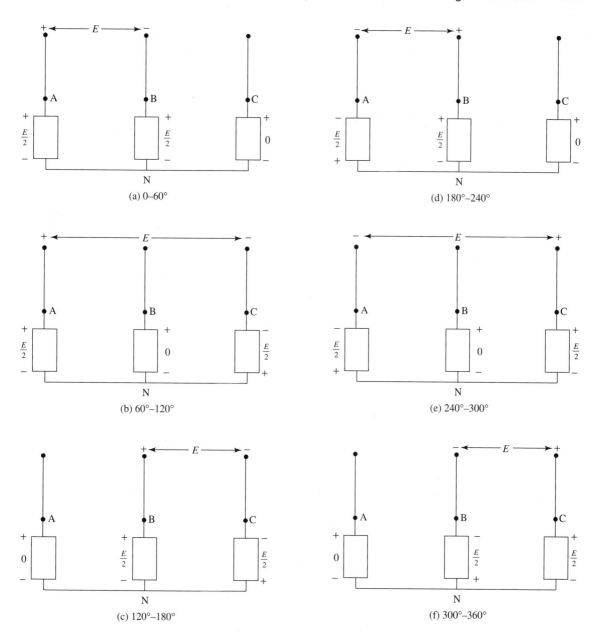

Figure 10.24
Three-phase bridge inverter equivalent circuits

Table 10.5

Interval	S_1	S_2	S_3	S_4	S_5	S_6	V_{AN}	V_{BN}	V_{CN}
0 to 60°	On	Off	Off	Off	Off	On	$+E/2$	$-E/2$	0
60 to 120°	On	On	Off	Off	Off	Off	$+E/2$	0	$-E/2$
120 to 180°	Off	On	On	Off	Off	Off	0	$+E/2$	$-E/2$
180 to 240°	Off	Off	On	On	Off	Off	$-E/2$	$+E/2$	0
240 to 300°	Off	Off	Off	On	On	Off	$-E/2$	0	$+E/2$
300 to 360°	Off	Off	Off	Off	On	On	0	$-E/2$	$+E/2$

Figure 10.25 shows the three phase voltages, v_{AN}, v_{BN}, and v_{CN}. The line voltages can be determined from the phase voltages using

$$v_{AB} = v_{AN} - v_{BN} \tag{10.9}$$

$$v_{BC} = v_{BN} - v_{CN} \tag{10.10}$$

$$v_{CA} = v_{CN} - v_{AN} \tag{10.11}$$

The three line voltages are also shown in Figure 10.25. The step-wave line voltage waveforms are identical in shape but are displaced from each other by 120°.

Each switch is on for a duration of 120° in sequence. When S_1 is on at $\omega t = 0$, terminal A is connected to the positive side of the DC source. When S_4 is on at $\omega t = \pi$, terminal A is connected to the negative side of the DC source.

When the load is inductive, the waveform will vary from those shown because the terminal voltage in the off period will be affected by the transient current behavior.

With a balanced Y-connected resistive load, the output power is given by

$$P_o = \frac{(E/2)^2}{R} + \frac{(E/2)^2}{R} \tag{10.12}$$
$$= E^2/2\,R$$

where R is the resistance per phase.

The RMS value of the phase voltage is

$$V_{ph(RMS)} = \frac{E\sqrt{2}}{2\sqrt{3}} = \frac{E}{\sqrt{6}} \tag{10.13}$$

$$\text{RMS value of line voltage} = V_{L(RMS)}\sqrt{3}\ V_{ph(RMS)} = \frac{E}{\sqrt{2}} \tag{10.14}$$

The RMS current through the switch is

$$i_{Switch(RMS)} = E/2\sqrt{3}R \tag{10.15}$$

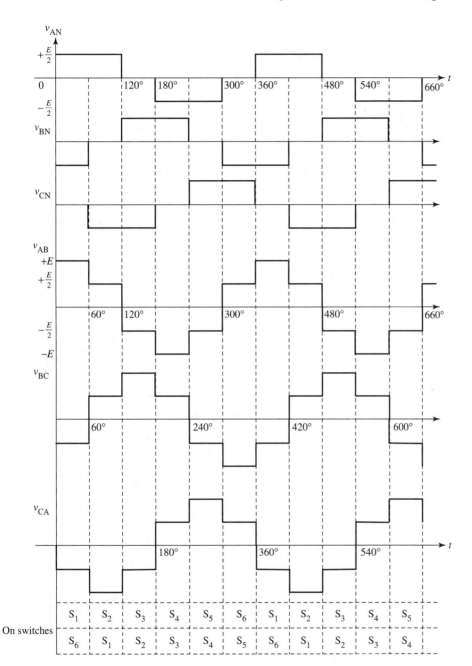

Figure 10.25
Circuit voltage waveforms for 120° conduction

The RMS value of the output current is

$$I_{O(RMS)} = \sqrt{2}\, I_{switch(RMS)} \qquad \textbf{10.16}$$

$$\text{reverse voltage rating for the switch} = E \qquad \textbf{10.17}$$

Example 10.6 For a three-phase bridge inverter with a Y-connected resistive load, plot the three output currents and the six switch currents for the 120° conduction mode.

Solution The waveforms of the output current are shown in Figure 10.26. The output currents are step-waves, with each switch conducting the load current for 120°. The three currents are displaced by 120°.

Figure 10.26
Waveforms for currents of a three-phase bridge inverter with 120° conduction and a resistive load

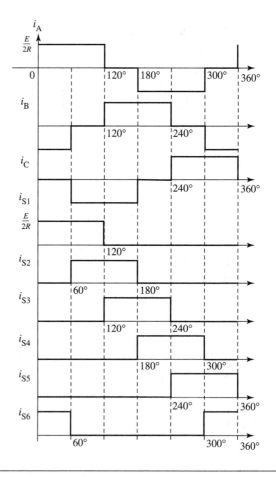

Example 10.7 For a three-phase bridge inverter with a Δ-connected resistive load, plot the three output voltages for 120° conduction mode.

Solution Figure 10.27 shows the switching sequence necessary to obtain the three-phase output voltage shown in Figure 10.28. In Figure 10.27(a), which shows the equivalent circuit from 0° to 60°, S₁ connects the positive terminal of E to A and

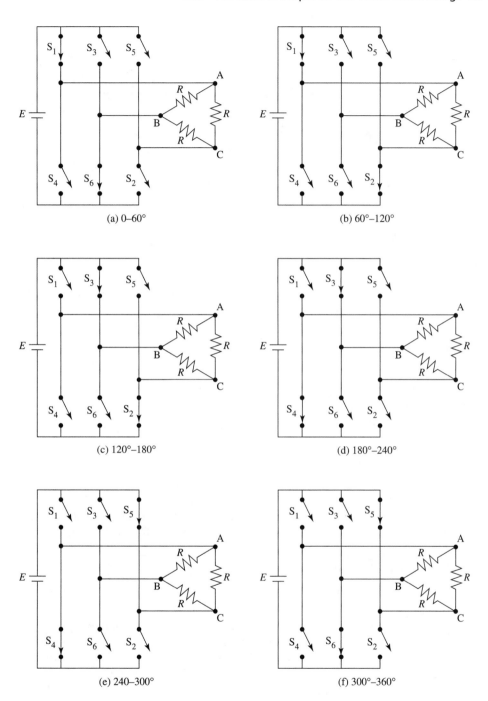

Figure 10.27
Equivalent circuits of the three-phase inverter bridge

Figure 10.28
The output voltage waveforms of the three-phase bridge inverter with a delta load

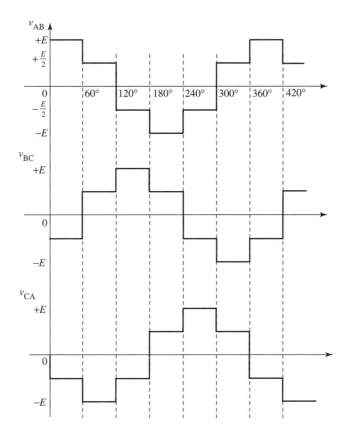

S_6 connects the negative terminal of E to B. Therefore, the output voltage v_{AB} equals $+E$. Voltages v_{AC} and v_{CB} are both equal to $+E/2$. The output voltages v_{CA} and v_{BC} are therefore equal to $-E/2$.

The other circuits (10.27 (b), (c), (d), (e), and (f)) can be analyzed in a similar way to obtain the three output voltage waveforms in Figure 10.28.

10.8.2 The 180° Conduction Type

Switching for this type is carried out without an off period, that is, each switch is always on either the positive or negative terminal, but the situation to avoid is having all three on the positive or negative terminal simultaneously. At any given instant, three switches are conducting, say S_1, S_2, and S_3. After a period of 60°, S_2, S_3, and S_4 will be conducting. The conduction period for each switch is 180°, so that no two switches in the same leg conduct simultaneously.

Six distinct 60° intervals exist for one output cycle, and the rate of sequencing these intervals specifies the inverter output frequency. The complete switching pattern for the six intervals is shown in Table 10.6. The six-step sequence creates a cyclic pattern: 1-2-3, 2-3-4, 3-4-5, 4-5-6, 5-6-1, 6-1-2, ... It can be seen that each switch conducts for a period of 180°.

Table 10.6

Interval	S_1	S_2	S_3	S_4	S_5	S_6
0 to 60°	On	Off	Off	Off	On	On
60 to 120°	On	On	Off	Off	Off	On
120 to 180°	On	On	On	Off	Off	Off
180 to 240°	Off	On	On	On	Off	Off
240 to 300°	Off	Off	On	On	On	Off
300 to 360°	Off	Off	Off	On	On	On

The three output voltage waveforms can now be derived by assuming a balanced, Y-connected resistive load R. The phase voltages for the various 60° intervals can be obtained by considering the equivalent circuit for each interval, as shown in Figure 10.29. From these equivalent circuits, the voltages associated with each phase of the load can be determined. A summary of the results is given in Table 10.7. The phase voltages waveforms shown in Figure 10.30 are identical but are displaced by 120°. They are not pure sine waves, but they do show some resemblance. The phase voltages have six discontinuities per cycle, corresponding to the six switching points per cycle.

The line voltages can be obtained from the following relationships:

$$v_{AB} = v_{AN} - v_{BN} \qquad\qquad\qquad \textbf{10.18}$$

$$v_{BC} = v_{BN} - v_{CN} \qquad\qquad\qquad \textbf{10.19}$$

$$v_{CA} = v_{CN} - v_{AN} \qquad\qquad\qquad \textbf{10.20}$$

Table 10.7

Interval	v_{AN}	v_{BN}	v_{CN}	v_{AB}	v_{BC}	v_{CA}
0 to 60°	$E/3$	$-2E/3$	$E/3$	$+E$	$-E$	0
60 to 120°	$2E/3$	$-E/3$	$-E/3$	$+E$	0	$-E$
120 to 180°	$E/3$	$E/3$	$-2E/3$	0	$+E$	$-E$
180 to 240°	$-E/3$	$+2E/3$	$-E/3$	$-E$	$+E$	0
240 to 300°	$-2E/3$	$E/3$	$E/3$	$-E$	0	$+E$
300 to 360°	$-E/3$	$-E/3$	$+2E/3$	0	$-E$	$+E$

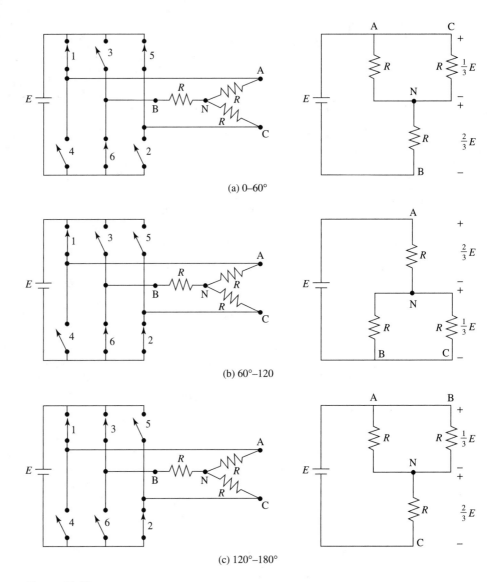

Figure 10.29
Three-phase bridge inverter equivalent circuits

The line voltages for the six intervals are also shown in Table 10.7 and plotted in Figure 10.30. The three line voltages are also displaced by 120°. Basically, the line or phase voltage waveforms are independent of the load characteristics, which may have any combination of resistance, inductance, and capacitance, balanced or unbalanced. If the load is resistive, the load current has the same waveform as the phase voltage.

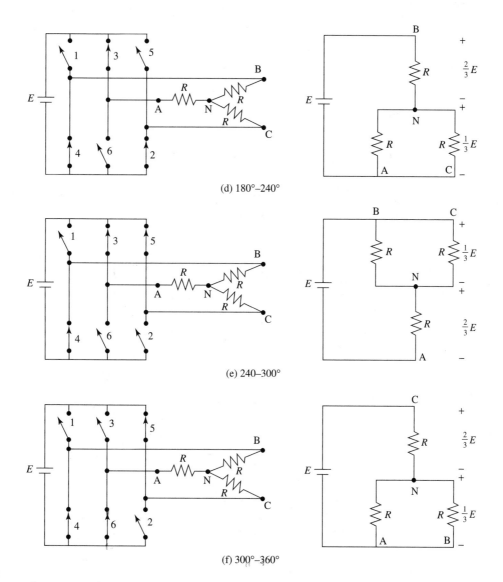

(d) 180°–240°

(e) 240–300°

(f) 300°–360°

Figure 10.29 (con't)
Three-phase bridge inverter equivalent circuits

For a balanced Y-connected load, the output power is given by

$$P_o = \frac{(E/3)^2}{R} + \frac{(E/3)^2}{R} + \frac{(2E/3)^2}{R}$$

$$= \frac{2\,E^2}{3\,R}$$

10.21

Figure 10.30
Output voltage waveforms

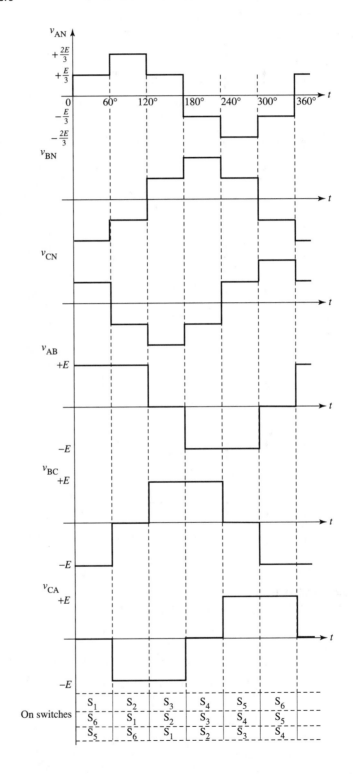

where R is the resistance per phase. Note that the power output here is 1.33 times the power output in the 120° conduction mode.

The RMS current through the switch is

$$I_{switch(RMS)} = E/3R \qquad\qquad \textbf{10.22}$$

The RMS value of the output current is

$$I_o = \sqrt{2}\, I_{switch(RMS)} \qquad\qquad \textbf{10.23}$$

reverse voltage rating for the switch $= E$ **10.24**

The RMS value of the output line voltage is

$$V_{L(RMS)} = \frac{\sqrt{2}\,E}{\sqrt{3}} \qquad\qquad \textbf{10.25}$$

The RMS value of the output phase voltage is

$$V_{ph(RMS)} = \frac{\sqrt{2}}{3}\,E \qquad\qquad \textbf{10.26}$$

With a PWM ratio of α,

$$V_{L(RMS)} = \frac{\sqrt{2}}{\sqrt{3}}\,\alpha\,E \qquad\qquad \textbf{10.27}$$

and

$$V_{ph(RMS)} = \frac{\sqrt{2}}{3}\,\alpha\,E \qquad\qquad \textbf{10.28}$$

The DC input current is

$$I_i = \frac{3\sqrt{2}\,I_o}{\pi}\cos\theta \qquad\qquad \textbf{10.29}$$

where θ is the load phase angle.

For PWM,

$$I_i = \frac{3\sqrt{2}}{\pi}\,\alpha\,I_o\cos\theta \qquad\qquad \textbf{10.30}$$

Example 10.8   For a three-phase bridge inverter with a Y-connected resistive load, plot the three output currents and the six switch currents for the 180° conduction mode.

Solution   The waveforms of the output currents, i_A, i_B and i_C, are shown in Figure 10.31. The output current waveforms are stepped, with each switch conducting the load current for 180°. The three currents are displaced by 120°. The switch currents can be obtained easily by referring to Fig. 10.29. For example, $i_{s1} = i_A$ for $0 - 180°$ and $i_{s1} = 0$ for the rest of the cycle.

Figure 10.31
Current waveforms for a three-phase bridge inverter with 180° conduction and a resistive load

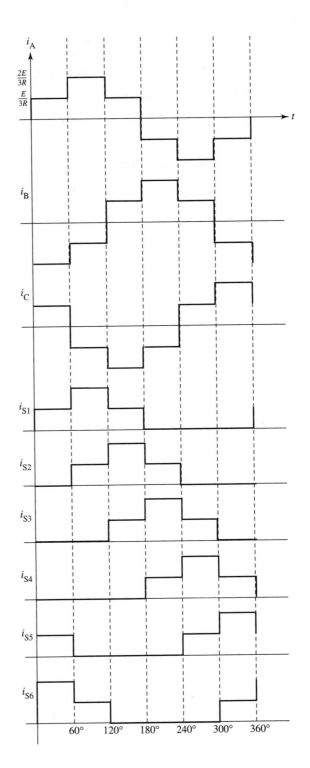

Example 10.9   A three-phase bridge inverter supplies a Y-connected load with 10 Ω resistance per phase. The DC source voltage is 440 V and the inverter operates in the 180° conduction mode. Find
a) the source current
b) the average power absorbed by the load
c) the RMS value of the output phase voltage
d) the RMS value of the output line voltage

Solution   a) At any instant the equivalent circuit of the three-phase bridge inverter is as shown in Figure 10.32. The total resistance seen by the source is:

$$R_T = (R \parallel R) + R = R/2 + R = 1.5\ R = 1.5\ (10) = 15\ \Omega$$

The source current is constant given by

$$I_s = E/R_T = 440/15 = 29.33\ \text{A}$$

Figure 10.32
Equivalent circuit
for Example 10.9

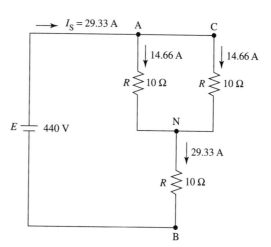

b) The power delivered to the load can be determined from Figure 10.32 by adding the power consumed by each leg:

$$P_L = (29.33^2 * 10) + (14.66^2 * 10) + (14.66^2 * 10)$$
$$= 8602.5 + 2150.6 + 2150.6$$
$$= 12904\ \text{W}$$

Note that the power input from the source ($P_s = E * I_s = 440 * 29.33 = 12905$ W) equals the power to the load.
The load power per phase is

$$P_{\text{ph.}} = P_L/3 = 4301\ \text{W}$$

c) The power per phase is

$$P_{ph.} = V^2_{ph.}/R$$
$$V_{ph} = \sqrt{P_{ph.} * R} = \sqrt{4301 * 10} = 207 \text{ V}$$

Therefore the RMS value of the three phase voltages is

$$V_{AN} = V_{BN} = V_{CN} = 207 \text{ V}$$

Figure 10.33
Output waveforms with a
Δ-connected resistive load
in 180° operation

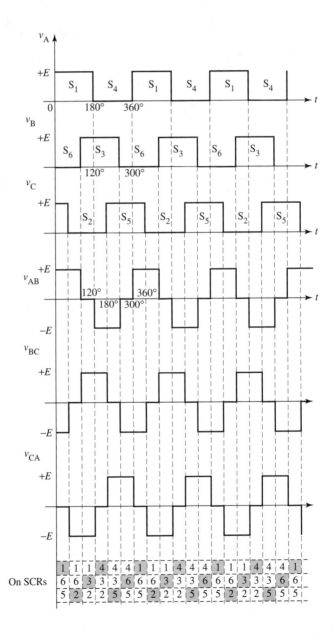

d) The RMS value of three line voltages is

$$V_{AB} = V_{BC} = V_{CA} = \frac{\sqrt{2}}{\sqrt{3}}E = 0.816\,(440) = 359\text{ V}$$

Note the $\sqrt{3}$ relationship between the line and phase voltages.

Example 10.10   Plot the three output voltages for the 180° conduction mode of a three-phase bridge inverter with a Δ-connected resistive load.

Solution   The voltage waveforms of the three output terminals, A, B, and C, are shown in Figure 10.33. From these three waveforms we can obtain the three output line voltages for the inverter.

Example 10.11   The reference sine wave waveform, $v_{R(A)}$, $v_{R(B)}$, and $v_{R(C)}$, for a three-phase bridge inverter with sinusoidal pulse-width modulation are shown in Figure 10.34(a). Plot the output line voltage v_{AB}.

Solution   In contrast to the single-phase bridge inverter, the output voltage of a three-phase bridge inverter cannot be controlled by varying the duty cycle. In this case, a pulse-width modulation technique in which the output voltage closely resembles a sine wave is commonly applied.

To implement sinusoidal pulse-width modulation, we need three reference sine waves, $v_{R(A)}$, $v_{R(B)}$, and $v_{R(C)}$, one for each leg of the inverter. The three sinusoidal reference voltages are displaced from each other by an angle of 120°. These voltages are compared with the carrier triangular wave to find the switching points for the switch pairs S_1 and S_4, S_3 and S_6, and S_5 and S_2.

One of the two switches in each leg is conducting at all times, connecting the output terminal to either the positive or the negative side of the DC source. For example, suppose i_A is positive and S_1 is conducting. Then when S_4 is turned on, S_1 turns off and the output current transfers to diode D_4. However, if i_A is negative, diode D_1 would be conducting and when S_4 is turned on, it would take the load current immediately.

The output voltages v_A, v_B, and v_C with respect to the DC source hypothetical midpoint are shown in Figure 10.34(b). The line voltages are:

$$v_{AB} = v_A - v_B$$
$$v_{BC} = v_B - v_C$$
$$v_{CA} = v_C - v_A$$

The line voltage v_{AB} is shown in Figure 10.34(c).

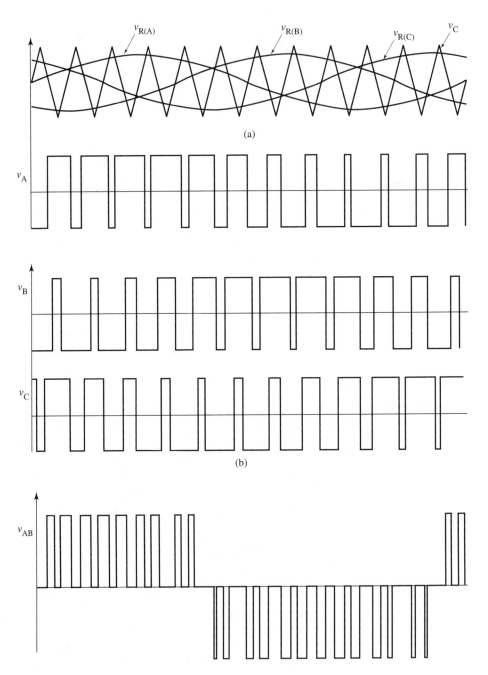

Figure 10.34
Sinusoidal pulse-width modulated waveforms for a three-phase bridge inverter

10.9   The Ideal Current Source Inverter (CSI)

A current source inverter is one in which the input current from the DC source is maintained at a constant level, regardless of the DC input voltage variation. In practice, this is achieved by inserting a large inductor in series with the DC voltage source. This arrangement prevents sudden changes in current and effectively maintains a constant level of supply current. The inverter converts the input DC current into a rectangular wave AC output current.

10.9.1   The Single-Phase Current Source Bridge Inverter

Figure 10.35(a) shows a single-phase current source bridge inverter. Unlike the voltage source inverter, it does not require freewheeling diodes, and the current flows unidirectionally through each SCR. The SCRs are turned on in pairs, SCR_1

Figure 10.35
Single-phase current source inverter (a) circuit diagram (b) load current waveform

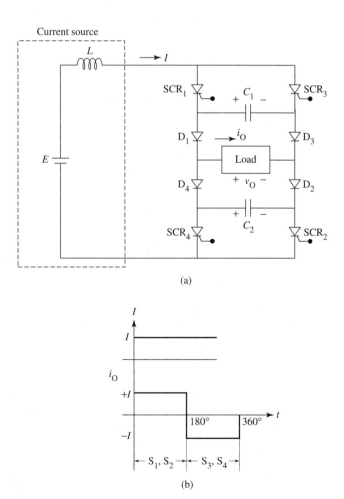

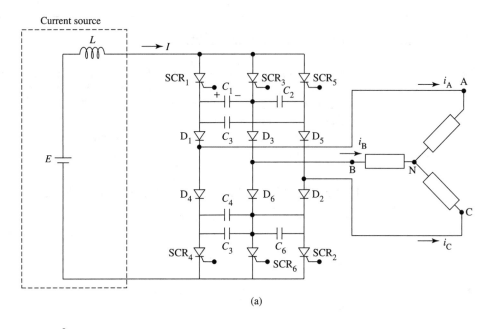

(a)

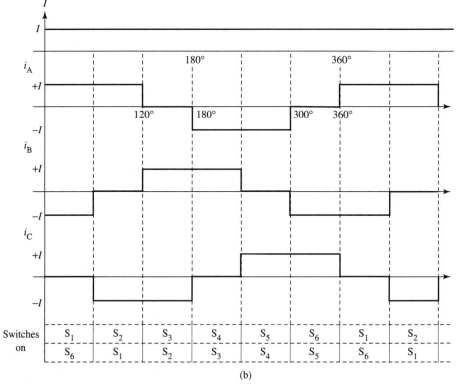

(b)

Figure 10.36
Three-phase current source inverter (a) circuit diagram (b) current waveforms

and SCR_2 and then SCR_3 and SCR_4, at a constant frequency. The resulting output current, shown in Figure 10.35(b), is an AC rectangular wave whose magnitude is equal to the DC input current. Each switch conducts for a period of 180°.

10.9.2 The Three-Phase CSI

A three-phase CSI is shown in Figure 10.36(a). A series inductor is present at the input side to provide a constant DC current source. The waveforms of the input and output currents are shown in Figure 10.36(b).

The upper group is SCR_1, SCR_3, and SCR_5, and the lower group is SCR_4, SCR_6, and SCR_2. Only two SCRs are on at any instant, with each SCR conducting for a period of 120°. The SCRs are numbered in the order they conduct, that is, after SCR_1 conducts, SCR_2 conducts, then SCR_3, SCR_4, SCR_5, SCR_6, SCR_1, ... When an SCR is turned on, it must instantly turn off the conducting SCR of the same group. For example, suppose that SCR_1 and SCR_2 are on. Input current I will flow from the positive side of the source, through SCR_1, the phase A load, the phase C load, and SCR_2 and back to the negative side of the source. If SCR_3 now turns on, it must immediately commute SCR_1. The current from SCR_1 will be transferred to SCR_3. The input current now flows through SCR_3, the phase B load, the phase C load, and SCR_2 and back to the source.

The three-phase output current is a 120° wide step wave of magnitude I. The frequency of the output current is adjusted by controlling the opening and closing of the switches in the required sequence.

The RMS value of the output line current is

$$I_{O(RMS)} = \frac{\sqrt{2}}{\sqrt{3}} I \qquad\qquad\qquad \textbf{10.31}$$

The RMS value of the output phase current is

$$I_{Oph(RMS)} = \frac{\sqrt{2}\, I}{3} \qquad\qquad\qquad \textbf{10.32}$$

The DC input voltage is

$$E = \sqrt{6}\, V_{ph} \cos \theta \qquad\qquad\qquad \textbf{10.33}$$

where θ is the load phase angle.

Example 10.12   Show the switching scheme for each 60° interval of a three-phase current source inverter supplying a balanced three-phase resistive load.

Solution   The operation of the circuit can be divided into six modes in each cycle. The switches are turned on in pairs for 60° intervals. Table 10.8 shows the conducting switches in each of the six modes.

Table 10.8

Interval	S_1	S_2	S_3	S_4	S_5	S_6	i_A	i_B	i_C
0 to 60°	On	Off	Off	Off	Off	On	+I	−I	0
60 to 120°	On	On	Off	Off	Off	Off	+I	0	−I
120 to 180°	Off	On	On	Off	Off	Off	0	+I	−I
180 to 240°	Off	Off	On	On	Off	Off	−I	+I	0
240 to 300°	Off	Off	Off	On	On	Off	−I	0	+I
300 to 360°	Off	Off	Off	Off	On	On	0	−I	+I

10.10   Problems

10.1   What is an inverter? Give some applications of inverters.

10.2   For the inverter circuits discussed in this chapter, a freewheeling diode is connected across each switch. What is the purpose of this diode?

10.3   For a half-bridge circuit with an inductive load, is it possible to obtain voltage control by implementing the voltage pattern in Figure 10.37? Verify your answer by analyzing the half-bridge circuit with an *RL* load.

Figure 10.37
See Problem 10.3

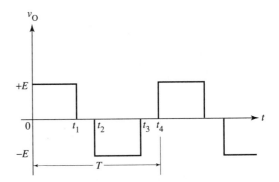

10.4   A single-phase full-bridge inverter produces a square-wave output across a purely inductive load with a freewheeling diode across each switch. Plot the waveforms for the output voltage, output current, switch currents, diode currents, and source current. What is the average power absorbed by the source?

10.5   In Problem 10.4, draw the waveform of the output current if the inverter bridge supplies an *RL* load.

10.6 A single-phase full-bridge inverter uses pulse-width modulation for voltage control. Plot the waveforms for the output voltage if the conduction interval of diagonally opposite switches has a phase angle of 120°.

10.7 A three-phase bridge inverter fed from a 120 V AC source supplies a Y-connected balanced resistive load. Using a 120° conduction scheme, plot the line and phase voltages. If the load resistance is 10 Ω per phase, calculate the RMS value of the output line and phase voltages.

10.8 A three-phase bridge inverter supplies a Y-connected load with 10 Ω resistance per phase. The DC source voltage is 440 V, and the inverter operates in the 120° conduction mode. Find
a) the source current
b) the average power absorbed by the load
c) the RMS value of the output phase voltage
d) the RMS value of the output line voltage

10.9 A three-phase bridge inverter is fed from a 500 V DC supply. The load is Y-connected with a per phase resistance of 10 Ω. For the 120° conduction scheme, find
a) the RMS value of the phase current
b) the RMS value of the switch current
c) the power delivered to the load

10.10 Repeat problem 10.9 for a 180° conduction scheme.

10.11 A three-phase bridge inverter supplies a balanced Δ-connected inductive load. If the inverter is operating in the 180° conduction mode, plot the phase and line current waveforms.

10.12 A three-phase bridge inverter supplies a balanced Δ-connected load consisting of a resistor in series with an inductor in each leg. If the inverter is operating in the 180° conduction mode, plot the waveforms of the three phase currents and the line currents.

10.11 Equations

$$f = 1/T \quad \text{(Hz)} \tag{10.1}$$

$$V_{o(avg.)} = 2\,E\,d \tag{10.2}$$

$$d = T_{ON}/T \tag{10.3}$$

$$V_{o(RMS)} = \sqrt{(2d)}\ E \tag{10.4}$$

$$I_{o(avg.)} = V_{o(avg.)}/R \tag{10.5}$$

$$\text{average current in the switch} = I_{o(avg.)}/2 \tag{10.6}$$

$$P_L = 2\,d\,\frac{E^2}{R} \tag{10.7}$$

$$V_{o(\text{avg.})} = E\left(1 - \frac{2\delta}{T}\right)$$ **10.8**

$$v_{AB} = v_{AN} - v_{BN}$$ **10.9**

$$v_{BC} = v_{BN} - v_{CN}$$ **10.10**

$$v_{CA} = v_{CN} - v_{AN}$$ **10.11**

$$P_o = E^2/2\,R$$ **10.12**

$$V_{ph.(\text{RMS})} = \frac{E}{\sqrt{6}}$$ **10.13**

$$V_{L(\text{RMS})} = \frac{E}{\sqrt{2}}$$ **10.14**

$$I_{\text{Switch(RMS)}} = E/2\,\sqrt{3}R$$ **10.15**

$$I_{O(\text{RMS})} = \sqrt{2}\,I_{\text{switch(RMS)}}$$ **10.16**

reverse voltage rating for the switch $= E$ **10.17**

$$v_{AB} = v_{AN} - v_{BN}$$ **10.18**

$$v_{BC} = v_{BN} - v_{CN}$$ **10.19**

$$v_{CA} = v_{CN} - v_{AN}$$ **10.20**

$$P_o = \frac{2\,E^2}{3\,R}$$ **10.21**

$$I_{\text{switch(RMS)}} = E/3R$$ **10.22**

$$I_{O(\text{RMS})} = \sqrt{2}\,i_{\text{switch(RMS)}}$$ **10.23**

reverse voltage rating for the switch $= E$ **10.24**

$$V_{L(\text{RMS})} = \frac{\sqrt{2}}{\sqrt{3}}\,E$$ **10.25**

$$V_{ph(\text{RMS})} = \frac{\sqrt{2}}{3}\,E$$ **10.26**

$$V_{L(\text{RMS})} = \frac{\sqrt{2}}{\sqrt{3}}\,\alpha\,E$$ **10.27**

$$V_{ph(\text{RMS})} = \frac{\sqrt{2}}{3}\,\alpha\,E$$ **10.28**

$$I_i = \frac{3\,\sqrt{2}}{\pi}I_o\,\cos\theta$$ **10.29**

$$I_i = \frac{3\,\sqrt{2}}{\pi}\alpha\,I_o\,\cos\theta$$ **10.30**

$$I_{o(\text{RMS})} = \frac{\sqrt{2}}{\sqrt{3}}I$$ **10.31**

$$I_{oph(\text{RMS})} = \frac{\sqrt{2}}{3}I$$ **10.32**

$$E = \sqrt{6}\,V_{ph}\,\cos\theta$$ **10.33**

AC Voltage Controller

11

Learning Objectives

After completing this chapter, the student should be able to

- define the term *AC voltage controller*
- describe methods of controlling AC power
- describe integral cycle control
- describe AC phase control with resistive and inductive loads

- explain the operation of a three-phase controller with resistive and inductive loads
- explain the operation of a half-controlled AC voltage controller
- explain the operation of single-phase and three-phase cycloconverters

353

11.1 Introduction

An alternating-current voltage controller or regulator converts a fixed AC voltage source to a variable AC voltage source. The output frequency is always equal to the input frequency. The simplest way to control the AC voltage to a load is by using an AC switch. This switch can be a bidirectional switch like a triac or a pair of SCRs connected in "antiparallel," as shown in Figure 11.1. Switching devices other than thyristors can also be used to implement bidirectional switches. For most purposes, the control result is independent of the switch used. The practical limitations of the presently available triac ratings often make it necessary to use SCRs in very high power applications for which triacs might otherwise be used.

Figure 11.1
Basic AC power controller
circuits (a) an SCR circuit
(b) a triac circuit

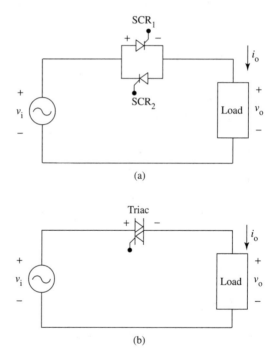

(a)

(b)

The major applications of AC voltage controllers include lighting control, industrial heating, spot (resistance) welding, on-load transformer tap changing, static var compensation, and speed control for induction motors.

11.2 AC Power Control

There are two basic methods for controlling the load power: *integral cycle control* or *on-off control* and *phase control*. The first method is suitable for systems with a large time constant, such as a temperature control system. The load power can be controlled by connecting the source to the load for a few complete cycles then disconnecting the source from the load for another number of cycles, and repeating the switching cycle. The relative duration of the on and off periods, i.e., the duty cycle d, is adjusted so that the average power delivered to the load meets some particular objective. Figure 11.2 shows a typical pattern. In ideal circumstances, the average power to the load can be controlled from 0% through 100%.

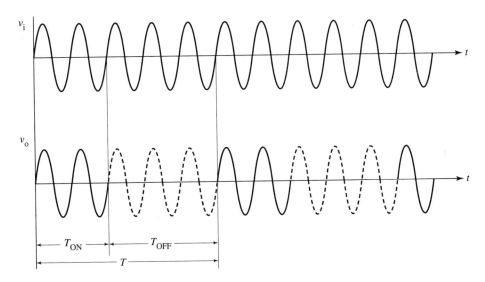

Figure 11.2
Integral cycle control

Integral cycle control is not suitable for loads with a short time constant. **Phase control** can be used in these situations. In phase control, the switch connects the load to the source for part of each cycle of input voltage. The graphs in Figure 11.3 illustrate waveforms for phase control with a resistive load. The voltage at the load can be varied by altering the firing angle for each half-cycle of a period. If $\alpha = 0$, the output voltage is maximum ($v_o = v_i$). When $\alpha = \pi$, the output voltage is minimum ($v_o = 0$). Therefore, the output voltage can be controlled to any value between zero and the source voltage. This process produces a phase-controlled alternating output that is suitable for applications such as lighting control and motor-speed control.

Figure 11.3
AC phase control waveforms
with a resistive load

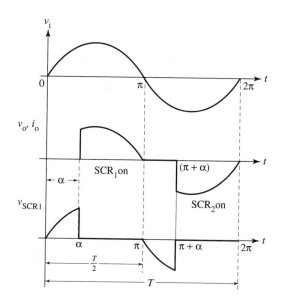

Example 11.1   For the following applications, choose integral cycle control or phase angle control and justify your choice.
a) lighting control
b) motor speed control
c) heating loads

Solution   a) Due to the system frequency (60 Hz), integral cycle control is not practical for the brightness control of an incandescent lamp. If we assume a minimum possible off time of one cycle (i.e., $T_{OFF} = 1$), the lamp will flicker once every two cycles with a frequency of 30 Hz. If we increase the off time for lighting control, the frequency will decrease. A flicker of less than 30 Hz is both noticeable and annoying. Therefore, integral cycle control is unsuitable for this application.

   With phase control, there is an off interval in every half-cycle of the load voltage. The fluctuation of current thus has a frequency of 120 Hz on a 60 Hz system. A flicker of 120 Hz is barely noticeable to a human eye. Lamp dimmers usually use phase control circuits.

b) With an assumed minimum off time of one cycle, integral cycle control could cause severe current variation in motors, resulting in pulsating torque and speed oscillations, especially if the system inertia is low. Consequently, integral cycle control is not suitable for the speed control of motors. Phase control is appropriate since the off time is much shorter—usually some fraction of a half-cycle. However, this method of motor speed control is suitable only for variable torque loads, such as fans and pumps, in which the torque varies as the square of the speed.

c) Heating loads that are required to be maintained at a certain temperature, for example, an electric furnace or a tank for heating liquid, are applications in which a long off time is acceptable. Such applications can be part of a closed-loop system. In these cases the time constant for load response is relatively long (seconds, not fractions of a second), so the load responds well to average power.

Integral cycle control is therefore suitable in applications where the switch on-off intervals are controlled to produce the desired temperature in a load. Once the desired temperature is reached, the intermittent nature of the current does not cause any major change in temperature unless T_{OFF} is very long.

11.3 Integral Cycle Control

In the AC voltage controller in Figure 11.1, the thyristors can be fired at $\alpha = 0°$ to allow complete cycles of source voltage to be applied to the load. If there is no firing signal in any cycle, then no voltage appears across the load. Thus it is possible to allow complete cycles of source voltage to be applied to the load followed by complete cycles of extinction. If the load voltage is turned on and off in this manner (Figure 11.2), the average power to the load can be varied. The ratio of on time to total cycle time (the period in which the conduction pattern repeats) controls the average load power. In Figure 11.2, T_{ON} is the number of cycles for which the load is energized and T is the number of cycles in the full period of operation. During the T_{ON} part of the cycle, the switch is on and the load power is maximum. During the remaining T_{OFF} ($T_{OFF} = T - T_{ON}$) cycles, the switch is off and the load power is zero.

For a resistive load R, the average load power is given by

$$P_{o(avg.)} = \frac{V_i^2\, T_{ON}}{R\, T} = \frac{V_i^2}{R}\, d = P_{o(max)}\, d \qquad \textbf{11.1}$$

The RMS value of the output voltage is given by

$$V_o = \frac{V_m}{\sqrt{2}} \sqrt{\frac{T_{ON}}{T}} = V_i\, \sqrt{d} \qquad \textbf{11.2}$$

where

V_m = maximum value of input voltage

V_i = RMS value of input voltage = $V_m/\sqrt{2}$

Because T_{ON} can be varied only as an integer, the average value of the load power is not a continuous function but has only discrete levels. The number of steps available for regulating the average power depends on the total number of cycles included in the repeat pattern.

Power conversion is the ratio of the average power output ($P_{o(avg.)}$) to the maximum possible power output ($P_{o(max)}$). $P_{o(avg.)}/P_{o(max)}$ is equal to the duty cycled

$$d = T_{ON}/(T_{ON} + T_{OFF}) = T_{ON}/T$$

where

$$T = \text{time period} = T_{ON} + T_{OFF}$$

The source current is always in time phase with the source voltage. However, this does not mean that an integral cycle control circuit operates at unity power factor—for part of the time, the source current is not present at all and therefore is not in phase with the source voltage.

The power factor is given by

$$PF = \sqrt{T_{ON}/T} = \sqrt{d} \qquad\qquad \textbf{11.3}$$

It is clear from Equation 11.3 that a power factor of one will result when $T_{ON} = T$, which would result in sinusoidal operation.

A closed-loop control system can be used to vary the value of T_{ON} to maintain some variable close to a selected set point. Such a system would depend on sufficient energy storage in the controlled system to smooth variations that result from the on-off nature of the control. Integral cycle control has the advantage of fewer switching operations and low radio frequency interference (RFI) due to control during the zero crossing of the AC voltage, that is, in this method, switching occurs only at zero voltage for resistive loads. The rate of change of the load current depends on the system frequency, which is small, so there is low electrical noise compared with other control methods.

Example 11.2   A single-phase 120 V AC source controls power to a 5 Ω resistive load using integral cycle control. Find
a) the average value of output current
b) the maximum switch current
c) the maximum power produced
d) the duty cycle and the value of T_{ON} to produce 1 kW power
e) the power factor for part (d)

Solution   a) The average value of the output current over any number of complete conduction cycles is 0.

b) $I_{o(RMS)} = 120/5 = 24$ A
$I_m = \sqrt{2}\,(24) = 33.9$ A

c) Maximum power will be produced when the switch is always on.

$$P_{o(max)} = 120 * 24 = 2880 \text{ W}$$

d) For $P_{o(avg.)} = 1000$ W,

$$d = \frac{T_{ON}}{T} = \frac{P_{o(avg.)}}{P_{o(max)}} = \frac{1000}{2880} = 0.35$$

If we choose $T = 15$ cycles, then

$$T_{ON} = 0.35 * 15 = 5 \text{ cycles}$$

e) $PF = \sqrt{\dfrac{T_{ON}}{T}} = \sqrt{\dfrac{5}{15}} = 0.58$

11.4 AC Phase Control

11.4.1 In Circuits with a Resistive Load

The basic circuit in Figure 11.1 can be used to control the power to a resistive load. As is done with a controlled rectifier, output voltage is varied by delaying conduction during each half-cycle by an angle α. The delay angle α is measured from the source voltage zero.

SCR_1, which is forward-biased during the positive half-cycle, is turned on at an angle α. It conducts from α to π, supplying power to the load. SCR_2 is turned on half a cycle later at $\pi + \alpha$. It conducts up to 2π, supplying power to the load. The waveforms in Figure 11.3 are identical to those of a full-wave rectifier with a resistive load. The difference here is that each second half-cycle has a negative current rather than a positive one. There is, however, no effect on the power, because power is a squared function.

The equation for the RMS value of the output voltage is

$$V_{o(RMS)} = V_i \left\{ 1 - \frac{\alpha}{\pi} + \frac{\sin 2\alpha}{2\pi} \right\}^{1/2} \qquad \textbf{11.4}$$

The equation for the RMS value of the output current with a resistive load is similar to Equation 11.4:

$$I_{o(RMS)} = \frac{V_i}{R} \left\{ 1 - \frac{\alpha}{\pi} + \frac{\sin 2\alpha}{2\pi} \right\}^{1/2} \qquad \textbf{11.5}$$

By varying the delay angle α, the output current of the load can be continuously adjusted between the maximum value of V_i/R at $\alpha = 0$ and zero at $\alpha = 180°$.

The RMS current rating of the triac is given by

$$I_{T(RMS)} = I_{o(RMS)} \qquad \textbf{11.6}$$

The RMS current rating of the SCRs is given by

$$I_{SCR(RMS)} = I_{o(RMS)}/\sqrt{2} \qquad \textbf{11.7}$$

Output power is given by

$$P_{o(avg.)} = I_{o(RMS)}^2 (R) \qquad \text{or} \qquad V_{o(RMS)}^2/R \qquad \textbf{11.8}$$

Examination of Equations 11.5 and 11.8 shows that the load power can be varied by changing α over the full range from zero to $180°$. Suitable trigger circuits exist to allow conduction to be adjusted essentially over this entire range.

Table 11.1

α (°)	$V_{o(RMS)}$ (V)	$P_{o(avg.)}$ (W)	$P_{o(avg.)}/P_{o(max)}$	$V_{o(RMS)}/V_i$
0	50.0	25.0	1.0	1.0
30	49.3	24.3	0.97	0.98
60	44.8	20.1	0.80	0.89
90	35.4	12.5	0.50	0.71
120	21.9	4.8	0.20	0.44
150	8.5	0.72	0.03	0.17
180	0.0	0.0	0.0	0.0

The **control characteristic** of a single-phase AC power controller can be calculated as a function of the delay angle. If we assume $V_i = 50$ V and load resistance $R = 100\ \Omega$, then at $\alpha = 0°$, using Equation 11.4, output voltage $V_{o(RMS)} = V_i = 50$ V and

$$P_{o(max)} = V_i^2/R = 50^2/100 = 25\ \text{W}$$

while

$$P_{o(avg.)} = V_{o(RMS)}^2/R$$

Evaluating output voltage and power for successive values of the delay angle gives the results shown in Table 11.1.

The control characteristic, $V_{o(RMS)}/V_i$ and $P_{o(avg.)}/P_{o(max)}$ versus α, for a resistive load is plotted in Figure 11.4.

Figure 11.4
Variation of output voltage and power with delay angle for a resistive load

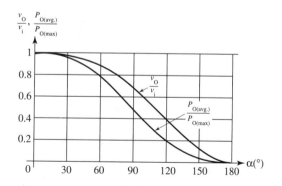

Because the current is nonsinusoidal, the power factor presented to the AC source is less than unity, although the load is resistive. Whatever the waveform, by definition the power factor is given by:

$$PF = \frac{\text{active power}}{\text{apparent power}}$$

$$= \frac{P}{V_i I_i}$$

$$= \frac{\{V_{o(RMS)}^2/R\}}{V_i \{V_{o(RMS)}/R\}}$$

$$= \frac{V_{o(RMS)}}{V_i}$$

Substituting Equation 11.4, we obtain

$$PF = \left\{1 - \frac{\alpha}{\pi} + \frac{\sin 2\alpha}{2\pi}\right\}^{1/2} \qquad \textbf{11.9}$$

The resulting power factor is unity only when α is zero; it becomes progressively smaller as α increases, becoming approximately zero for $\alpha = \pi$.

The switch current becomes zero just when the source voltage is zero, because the load is resistive. Therefore, when the switch begins blocking at the time of the current zero, negligible source voltage is present. The problem of dv/dt being large at turnoff does not exist, and no snubber is required to reduce the rate of voltage buildup across the device terminals.

For values of $\alpha > 90°$, the switch blocks the peak source voltage before it turns on. The minimum switch voltage capability therefore is the peak value of the source voltage. This blocking capability is of course necessary in both directions for either the SCR or the triac implementation of the switch.

$$PIV \geq V_{i(m)} \qquad \textbf{11.10}$$

Example 11.3   A single-phase 120 V AC source controls power to a 5 Ω resistive load using integral cycle control. If $T_{ON} = 2$ cycles and $T = 4$ cycles, find
a) the output power
b) the delay angle required if the phase control method is used to produce the same power
c) the output power, if the load is always connected to the source

Solution   a) From Equation 11.1,

$$P_{o(avg.)} = \frac{V_i^2 T_{ON}}{R\, T} = \frac{120^2}{5}\frac{2}{4} = 1440 \text{ W}$$

b) $\qquad P_{o(avg.)} = I_{o(RMS)}^2 R$

From Equation 11.5,

$$I_{o(RMS)} = \frac{V_i}{R}\left\{1 - \frac{\alpha}{\pi} + \frac{\sin 2\alpha}{2\pi}\right\}^{1/2}$$

Therefore,

$$P_{o(avg.)} = \frac{V_i^2}{R}\left\{1 - \frac{\alpha}{\pi} + \frac{\sin 2\alpha}{2\pi}\right\} = 1440 \text{ W}$$

The required value is $\alpha = 90°$.

c) $P_{o(avg.)} = \dfrac{V_i^2}{R} = \dfrac{120^2}{5} = 2880 \text{ W}$

Figure 11.5
AC phase control waveforms for a resistive load, for delay angles varying from 30° to 150°

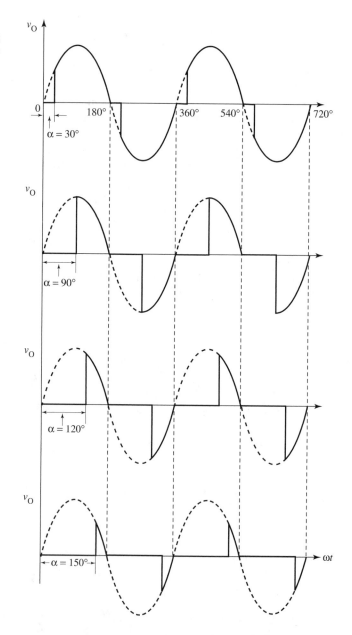

Example 11.4 A single-phase power controller as shown in Figure 11.1(a) is supplying a resistive load. Plot the waveform of the output voltage if the delay angle is
a) 30°
b) 90°
c) 120°
d) 150°

Solution The voltage waveforms for a resistive load and the given delay angles are shown in Figure 11.5.

Example 11.5 A single-phase power controller fed from a 120 V supply is supplying a 25 Ω resistive load. Find the RMS value of the output current and the power factor if the delay angle is:
a) 0°
b) 30°
c) 60°
d) 90°
e) 120°
f) 150°
g) 180°

Solution The RMS value of output current is given by Equation 11.5 as

$$I_{o(RMS)} = \frac{V_i}{R}\left\{1 - \frac{\alpha}{\pi} + \frac{\sin 2\alpha}{2\pi}\right\}^{1/2}$$

Here,

$$V_i/R = 120/25 = 4.8 \text{ A}$$

Table 11.2

α	$I_{o(RMS)}$ (A)	PF
(a) 0°	4.80	1.0
(b) 30°	4.73	0.98
(c) 60°	4.30	0.90
(d) 90°	3.39	0.71
(e) 120°	2.13	0.44
(f) 150°	0.83	0.17
(g) 180°	0.0	0

The power factor is given by Equation 11.9:

$$PF = \left\{ 1 - \frac{\alpha}{\pi} + \frac{\sin 2\alpha}{2\pi} \right\}^{1/2}$$

Substituting values of α into the equations for $I_{o(RMS)}$ and PF gives the results shown in Table 11.2.

Example 11.6   A single-phase power controller as shown in Figure 11.1(a) is supplying a 100 Ω resistive load through a 50 V source. Plot the waveforms for output voltage, output current, voltages across SCR$_1$ and SCR$_2$, and current through SCR$_1$ and SCR$_2$ if the delay angle is 60°.

Solution   The waveforms are shown in Figure 11.6.

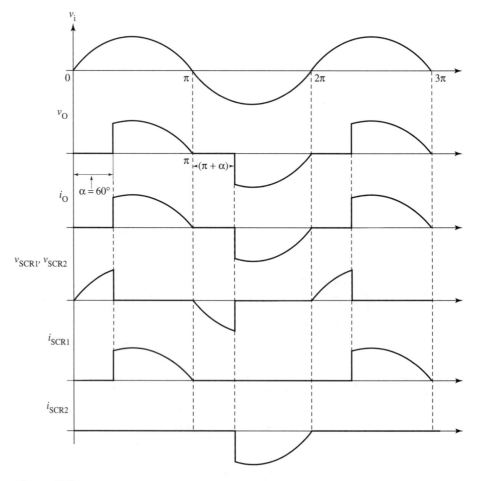

Figure 11.6
Waveforms for resistive load and a delay angle of 60°

Example 11.7   A single-phase 110 V AC source controls power to a 10 Ω resistive load, using the circuit shown in figure 11.1(a). If $\alpha = 30°$, find
a) the maximum output current
b) the average output power
c) the peak reverse voltage
d) the RMS value of the input current
e) the circuit power factor
f) the RMS value of the SCR current

Solution   a) $V_m = \sqrt{2}\,(110) = 155.5$ V

$$I_m = \frac{V_m}{R} = \frac{155.5}{10} = 15.5 \text{ A}$$

b) From Equation 11.5,

$$I_{o(RMS)} = 10.8 \text{ A}$$
$$P_{o(avg.)} = I_{o(RMS)}^2\,R = (10.8)^2\,10 = 1175 \text{ W}$$

c) The switch must block the maximum source voltage V_m of 155.5 V.
d) The RMS value of the input current is

$$I_{i(RMS)} = I_{o(RMS)} = 10.8 \text{ A}$$

e) P F $= \dfrac{P_{o(avg.)}}{V_i I_i} = \dfrac{1175}{110 * 10.8} = 0.99$

f) The RMS value of the SCR current is

$$I_{SCR(RMS)} = \frac{I_{o(RMS)}}{\sqrt{2}} = 7.64 \text{ A}$$

Example 11.8   A 120 V source controls power to a 5 Ω resistive load using a phase-control switch. If the load power required is 1 kW, find
a) the maximum load current
b) the RMS value of the load current
c) the delay angle α
d) the RMS value of the switch current, if the switch is a triac
e) the average current in each of the two SCRs if the switch is like that in Figure 11.1(a)
f) the peak reverse voltage rating of the switch
g) the power factor

Solution   a) $V_m = \sqrt{2}\,(120) = 170$ V

$$I_m = \frac{V_m}{R} = \frac{170}{5} = 34 \text{ A}$$

b) For $P_{o(avg.)} = 1000$ W,

$$1000 = I_{o(RMS)}^2 * 5$$
$$I_{o(RMS)} = 14.14 \text{ A}$$

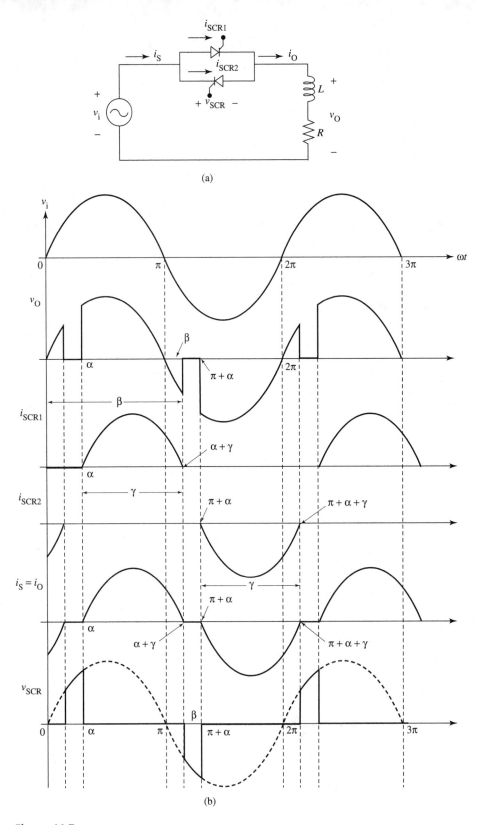

Figure 11.7
(a) AC phase-control circuit with an *RL* load (b) voltage and current waveform with
an *RL* load

c) Using Equation 11.5 to solve for α,

$\alpha = 105°$

d) $I_{T(RMS)}$ is the same as the load current, 14.14 A
e) The SCR current is a half-wave controlled waveform. The average value of each SCR current can be found by using Equation 6.2:

$$I_{SCR(avg.)} = \frac{I_m}{2\pi}(1 + \cos \alpha) = \frac{34}{2\pi}(1 + \cos 105°) = 4 \text{ A}$$

f) The switch must block at least the maximum source voltage, 170 V.

g) $PF = \left\{1 - \dfrac{105}{180} + \dfrac{\sin 2 * 105}{2\pi}\right\}^{1/2} = 0.58$

11.4.2 In Circuits with an Inductive (*RL*) Load

Consider the AC voltage controller circuit in which the load now consists of resistor R in series with an inductor L. The circuit is shown in Figure 11.7(a) and the corresponding waveforms in Figure 11.7(b). SCR$_1$ is triggered at α and SCR$_2$ is triggered at $\pi + \alpha$. When SCR$_1$ turns on, the source voltage is connected to the load, making the output voltage $v_o = v_i$. The output current i_o builds up at α. However, it does not become zero at π but continues to flow until β, which is known as the extinction angle. The interval during which SCR$_1$ conducts is called the conduction angle γ ($\gamma = \beta - \alpha$). When SCR$_2$ turns on, a reverse current flows in the load.

Note in the graph that the onset of output current coincides with the firing angle, that is, the load phase angle Φ ($\Phi = \tan^{-1} X_L/R$), the angle by which the output current lags the voltage, is equal to α. Under this condition, full output voltage is obtained. Furthermore, due to the load inductance, current flow is maintained through the SCR even after the input voltage has reversed polarity and goes negative. At the time when the output current decays to zero, the voltage across the switch has an ideal discontinuity. The output voltage is equal to the source voltage when either SCR conducts. The output voltage waveform has the shape of a sinusoid with a vertical portion removed. The missing portion of the output voltage waveform forms the voltage drop across the SCR switch.

The RMS value of the output current is given by

$$I_{o(RMS)} = \frac{V_i}{R}\left\{4\left(1 - \frac{\alpha}{\pi}\right)\left(\cos^2\alpha + \frac{1}{2}\right) + \frac{6}{\pi}\sin \alpha \cos \alpha\right\}^{1/2} \qquad \textbf{11.11}$$

where α is in the range from $\pi/2$ to π.

Example 11.9 A single-phase power controller as shown in Figure 11.7(a) is supplying an inductive load. Plot the waveforms of the output voltage and current if the delay angle is
a) 30°
b) 90°

Figure 11.8
Voltage and current
waveforms for inductive
load with delay angles of
(a) 30° (b) 90° (c) 120°
(d) 150°

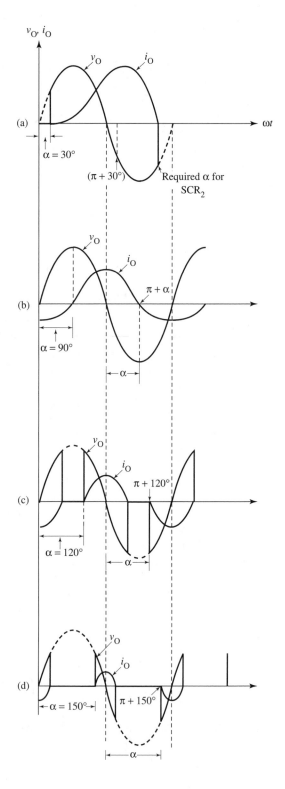

c) 120°

d) 150°

Solution For an inductive load, the output current lags the output voltage. If the load is purely inductive, the phase angle is 90°. Therefore, if the delay angle is less than 90°, the current will not be symmetrical. With a delay angle of 30°, the waveform shown in Figure 11.8(a) results, where conduction in SCR_1 lasts for more than 180° and SCR_2 does not conduct at all because it does not experience forward voltage when it receives its firing pulse at $\pi + 30°$. The output current is therefore unidirectional. To avoid this condition, the firing angle should be at least 90° (see Figure 11.8(b)). When π is between 90° and 180°, the waveforms are of the form shown in Figure 11.8(c) and (d). Thus for inductive loads, α is limited to the range 90° to 180°.

Example 11.10 A single-phase voltage controller with an RL load is connected to a 110 V source. If $R = 10$ Ω, $L = 20$ mH, and $\alpha = 80°$, find

a) the RMS output current

b) the SCR RMS current

c) the power delivered to the load

d) the power factor

Solution a) From Equation 11.11,

$$I_{o(RMS)} = \frac{110}{10} \left\{ 4\left(1 - \frac{80}{180}\right) \left(\cos^2 80 + \frac{1}{2}\right) + \frac{6}{180} \sin 80 \cos 80 \right\}^{1/2} = 5.5 \text{ A}$$

b) $\quad I_{SCR(RMS)} = I_{o(RMS)}/\sqrt{2} = 3.9$ A

c) $\quad P_{o(avg.)} = I_{o(RMS)}^2 R = 5.5^2(10) = 302.5$ W

d) $\quad \text{PF} = \frac{P}{S} = \frac{P_{o(avg.)}}{V_i I_i} = \frac{P_{o(avg.)}}{V_i I_{o(RMS)}} = 0.5$

11.5 Three-Phase AC Phase Control

11.5.1 In Circuits with a Resistive Load

The phase-control methods applied to single-phase loads can also be applied to three-phase systems. A three-phase AC power controller consists of three single-phase bidirectional connections using the phase-control principle. The circuits shown in Figure 11.9 can be used to vary the power supplied to a three-phase Y- or Δ-connected resistive load. As shown, the switch in each line is implemented by using two SCRs in an inverse-parallel arrangement.

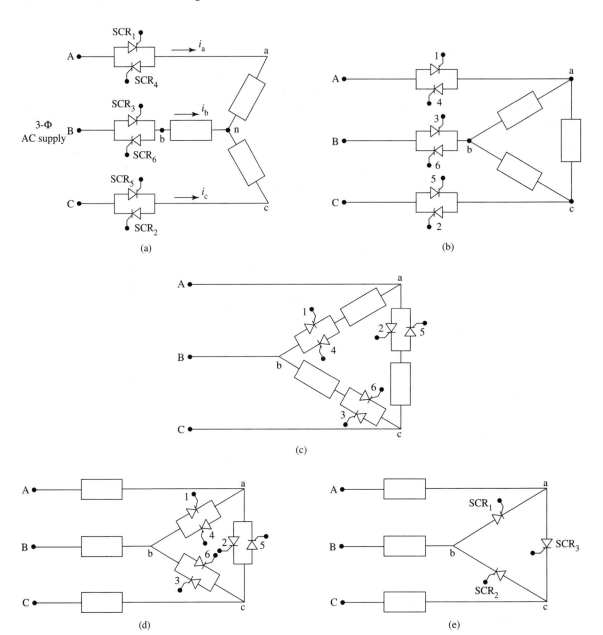

Figure 11.9
AC phase control, three-phase switch (a) line-controlled Y connection (b) line-controlled Δ connection (c) branch Δ connection (d) neutral point switching using six SCRs (e) neutral point switching using three SCRs

The primary considerations in the selection of circuits shown in Figure 11.9 are:

1. For a given power, circuits 11.9(a) and (b) give lower voltages ($\sqrt{3}/2$ times the supply phase voltage or half the line voltage) and higher currents in the SCR. Two SCR pairs are always required in series to block voltage or conduct current.

2. Circuit 11.9(c) gives higher voltages and lower currents in the SCRs. Each SCR can conduct current independently of the other.

3. Circuit 11.9(d) is functionally similar to Figure 11.9(a). It produces identical output voltage waveforms, but since each SCR is part of only one current path instead of two, the average SCR current is halved. In addition, as in circuit 11.9(c), each SCR can conduct current independently of other.

4. For circuit 11.9(e), control of 3Φ output voltage is also possible by using 3 SCRs instead of six. The waveform corresponds to the six SCR current of Fig. 11.9(d), however the SCR current ratings must be doubled.

To illustrate the method of analysis of three-phase AC voltage controllers. We will use circuit shown in Figure 11.9(a) as an example. The SCRs' turn-on is delayed by the angle α beyond the normal beginning of conduction. For symmetrical operation of the circuit, the gate trigger pulses of the thyristors in the three branches must have the same sequence and phase displacement as the supply voltage. If SCR_1 is triggered at α, SCR_3, must be turned on at $\alpha = 120°$ and SCR_5 at $\alpha = 240°$. The inverse parallel SCRs are triggered 180° from their partners. Therefore, SCR_4 (which is across SCR_1) is triggered at $\alpha = 180°$, SCR_6 at $\alpha + 300°$, and finally SCR_2 at $\alpha + 420°$(or $\alpha + 60°$). The SCR conduction order is therefore, SCR_1, SCR_2, SCR_3, SCR_4, SCR_5, SCR_6, SCR_1, . . . , with a phase displacement of 60°.

The output voltage waveform can be obtained by first considering the various output voltage constituents resulting from different SCR conduction patterns. There are four such configurations, which are shown in shown Figure 11.10. Table 11.3 shows the output voltages for each.

Figure 11.11 shows the phase and line voltage waveforms for the circuit in Figure 11.9(a) for different delay angles. Figure 11.11(a) shows the maximum output condition, which occurs when $\alpha = 0$. Note that the delay angle α for each SCR is measured from the reference point where current starts to flow through a purely resistive load. When the delay angle is small, as in Figure 11.11(b), where $\alpha = 30°$, conduction in each phase stops 180° after the reference point. All three lines start conducting again as each SCR is turned on. When α becomes 60° (see Figure 11.11(c)), the turning on of one SCR causes another SCR that was previously conducting to turn off, so that only two lines are always conducting. For $\alpha > 90°$, the conduction period is reduced to the point where it becomes necessary to fire pairs of SCRs simultaneously to establish

Figure 11.10
Circuit configurations to
obtain the output voltage in
Figure 11.9(a) with a
balanced load

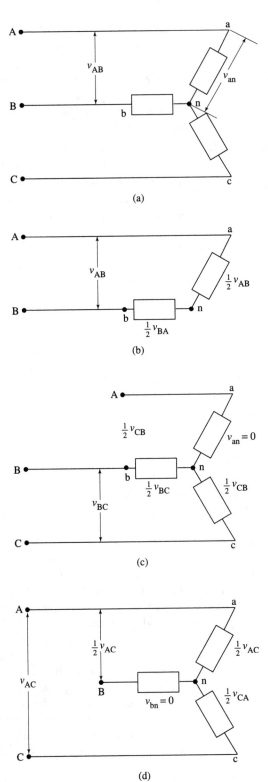

(a)

(b)

(c)

(d)

Table 11.3

Circuit configuration	Conducting lines	Nonconducting lines	Output voltage Phase (v_{an})	Output voltage Line (v_{ab})
a	All	None	$v_{an} = v_{AB}/\sqrt{3}$	v_{AB}
b	A, B	C	$(1/2)v_{AB}$	v_{AB}
c	B, C	A	0	$(1/2)v_{CB}$
d	C, A	B	$(1/2)v_{AC}$	$(1/2)v_{AC}$
	None	All	0	v_{AB}

conducting paths. This means that each SCR must receive two firing pulses separated by 60° in each cycle, as shown in Figure 11.11(d) and (e). If α reaches 150°, the current in each line falls to zero, giving zero output. Thus, the operating range for the delay angle is from 0° to 150°.

The preceding analysis can be summarized into the following three possible modes of operation for the circuit in Figure 11.9(a):

Mode I (0° ≤ α ≤ 60°) One device in each line conducts, in other words, three devices conduct simultaneously, and normal three-phase theory applies. Full output occurs when $\alpha = 0$. When $\alpha \leq 60°$ and all three devices are in conduction, the load currents are the same as for an uncontrolled three-phase resistive load. The RMS value of the output current is given by:

$$I_{o(RMS)} = \frac{V_i}{R}\left\{\frac{1}{3} - \frac{\alpha}{2\pi} + \frac{\sin 2\alpha}{4\pi}\right\}^{1/2} \qquad \textbf{11.12}$$

Mode II (60° ≤ α ≤ 90°) One device conducts in each of two AC lines, that is, a total of only two SCRs are conducting, and two lines act as a single-phase supply to the load.

During the intervals when one of the line currents is zero, the remaining two phases are effectively in series and form a single-phase load connected to two of the three lines of the voltage source. The phase voltage is equal to half the line voltage. The conduction pattern during any 60° interval is repeated during the following 60° interval with a permutation in phases and the sign of the current. For example, the current variation for Phase A during a given 60° interval is repeated during the next 60° for Phase C, except for a change in the algebraic sign of the current.

The RMS value of the output current is given by

$$I_{o(RMS)} = \frac{V_i}{R}\left\{\frac{1}{6} - \frac{3\sin 2\alpha}{8\pi} + \frac{\sqrt{3}\cos 2\alpha}{8\pi}\right\}^{1/2} \qquad \textbf{11.13}$$

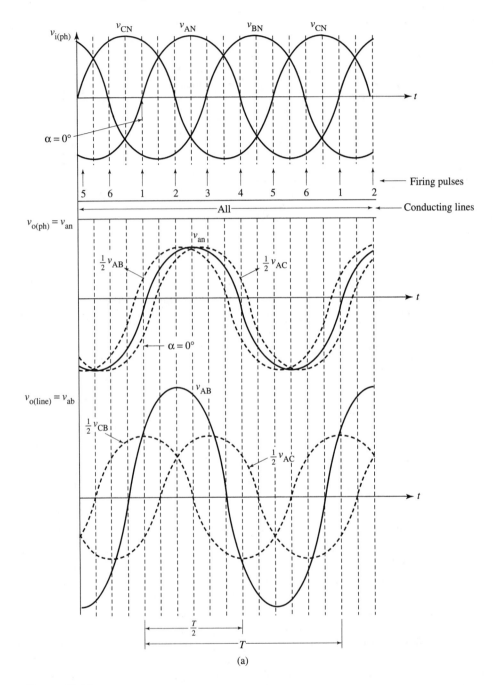

Figure 11.11
Output voltage waveforms for the circuit in Figure 11.9(a) with delay angles of (a) 0°

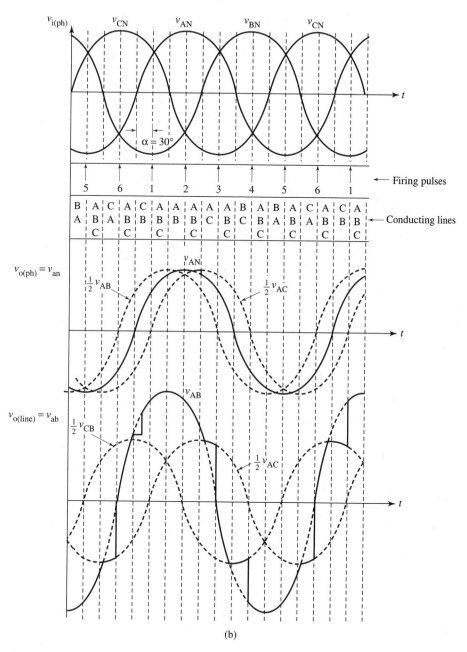

Figure 11.11
(b) 30°

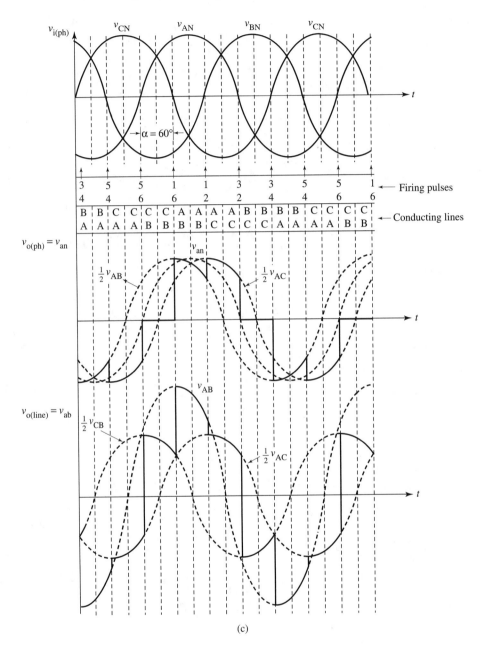

Figure 11.11
(c) 60°

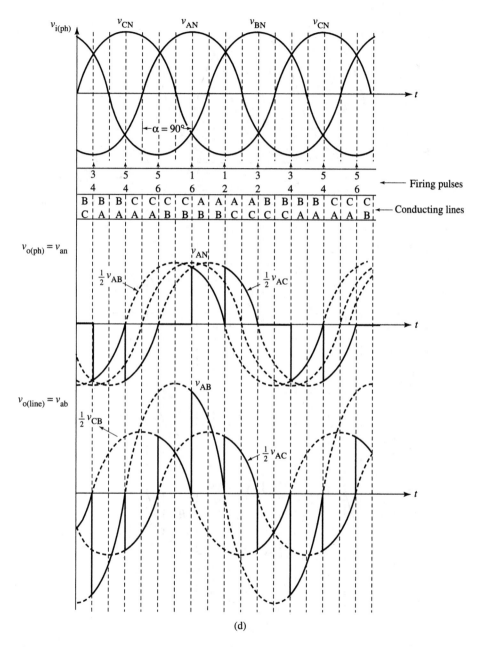

Figure 11.11
(d) 90°

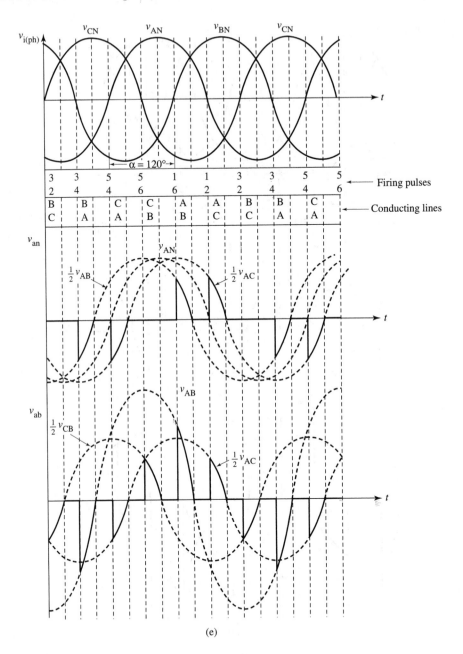

Figure 11.11
(e) 120°

Mode III ($90° \leq \alpha \leq 150°$)   No more than two SCRs conduct at any one time. At times none of the devices conduct. For $\alpha \geq 90°$, when all three devices are off, a zero output period develops. The output voltage becomes zero for $\alpha = 150°$.

The equation for the RMS value o the output current is

$$I_{o(RMS)} = \frac{V_i}{R}\left\{\frac{5}{12} - \frac{\alpha}{2\pi} + \frac{\sqrt{3}\cos 2\alpha}{8\pi} + \frac{\sin 2\alpha}{8\pi}\right\}^{1/2} \qquad \textbf{11.14}$$

Note that for all three modes (Equations 11.12–11.14),

$$V_{o(RMS)} = \sqrt{3}\,I_{o(RMS)}\,R \qquad \textbf{11.15}$$

The waveforms in Figure 11.11 suggest that the maximum switch current may be less than in the case without phase control. For $\alpha < 30°$, maximum current is not affected, whereas for $\alpha > 30°$, maximum current is reduced. If we operate the controller with a zero value of α, the current rating of the switching device should be selected based on full-conduction conditions.

The circuit shown in Figure 11.12 can be used to find the voltage rating of the switching devices. During the interval when Phase A is not conducting, the voltage across the switch can be determined by writing KVL in the upper loop:

$$v_{AB} - v_{SW} + v_{bn} = 0$$
$$v_{SW} = v_{AB} + v_{bn}$$

Now

$$v_{bn} = \frac{v_{BC}}{2}$$

Therefore,

$$v_{SW} = v_{AB} + (v_{BC}/2) = 1.5v_{phase} \qquad \textbf{11.16}$$

A suitable rating for the switching device would therefore be at least equal to $V_{L(max)}$.

Figure 11.12
Voltage rating of the switching device

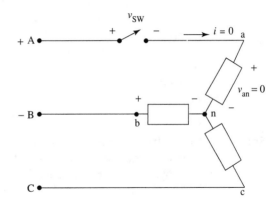

Example 11.11 The three-phase power controller shown in Figure 11.13 is supplying a balanced Δ-connected resistive load. If the delay angle is 45°, plot the waveform of the output voltage across any one phase and the voltage across any pair of SCRs.

Figure 11.13
See Example 11.11

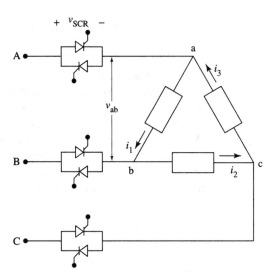

Solution The waveforms are shown in Figure 11.14.

Example 11.12 Repeat Example 11.11 for a balanced Y-connected resistive load.

Solution The waveforms are shown in Figure 11.15.

Example 11.13 The three-phase power controller shown in Figure 11.9(c) supplies a balanced resistive load. Plot the waveforms of the output voltage, the output current, and the voltage across the SCR, for the following delay angles:
a) 0°
b) 30°
c) 60°
d) 90°
e) 120°
f) 150°
g) 180°

Solution The voltage waveforms are shown in Figure 11.16. The output current waveforms are identical to the output voltage waveforms, since the current follows the voltage for a resistive load.

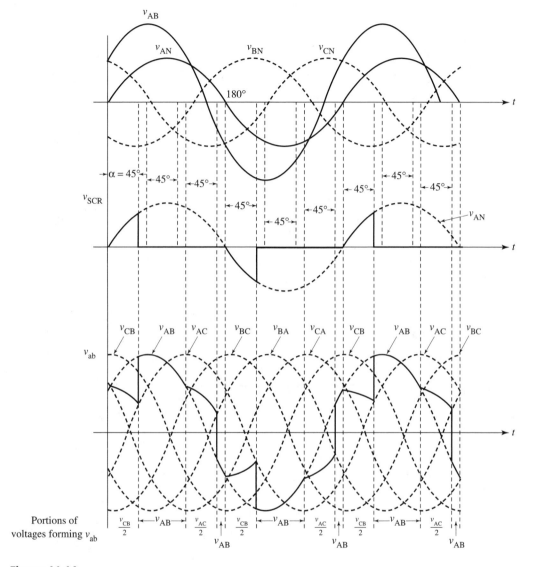

Figure 11.14
Voltage waveforms for a Δ-connected resistive load

Example 11.14   The three-phase power controller shown in Figure 11.9(a) supplies a balanced resistive load. Plot the waveform of the three output phase voltages for the following delay angles:

a) $\alpha < 60°$

b) $60° < \alpha < 90°$

c) $\alpha > 90°$

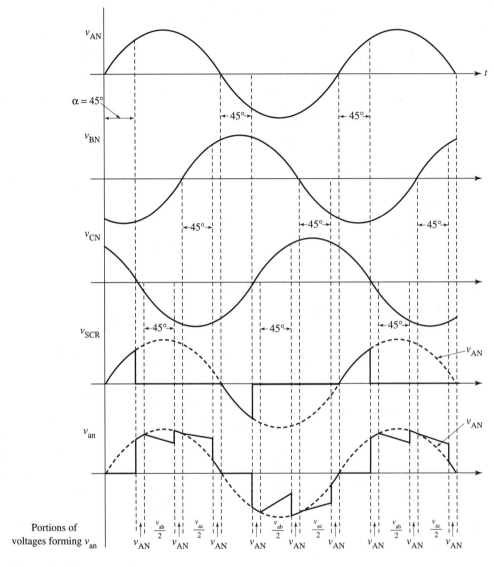

v_{AN}

$\alpha = 45°$

v_{BN}

v_{CN}

v_{SCR}

v_{an}

Portions of
voltages forming v_{an}

Figure 11.15
Voltage waveforms for a Y-connected resistive load

Solution   Figure 11.17 shows the phase voltage waveforms for different delay angles.

Example 11.15   The three-phase power controller shown in Figure 11.9(a) supplies a balanced resistive load. Plot the waveforms of the line current i_a, the output line voltage v_{ab}, and the voltage across the SCR when the delay angle α is 100°.

Solution   Figure 11.18 shows the waveforms for resistive load at a delay angle of 100°.

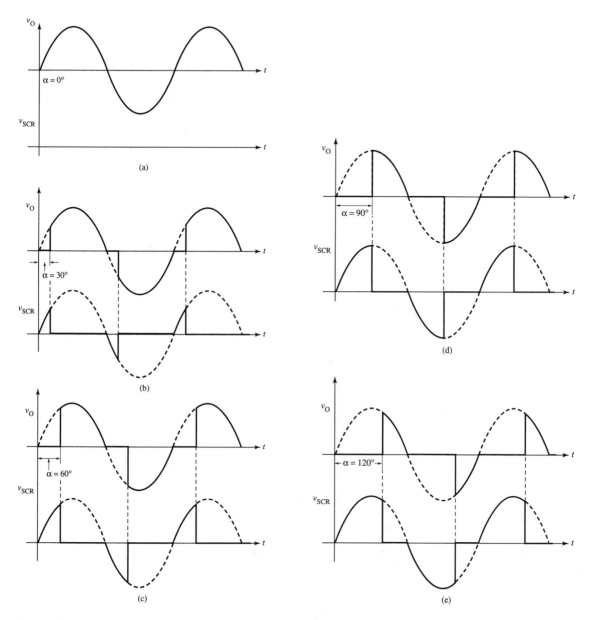

Figure 11.16
Waveforms for a balanced resistive load

Figure 11.16
Waveforms for a balanced
resistive load

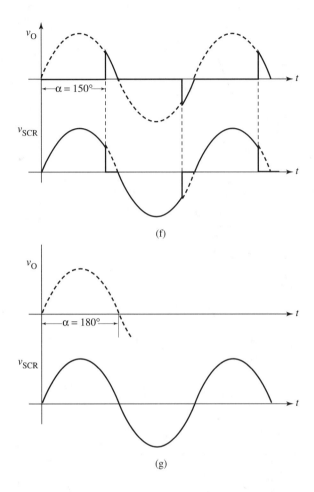

11.5.2 In Circuits with an Inductive (*RL*) Load

With an *RL* load, the waveforms in Figure 11.11 are slightly different because the current is no longer continuous at the points where it switches. The voltages and the currents cannot be determined easily since each depends not only on the present value but also on previous conditions.

The waveforms of the load currents shown in Figure 11.19 are drawn for an inductive load with a delay angle of 100°. As shown, the current waveform in one phase is identical to the current waveform of another phase, except for a 60° phase shift and a reversal of sign.

The voltage rating of the switches should be at least equal to the maximum line voltage of the source. Since the load is inductive, each switch is subjected to a rapid change in voltage as its current becomes zero. A snubber circuit across the switch is usually used to prevent unscheduled firing. The current rating of the switching devices is determined by the current at $\alpha = 0°$.

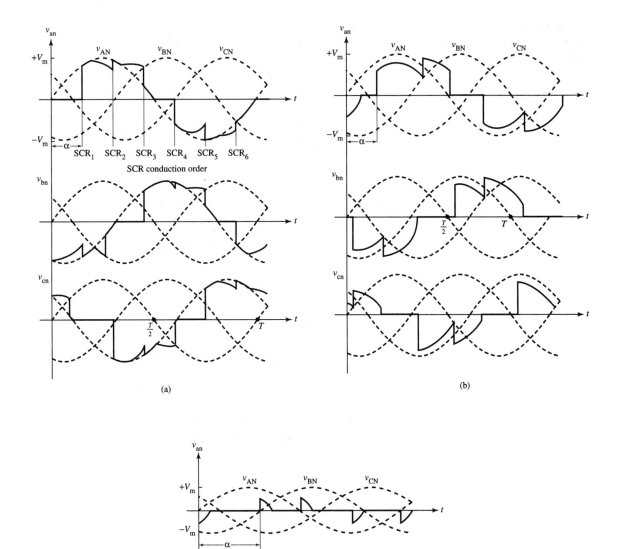

Figure 11.17
Phase voltage waveforms across the load (a) $\alpha < 60°$ (b) $60° < \alpha < 90°$ (c) $\alpha > 90°$

Figure 11.18
Waveforms for resistive load
at a delay angle of 100°

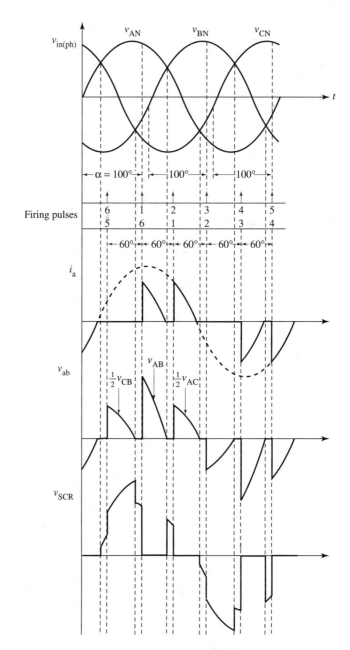

Example 11.16   The three-phase power controller shown in Figure 11.9(c) supplies a balanced inductive load. Plot the waveform of the output voltage, the voltage across the SCR, and the phase and line currents for the following delay angles:

a) 90°

b) 120°

c) 150°

Figure 11.19
Waveforms for an inductive load with a delay angle of 100°

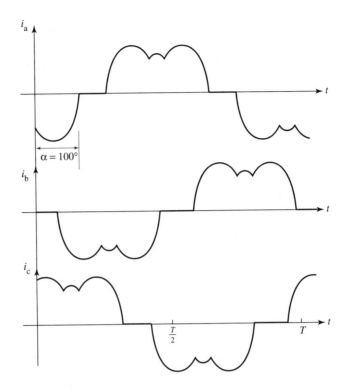

d) 165°

e) 180°

Solution   The waveforms are shown in Figure 11.20.

11.6 Half-Controlled AC Voltage Controllers

The half-controlled three-phase controller is simpler, as it requires only three SCRs. The return path is through the diodes. Figure 11.21 shows the circuit with balanced Y- and Δ-connected resistive loads. The phase and line voltage waveforms for three different delay angles are shown in Figure 11.22.

There are three different modes of operation.

Model I $(0 \leq \alpha \leq 60°)$

Before turn-on, one SCR and one diode conduct in the other two phases. After turn-on, two SCRs and one diode conduct and the three-phase AC source appears across the output.

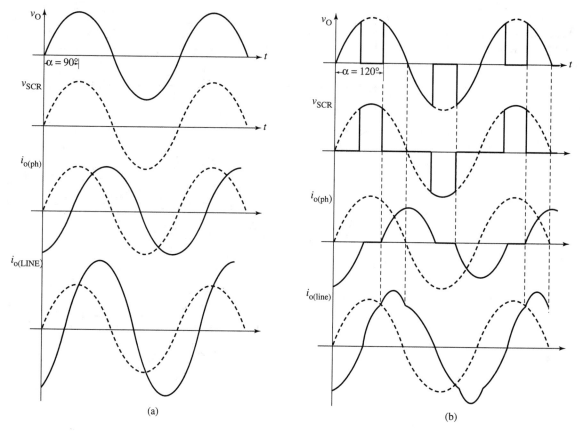

Figure 11.20
Waveform for a balanced *RL* load (a) $\alpha = 90°$ (b) $\alpha = 120°$

The equation for the RMS value of the output current is

$$I_{o(RMS)} = \frac{V_i}{R}\left\{\frac{1}{3} - \frac{\alpha}{4\pi} + \frac{\sin 2\alpha}{8\pi}\right\}^{1/2} \qquad \textbf{11.17}$$

Mode II $(60° \leq \alpha \leq 120°)$

Only one SCR at a time conducts, and the return current is shared at different intervals by one or two diodes.

The equation for the RMS value of the output current is

$$I_{o(RMS)} = \frac{V_i}{R}\left\{\frac{11}{24} - \frac{\alpha}{2\pi}\right\}^{1/2} \qquad \textbf{11.18}$$

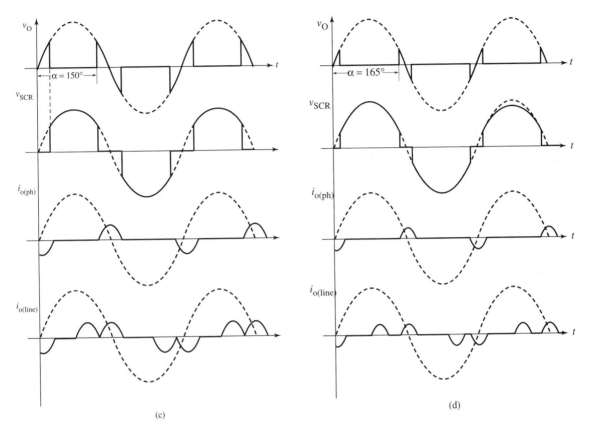

Figure 11.20
Waveforms for a balanced RL load (c) $\alpha = 150°$ (d) $\alpha = 165°$

Mode III $(120° \leq \alpha \leq 210°)$

Only one SCR and one diode conduct, and at 210° the power delivered to the load is zero.

The equation for the RMS value of the output current is

$$I_{O(RMS)} = \frac{V_i}{R} \left\{ \frac{7}{24} - \frac{\alpha}{4\pi} - \frac{\sqrt{3} \cos 2\alpha}{16\pi} + \frac{\sin 2\alpha}{16\pi} \right\}^{1/2} \qquad \textbf{11.19}$$

Example 11.17   The three-phase half-wave power controller shown in Figure 11.21(a) supplies a balanced resistive load. Plot the waveform of the output phase voltage for the following delay angles:
a) 45°
b) 75°

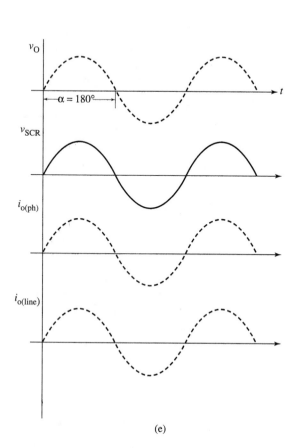

(e)

Figure 11.20
Waveforms for a balanced *RL* load
(e) α = 180°

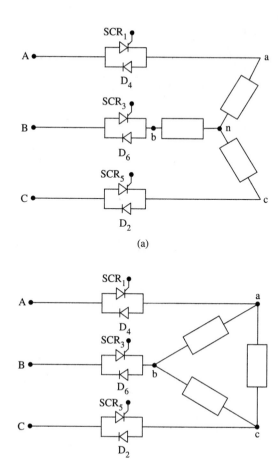

(a)

(b)

Figure 11.21
Three-phase half-wave AC voltage
controller circuit diagrams (a) Y-connected
load (b) Δ-connected load

c) 105°
d) 135°

Solution The waveforms are shown in Figure 11.23.

Example 11.18 A three-phase half-wave power controller as shown in figure 11.21(a) supplies a balanced resistive load. Plot the waveform of the line current for the following delay angles:
a) 20°
b) 170°

Solution The line current waveforms for a half-controlled load are shown in Figure 11.24.

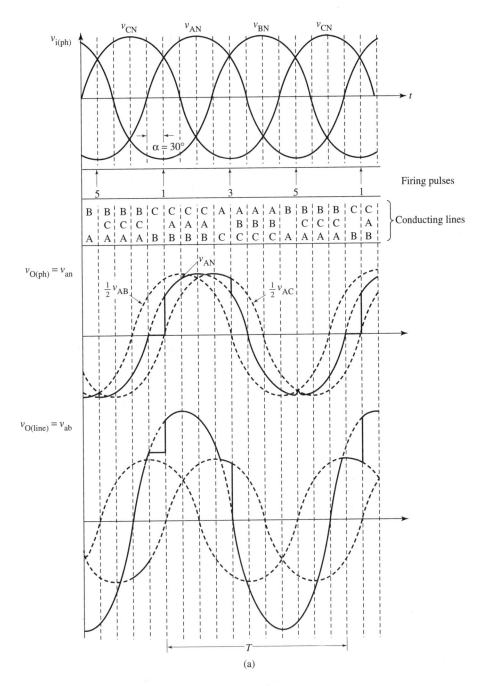

Figure 11.22
Three-phase half-wave AC voltage controller phase voltage and line voltages for delay angles of (a) 30°

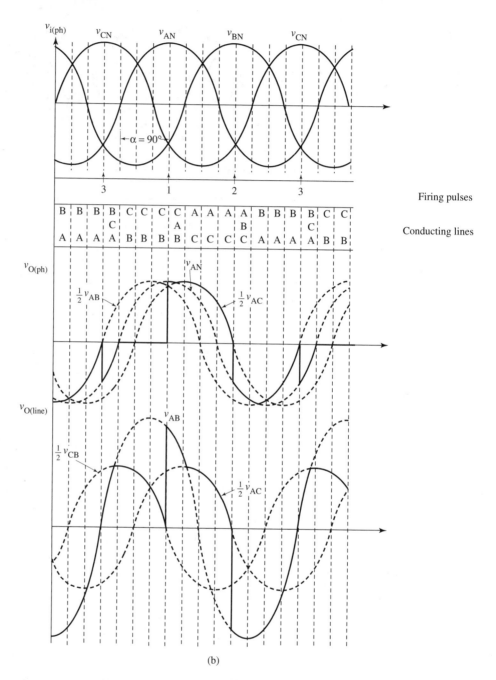

Firing pulses

Conducting lines

Figure 11.22
(b) 90°

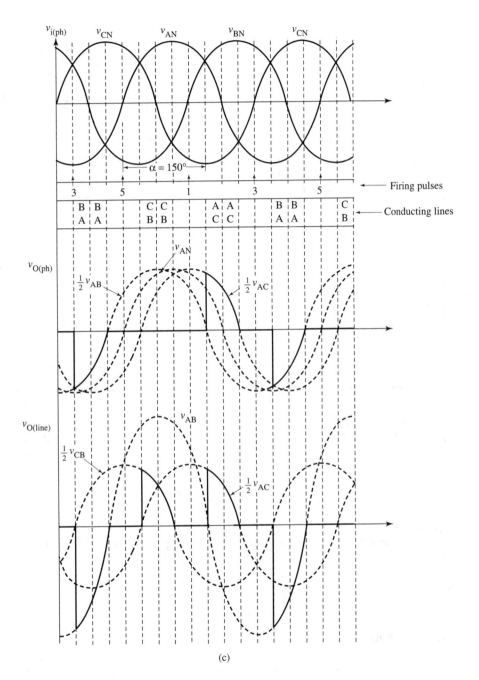

Figure 11.22
(c) 150°

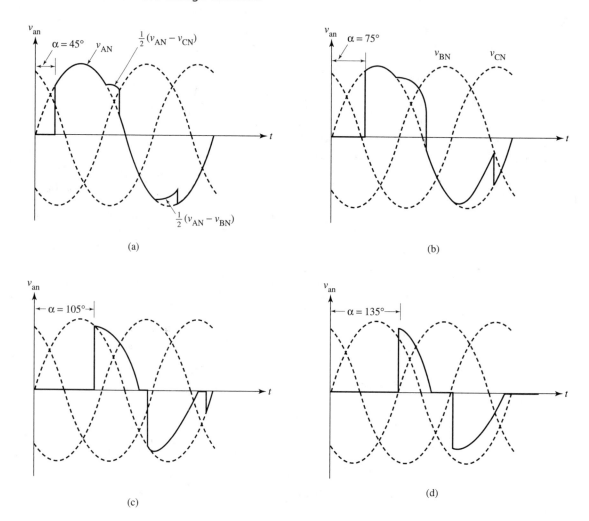

Figure 11.23

Phase voltage waveforms for four different delay angles (a) 45° (b) 75° (c) 105° (d) 135°

11.7 Cycloconverters

A cycloconverter is a frequency changer that converts AC power at one input frequency to AC output power at a different (normally lower) frequency. Variable frequency output can be obtained by using either of two methods. The most obvious method is to convert the given AC to DC using rectification, followed by inversion to obtain the desired frequency output. This arrangement, called a DC-link converter, is implemented by using phase-controlled converters. A more

Figure 11.24
Line current waveforms for
two different delay angles
(a) 20° (b) 170°

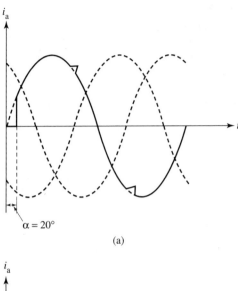

(a)

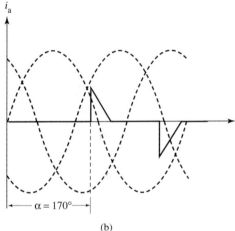

(b)

direct method is to produce the desired output frequency in one stage by using
a cycloconverter. A cycloconverter is basically a dual converter (see Section 6.6)
operated in such a way as to produce an alternating output voltage. For efficient operation, cycloconverters are usually operated in the frequency range of
zero to less than one-third the source frequency.

Cycloconverter systems can provide reverse operation and regeneration.
They are used in low-speed AC drives where the motor drive is started by
reducing the input voltage and frequency. They are also used in variable-speed
constant-frequency (VSCF) generator systems in aircraft to provide a regulated
output voltage at constant frequency regardless of speed changes in the prime
mover.

Figure 11.25
Single-phase to single-phase cycloconverter (a) circuit diagram with a resistive load (b) output voltage and input current waveforms with full conduction of each SCR

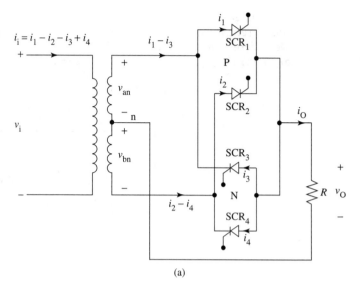

(a)

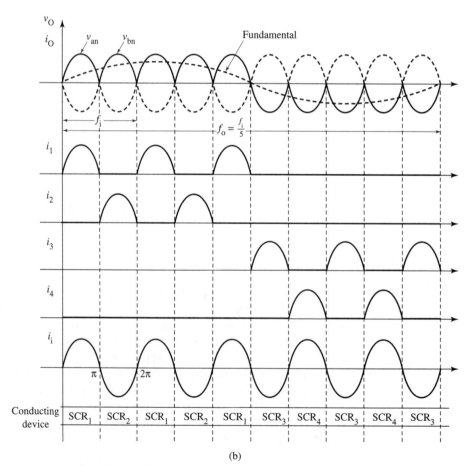

(b)

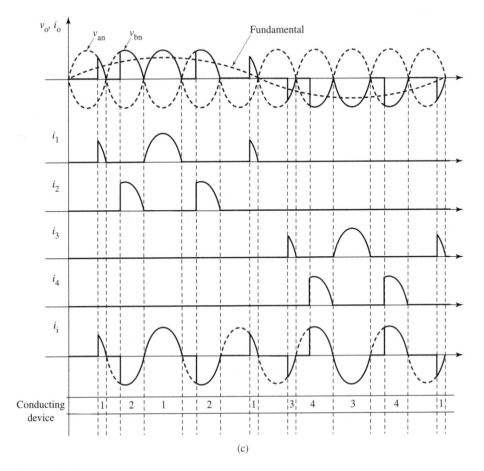

(c)

Figure 11.25
(c) waveforms when phase control is used on each SCR

11.7.1 The Single-Phase to Single-Phase Cycloconverter

The simplest cycloconverter is a single-phase input to single-phase output converter. It consists of a dual-converter, that is, two converters connected back to back as shown in Figure 11.25(a). Such a converter can supply a DC output voltage of either polarity. SCR_1 and SCR_2 form the positive group (P), while SCR_3 and SCR_4 form the negative group (N). If the SCRs are fired without delay, the converters will function as diode rectifiers. Figure 11.25(b) shows the low-frequency output waveform synthesized from selected portions of the high-frequency input voltage. The P group conducts for the first five half-cycles, then the N group conducts for the next five half-cycles. The output voltage waveform clearly shows an output frequency that is one-fifth of the input frequency.

The shape of the output voltage waveform in Figure 11.25(b) is closer to a square wave than a sine wave. A better approximation of a sine wave can be obtained by phase-delaying the firing of the SCR to produce the output voltage. As can be seen in Figure 11.25(c), the middle pulse of voltages is obtained at a lower value of the delay angle, while the side pulses are obtained at increasingly higher values of the delay angles.

An alternative arrangement for a single-phase cycloconverter with a resistive load is shown in Figure 11.26(a). The positive group converter P is a controlled bridge rectifier that produces the positive half-cycle of the output frequency, while the negative group converter yields the negative half-cycle of the output voltage. If the firing angles of the two converters are zero, the output voltage v_o is maximum. During the positive half-cycle of source voltage, converter P is fired, and during the negative half-cycle, converter N is fired. The output voltage waveform v_o is shown in Figure 11.26(b), where each half-cycle is made up of a whole number of half-cycles of the single-phase source waveform. Note that the output frequency is one-third the input frequency.

If the firing angle is increased, the output voltage v_o is reduced. The waveform for a firing angle of 60° is shown in Figure 11.26(c). If we phase-delay the firing of the SCR so that the firing angle changes during the half-cycle, the waveform shown in Figure 11.26(d) is obtained. The output receives a full half-cycle of input voltage at its peak period, while it is reduced as the output voltage zero is approached.

11.7.2 The Three-Phase Cycloconverter

The single-phase cycloconverter discussed in the previous section produces a nonsinusoidal output voltage. A sinusoidal AC voltage can be generated from three-phase input voltages by using three-phase controlled rectifiers as shown in Figure 11.27(a). The average output voltage of this circuit varies as the cosine of the firing angle (recall Equation 8.1, $V_{o(avg.)} = 0.827\ V_m \cos \alpha$). If the firing angle α is zero, the output voltage is maximum. If the firing angle is increased, the output voltage is reduced. When α becomes 90°, the output voltage becomes zero. Further increases in α will make the output voltage negative, reaching a negative maximum at $\alpha = 180°$. By varying the firing angle with respect to time in this manner, an average output voltage that changes sinusoidally can be obtained.

Figure 11.27(b) shows how to obtain a sinusoidal output voltage from the three-phase input voltages. The average (DC) output voltage is a sinusoidally varying voltage shown as $V_{o(avg.)}$. At any instant, it corresponds to the existing firing angle. The firing angle required at various instants is also shown in Figure 11.27(b). As shown, if the firing angle of successive pulses is varied from 90° to 0°, back to 90°, from 90° to 180°, and back again to 90° in appropriate steps, it is possible to vary the output voltage successively from 0 to its positive maximum to 0 to its negative maximum and back to 0, through a complete AC cycle.

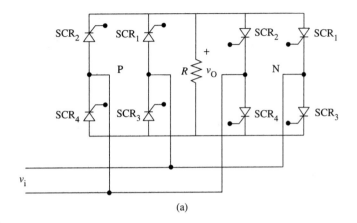

(a)

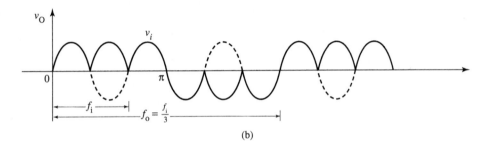

(b)

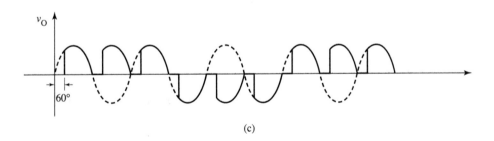

(c)

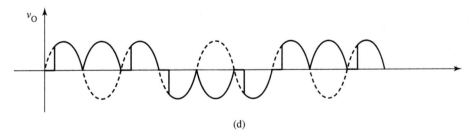

(d)

Figure 11.26
Single-phase cycloconverter (a) circuit diagram (b) waveforms for $\alpha = 0°$
(c) waveforms for $\alpha = 60°$ (d) waveforms at different values of α

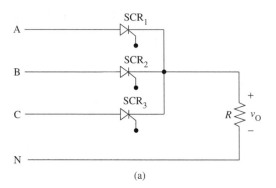

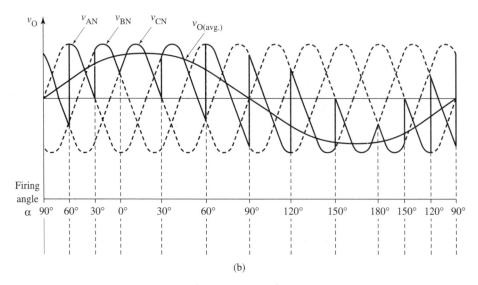

Figure 11.27
Synthesis of sinusoidal output voltage (a) controlled rectifier circuit (b) output voltage synthesis

11.7.2.1 The Three-Phase to Single-Phase Cycloconverter

The three-phase to single-phase cycloconverter is shown in Figure 11.28. It consists of two groups of SCRs connected to a three-phase source supplying a resistive load. The converter P formed by SCR_1, SCR_2, and SCR_3 conducts during the positive half-cycle. The converter N formed by SCR_4, SCR_5, and SCR_6 conducts during the negative half-cycle of the cycloconverter. To provide an alternating voltage, only one converter operates at any instant to supply current to the load. SCR gate firing is provided only to the group that needs to conduct the load current, while the gate trigger to the other group is blocked until the current in the first group has ceased. When the SCRs are triggered successively, a low-frequency AC voltage appears across the load. Figure 11.28(b)

Figure 11.28
The three-phase to single-phase cycloconverter
(a) circuit

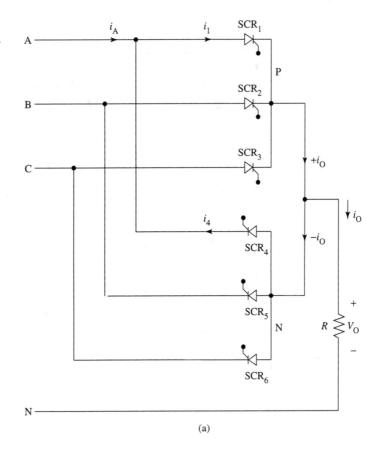

(a)

shows the output voltage waveform for a resistive load. The SCRs are triggered at such angles as to follow the fundamental sine wave as closely as possible. Note that the output voltage has the same maximum value as the source voltage and therefore has the same RMS value.

The maximum output voltage is the same as the average (DC) voltage, so

$$V_{o(max.)} = V_{o(avg.)} = \frac{3}{\pi} \sin 60° \; V_m \cos \alpha = 0.827 \; V_m \cos \alpha \qquad \textbf{11.20}$$

11.7.2.2 The Three-Phase to Three-Phase Cycloconverter

The cycloconverters we have considered thus far provide only single-phase output. To obtain three-phase AC output, three cycloconverters like the one in Figure 11.28 are required. Figure 11.29 shows a three-phase to three-phase cycloconverter using SCRs for switching. The input is a three-phase AC supply at a

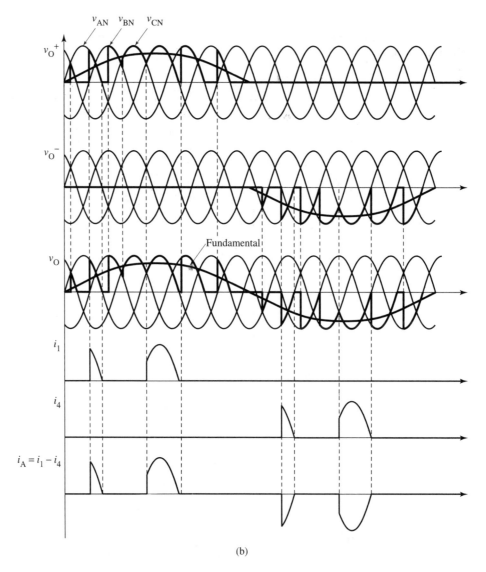

Figure 11.28
The three-phase to single-phase cycloconverter (b) output voltage waveform

frequency f_i (typically 60 Hz). The load is shown to be Y-connected. The cyclo-converter consists of a total of eighteen SCRs arranged into three groups of three-phase to single-phase cycloconverters. Each group is made up of two three-phase bridge rectifiers that function as one phase of AC output. Half of each group (A+, B+, and C+, the positive group (P)) produces output voltage during the positive half-cycle. The other half (A−, B−, and C−, the negative group

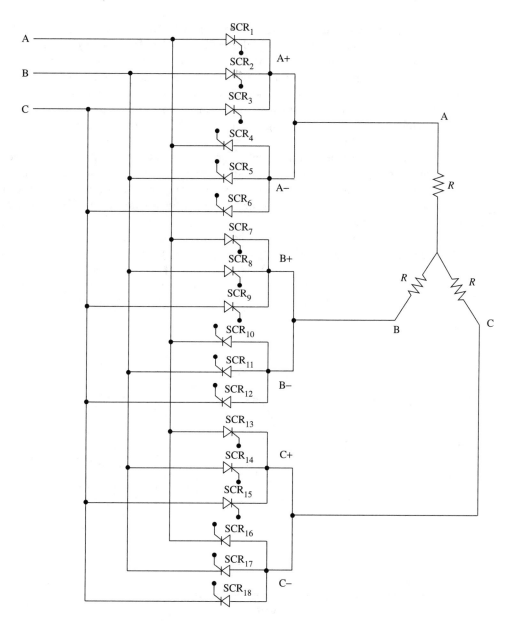

Figure 11.29
Three-pulse to three-phase cycloconverter

(N)) produces output voltage during the negative half-cycle. In this way, the two converters can provide an alternating output. The circuit converts the input frequency to a lower output frequency. The output frequency depends on the length of time during which the positive and negative groups operate. The switches are opened and closed at various instants to generate the output frequency out of selected small segments of the input frequency. The segments are determined by the gate firing of the SCRs. The output frequency can then be filtered to provide a smooth waveform.

Care must be taken to make sure that the SCRs from the positive and negative group are not conducting simultaneously, as this would short-circuit the source. To avoid this possibility, the gate-firing circuitry should contain an interlocking function that makes it impossible to give gate pulses to both converters simultaneously.

Figure 11.30 shows the voltage waveform for a resistive load across one phase of the output. The other two voltages have identical waveforms except for the usual 120° phase shift. The output frequency is one-third of source frequency, or 20 Hz from a 60-Hz source. The output voltage is determined by the firing angle α and is given by Equation 11.20.

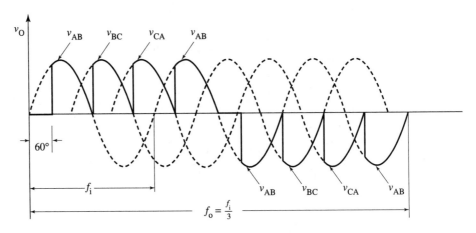

Figure 11.30
Output voltage waveform across one phase
for a three-phase cycloconverter

Example 11.19 A cycloconverter is supplied from a three-phase 440 V, 60 Hz AC source. If the cycloconverter supplies a resistive load with a current of 16 A, determine the maximum and RMS values of the load voltage for firing angles of
a) 0°
b) 45°
c) 60°

Solution

$$V_m = \sqrt{2} * 440 = 622.3 \text{ V}$$
$$V_{o(max.)} = 0.827 \ V_m \cos \alpha$$

a) $\alpha = 0°$

$$V_{o(max.)} = 0.827 * 622.3 * \cos 0 = 514.6 \text{ V}$$
$$V_{o(RMS)} = \frac{V_{o(max.)}}{\sqrt{2}} = \frac{514.6}{\sqrt{2}} = 364 \text{ V}$$

b) $\alpha = 45°$

$$V_{o(max.)} = 0.827 * 622.3 * \cos 45 = 364 \text{ V}$$
$$V_{o(RMS.)} = \frac{V_{o(max.)}}{\sqrt{2}} = \frac{364}{\sqrt{2}} = 257 \text{ V}$$

c) $\alpha = 60°$

$$V_{o(max.)} = 0.827 * 622.3 * \cos 60 = 257 \text{ V}$$
$$V_{o(RMS)} = \frac{V_{o(max.)}}{\sqrt{2}} = \frac{257}{\sqrt{2}} = 182 \text{ V}$$

11.8   Problems

11.1   Discuss the principle of the phase control method.

11.2   Discuss the principle of the integral cycle control method.

11.3   Which method of power control has low RFI?

11.4   Which method of power control would you recommend for lamp dimmers and variable-speed drills?

11.5   A single-phase half-wave AC controller is shown in Figure 11.31. Draw the output voltage waveform if the SCR is fired at 30°.

Figure 11.31
Single-phase half-wave AC
voltage controller

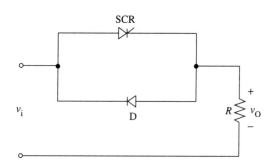

11.6 For the output voltage waveforms shown in Figure 11.32, find the output power if the source voltage is 120 V and the load resistance is 20 Ω.

Figure 11.32
Integral cycle control using
a triac

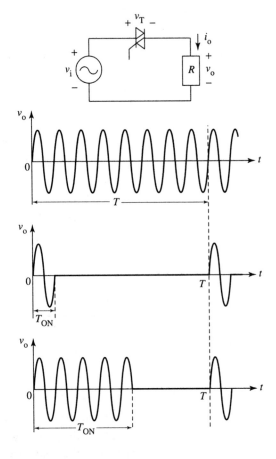

11.7 A single-phase full-wave AC controller is shown in Figure 11.33. Draw the output voltage waveform if the SCR is triggered at $\alpha = 30°$ during the positive and negative half-cycles.

Figure 11.33
Single-phase full-wave AC
voltage controller

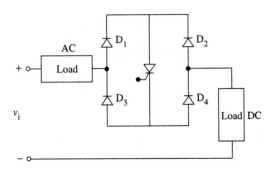

11.8 A single-phase full-wave AC controller is shown in Figure 11.34. SCR_1 and diode D_1 are triggered simultaneously during the positive half-cycle of the input voltage, while SCR_2 and D_2 are turned on during the negative half-cycle. Draw the output voltage waveform if the firing angle is 30°.

Figure 11.34
Single-phase full-wave AC voltage controller

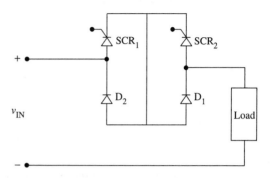

11.9 A single-phase 110 V 60 Hz AC source controls power to a 10 Ω resistive load using integral cycle control. If the total period T is 24 cycles, find the on time (T_{ON}) that will provide an output power that is 50% of the maximum value.

11.10 A single-phase 120 V 60 Hz AC source controls power to a 5 Ω resistive load using phase control. Find the firing angle if the output power is 40% of its maximum value. Find the range of output power if integral cycle control is used with a period of $T = 0.2$ s.

11.11 A single-phase resistive load is controlled from a single-phase 60 Hz AC source by means of an inverse parallel pair of SCRs. Find the firing angle of the SCRs if the output power is 60% of its maximum value.

11.12 A single-phase full-wave power controller as shown in Figure 11.1(a) supplies a resistive load of 10Ω. If the source voltage is 120V at 60 Hz and the firing angle for SCR_1 and SCR_2 is $\alpha_1 = \alpha_2 = 90°$, find
a) the RMS value of the output voltage
b) the RMS value of the SCR current

11.13 A single-phase power controller as shown in Figure 11.1(a) supplies a purely inductive load.
a) Plot the waveforms of the output voltage and the output current if the delay angle is 90°.
b) Repeat (a) if the delay angle is increased to 100°.

11.14 A single-phase power controller as shown in Figure 11.1(a) supplies an RL load. If the power factor angle is θ ($= \tan^{-1} X_L/R$), draw the waveform of the output current for delay angles of
a) $\alpha < \theta$
b) $\alpha = \theta$
c) $\alpha > \theta$

11.15 An SCR with a diode bridge is shown in Figure 11.35. Draw the wave-form of the voltage across the load if the firing angle is 30°.

Figure 11.35
SCR with diode bridge

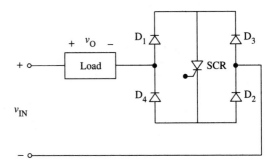

11.16 The single-phase cycloconverter of Figure 11.26(a) is connected to a 110 V, 60 Hz source and supplies a resistive load. If the output frequency is 15 Hz, determine the RMS value of the output voltage at α = 0° and α = 90°. Also draw the output voltage waveforms.

11.17 Draw the output voltage waveforms for a single-phase cycloconverter with the following output frequencies
a) 1/2 input frequency
b) 1/3 input frequency
c) 1/4 input frequency
d) 1/5 input frequency

11.9 Equations

$$P_{o(avg.)} = \frac{V_i^2\, T_{ON}}{R\, T} = \frac{V_i^2}{R\, d} = P_{o(max)}\, d \qquad \textbf{11.1}$$

$$V_o = \frac{V_m}{\sqrt{2}}\sqrt{\frac{T_{ON}}{T}} = V_i\,\sqrt{d} \qquad \textbf{11.2}$$

$$PF = \sqrt{T_{ON}/T} = \sqrt{d} \qquad \textbf{11.3}$$

$$V_{o(RMS)} = V_i\left\{\frac{1-\alpha}{\pi} + \frac{\sin 2\alpha}{2\pi}\right\}^{1/2} \qquad \textbf{11.4}$$

$$I_{o(RMS)} = \frac{V_i}{R}\left\{1 - \frac{\alpha}{\pi} + \frac{\sin 2\alpha}{2\pi}\right\}^{1/2} \qquad \textbf{11.5}$$

$$I_{T(RMS)} = I_{o(RMS)} \qquad \textbf{11.6}$$

$$I_{SCR(RMS)} = I_{o(RMS)}/\sqrt{2} \qquad \textbf{11.7}$$

$$P_{o(avg.)} = (I_{orms})^2\,(R) \quad \text{or} \quad V_{o(RMS)}^2/R \qquad \textbf{11.8}$$

$$PF = \left\{ 1 - \frac{\alpha}{\pi} + \frac{\sin 2\alpha}{2\pi} \right\}^{1/2} \qquad \textbf{11.9}$$

$$PIV \geq V_{i(m)} \qquad \textbf{11.10}$$

$$I_{o(RMS)} = \frac{V_i}{R} \left\{ 4 \left(1 - \frac{\alpha}{\pi} \right) \left(\cos^2\alpha + \frac{1}{2} \right) + \frac{6}{\pi} \sin \alpha \cos \alpha \right\}^{1/2} \qquad \textbf{11.11}$$

$$I_{o(RMS)} = \frac{V_i}{R} \left\{ \frac{1}{3} - \frac{\alpha}{2\pi} + \frac{\sin 2\alpha}{4\pi} \right\}^{1/2} \qquad \textbf{11.12}$$

$$I_{o(RMS)} = \frac{V_i}{R} \left\{ \frac{1}{6} - \frac{3 \sin 2\alpha}{8\pi} + \frac{\sqrt{3} \cos 2\alpha}{8\pi} \right\}^{1/2} \qquad \textbf{11.13}$$

$$I_{o(RMS)} = \frac{V_i}{R} \left\{ \frac{5}{12} - \frac{\alpha}{2\pi} + \frac{\sqrt{3} \cos 2\alpha}{8\pi} + \frac{\sin 2\alpha}{8\pi} \right\}^{1/2} \qquad \textbf{11.14}$$

$$V_{o(RMS)} = \sqrt{3} \, I_{o(RMS)} \, R \qquad \textbf{11.15}$$

$$v_{SW} = 1.5 \, v_{phase} \qquad \textbf{11.16}$$

$$I_{o(RMS)} = \frac{V_i}{R} \left\{ \frac{1}{3} - \frac{\alpha}{4\pi} + \frac{\sin 2\alpha}{8\pi} \right\}^{1/2} \qquad \textbf{11.17}$$

$$I_{o(RMS)} = \frac{V_i}{R} \left\{ \frac{11}{24} - \frac{\alpha}{2\pi} \right\}^{1/2} \qquad \textbf{11.18}$$

$$I_{o(RMS)} = \frac{V_i}{R} \left\{ \frac{7}{24} - \frac{\alpha}{4\pi} - \frac{\sqrt{3} \cos 2\alpha}{16\pi} + \frac{\sin 2\alpha}{16\pi} \right\}^{1/2} \qquad \textbf{11.19}$$

$$V_{o(m)} = V_{o(avg.)} = \frac{3}{\pi} \sin 60° \, V_m \cos \alpha = 0.827 \, V_m \cos \alpha \qquad \textbf{11.20}$$

Static Switches

12

Chapter Outline

Learning Objectives

After completing this chapter, the student should be able to

- define the term *static switch*
- list the advantages and disadvantages of a static switch compared to a mechanical switch
- explain the operation of single-phase and three-phase static switches
- describe solid-state relay
- describe the principle of static tap changing control
- discuss the operation of static VAR Controller

12.1 Introduction

A static switch turns the power to a load on or off but does not vary it in any other way. The bistable characteristic of semiconductor devices (like thyristors), that is, the existence of two stable states (conducting and nonconducting), suggests that these devices can be used as a contactless switch. Their applications in the field of static switching include on-off switches, circuit breakers, contactor, solid state relays, and so forth.

12.2 Comparison of Semiconductor and Mechanical Switches

A semiconductor switch offers several advantages over other switching devices:

1. It provides extremely high switching speeds, because the switch turns on immediately.
2. Operation is quiet because there are no moving parts and no arcing.
3. Radio frequency interference (RFI) is eliminated by using zero voltage switching.
4. No routine maintenance is required because there are no contacts or moving parts that wear.
5. Operational life is much longer.
6. It is completely safe in an explosive environment.
7. Immune from vibration and shock.
8. It can be installed in any position or location.
9. There is no switch contact bounce when closing.
10. It is small and lightweight.
11. It is easily tailored to electronic control.
12. The cost is low.
13. It offers greater reliability.
14. In addition to turning a load on or off, it can also be used to control the load power from zero to maximum.
15. The control circuit can be easily isolated from the power circuit.
16. It is easy to control remotely.

Some of the disadvantages of a semiconductor switch are:

1. Due to reverse leakage current when off, it does not allow the load to be completely isolated from the source.
2. It is likely to fail when subjected to overvoltage and overcurrent situations unless protected by an RC snubber circuit.

3. It has higher power losses in the on-state condition, so cooling is required.

4. The on-state voltage drop across the device may not be permissible in some applications.

5. Due to the higher cost of the device and the complexity of control, its use is normally limited to single-phase circuits.

6. The same switch cannot be used in both AC and DC circuits, since the AC switch turns off naturally while the DC switch needs additional force commutation circuitry to turn it off.

7. Continuous firing pulses are required to maintain the switch in the on state.

8. It can cause false triggering as a result of voltage transients caused by switching inductive loads on neighboring lines.

9. To prevent false triggering, the firing circuits must be completely isolated or shielded from power circuits.

10. Protection circuits are necessary to safely turn off the device before the surge current or fault current ratings are exceeded.

11. Overload capability is limited by the maximum current of the semiconductor device.

Static switches are used in both AC and DC switching operations. AC switching requires bidirectional control, which is usually implemented by using a triac or two SCRs connected in antiparallel. Since the device turns off naturally, the upper frequency limit is determined by the type of device used. For low-frequency switching applications, a single triac can be used. For high-frequency applications a configuration of two antiparallel SCRs is employed. DC switching requires control for only one direction of current flow, and the switching device is usually an SCR. The switching speed is limited by the commutation circuit and the reverse recovery time of the SCR.

■───

12.3 Static AC Switches

A static switch or contactor is a switching device that connects (on) or disconnects (off) a load from the supply. A static AC switch incorporates three main circuits: the power circuit, the firing circuit, and the protection circuit. The firing and protection circuits were discussed in Chapter 4.

12.3.1 Antiparallel SCR Connection

For high-power switching applications, the simplest single-phase AC switch is constructed from two SCRs in antiparallel, as shown in Figure 12.1 To turn on the power to the load, SCR_1 must be turned on at the beginning of the positive half-cycle. SCR$_2$ should be turned on at the beginning of the negative half-cycle for negative currents. The SCR switch will turn on immediately if a firing pulse is received at the gate. To maintain the switch in its on state, continuous firing pulses are required at the gate, since the thyristor turns off naturally at the zero crossing.

To switch off the power, it is sufficient to suppress further firing pulses. The AC current continues to flow until the SCR turns off naturally at zero current. The device will actually turn off within half a cycle of source frequency after the first zero crossing following removal of the firing pulses.

Voltage transients can occur at switching, especially with an inductive load, and they require suppression using an *RC* snubber circuit connected in parallel with the SCRs.

Figure 12.1
Single-phase static AC switch using two SCRs in antiparallel connection

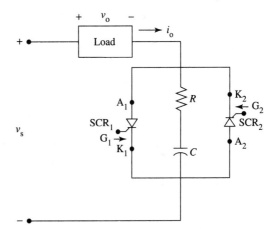

The RMS value of the output voltage is given by

$$V_{o(\text{RMS})} = \frac{V_m}{\sqrt{2}} \left[1 - \frac{\alpha}{\pi} + \frac{\sin 2\alpha}{2\pi} \right]^{1/2} \qquad \textbf{12.1}$$

where

V_m = maximum value of source voltage.

α = firing delay angle (0° for maximum output)

The RMS line current is the same as the RMS load current and is given by:

$$I_{\text{RMS}} = I_m/\sqrt{2} \qquad \textbf{12.2}$$

where I_m is the maximum value of line current.

Since each SCR carries the line current for only one half-cycle, the average current carried by each SCR is

$$I_{\text{SCR(avg.)}} = \frac{I_m}{\pi} \qquad \textbf{12.3}$$

Substituting I_m from Equation 12.2, we get

$$I_{SCR(avg.)} = \sqrt{2}\, I_{RMS}/\pi = 0.45\, I_{RMS} \qquad\qquad \textbf{12.4}$$

Therefore, it is essential to select SCRs with current ratings at least 0.45 times the load current.

The RMS value of the SCR current is

$$I_{SCR(RMS)} = \frac{I_m}{2} \qquad\qquad \textbf{12.5}$$

The forward blocking voltage V_{FB} must be at least equal to the maximum value of the source voltage, that is,

$$V_{FB} \geq V_{S(m)} = \sqrt{2}\, V_s \qquad\qquad \textbf{12.6}$$

where V_s is the RMS value of the source voltage.

The output power can be obtained from:

$$P_o = \frac{V_{o(RMS)}^2}{R} \qquad\qquad \textbf{12.7}$$

To reduce distortion of the load waveforms due to switching, the SCRs must be turned on at the zero crossing of each half-cycle of the AC source voltage. Figure 12.2(a) shows the load voltage and current waveforms for a pure resistive load where the gate signal is applied at $\alpha = 0$.

With an *RL* load, the current lags the voltage by a phase angle Φ due to the inductance. For highly inductive loads, Φ becomes greater than the width of the trigger pulse and the SCR will not trigger. To ensure reliable triggering, the gate signal needs a wide pulse width from 0 to π. The load voltage and current waveforms for a pure inductor are shown in Figure 12.2(b).

12.3.2   Triac Connection

Triacs are not yet available with voltage, current, and frequency ratings as high as those already available in SCRs. However, for low-power applications, provided the current rating is not exceeded, a single triac is more economical to use than a pair of SCRs. A triac also eliminates the need for a second firing circuit. Figure 12.3 shows an AC switch using a triac. The triac can be turned on by either a positive or a negative trigger pulse and can carry current in both directions. Like an SCR, once the triac has turned on, the gate loses control over the switching. It regains control after the current has fallen to zero, and another gate pulse is needed to turn the device on again.

12.3.3   Alternative Switch Implementation

Figure 12.4 shows two other ways of implementing single-phase AC static switches using combinations of SCRs and diodes. Diodes are less expensive than SCRs and do not require a firing circuit. Figure 12.4(a) shows an arrangement that

Figure 12.2
Load voltage and current
waveforms (a) resistive load
(b) inductive load

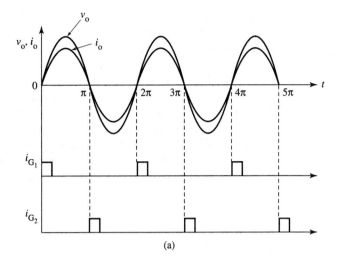

(a)

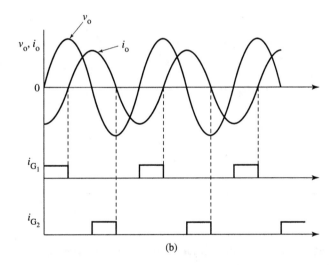

(b)

Figure 12.3
Single-phase AC static switch
using a triac

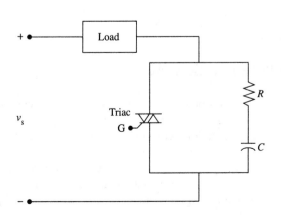

requires only one SCR to control both halves of the AC cycle. The SCR is con-nected across the DC terminals of a full-wave diode bridge rectifier. The positive current path is through diodes D_1, D_4, and the SCR, while negative current flows through D_2, D_3, and the SCR. The current is AC through the load and DC through the SCR. The load may be resistive or inductive. The gate-firing circuit is simple since there is only one SCR to be fired for both the positive and the negative half-cycle. However, since the current has to flow through three devices, the power loss is higher due to the forward voltage drop of the devices.

In Figure 12.4(b), SCR_1 and diode D_2 conduct for the positive half-cycle, and SCR_2 and diode D_1 conduct for the negative half-cycle. Since the two SCR cathodes are common in this scheme, the two gates can be tied together to pro-vide a common gate-firing control. The current has to flow through an SCR and a diode, causing added power loss.

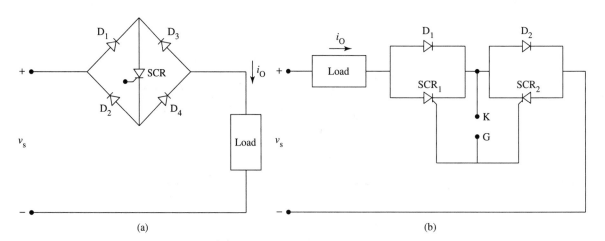

Figure 12.4
Single-phase AC static switch alternative arrangement (a) using a single SCR (b) using an SCR-diode bridge

12.4 Three-Phase Static Switches

Fundamentally, three-phase AC circuits can be switched on and off in the same way as single-phase AC circuits. Figure 12.5 shows various three-phase switch configurations controlling three-phase loads.

In Figure 12.5(a) and (b), the semiconductor switch constructed from SCRs in antiparallel is connected in series with each AC supply line of a Y- or Δ-connected load. With a Δ-connected load, the alternative arrangements shown in Figure 12.5(c) can be used if the individual load phase can be accessed. In

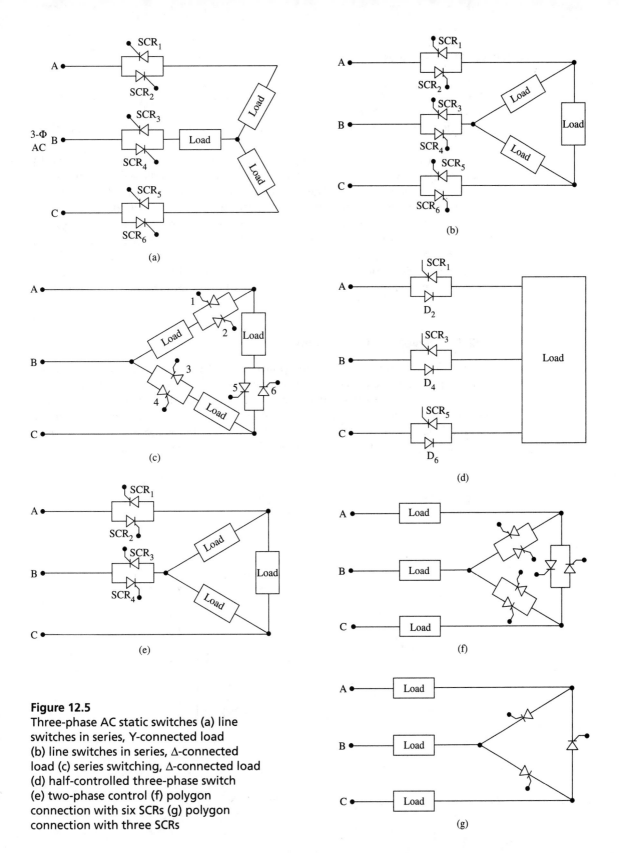

Figure 12.5
Three-phase AC static switches (a) line switches in series, Y-connected load (b) line switches in series, Δ-connected load (c) series switching, Δ-connected load (d) half-controlled three-phase switch (e) two-phase control (f) polygon connection with six SCRs (g) polygon connection with three SCRs

this circuit, the switch is in series with the load, and this combination is connected in Δ. The current in each switch is $\sqrt{3}$ times less than the line current. Therefore, the current rating of the SCR is less than that required for the previous two circuits.

In a three-phase system without the neutral, we can reduce the cost by substituting a diode for an SCR in each of the antiparallel-connected circuits of Figure 12.5(a) and (b). The circuit shown in Figure 12.5(d) is called a half-controlled switch. If unbalanced voltage operation is permitted, costs can be further reduced by removing one pair of SCRs, as shown in Figure 12.5(e).

The polygon connection of Figure 12.5(f) uses six SCRs connected in Δ through the three-phase load. The six SCRs can be reduced to three as shown in Figure 12.5(g). However, the voltage and current duty of each SCR would be more severe. The current rating of the SCRs is 1.5 times greater than that of the other connections in Figure 12.5.

12.4.1 Reversing Connections

Static switches are often used in applications where high switching frequencies are required, for example, in reversing connections where AC machines are switched from one direction of rotation to the other by reversing the rotating field. The circuit shown in Figure 12.6 reverses the three-phase power supplied to the load. When switches S_1, S_2, and S_3 are on, line A feeds a, line B feeds b, and line C feeds c for one direction of rotation. Turning off switches S_2 and S_3

Figure 12.6
Three-phase reversing contactor

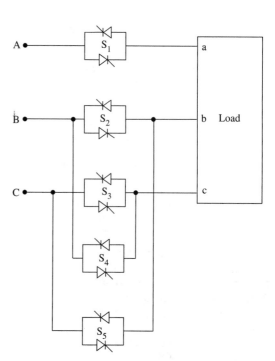

and turning on S_4 and S_5 results in line A feeding a, line B feeding c, and line C feeding b, thus achieving a reversal of phase sequence of the voltages supplied to the load. Care must be taken to ensure that S_2 and S_5 are never turned on simultaneously, or else a short circuit will occur across lines B and C. The same applies to switches 3 and 4. Also note that in this circuit all switches must be SCRs, because diodes would cause short circuits between main phases.

12.5 Hybrid Switches

By connecting a static switch and a mechanical switch in parallel, some of the best features of both can be obtained. Switching is performed by static switches, but in between times, they are shorted out by the mechanical switch, as shown in Figure 12.7. The mechanical switch carries the load current during this time, thus avoiding the power loss that occurs in the static switch due to the on-state voltage drop. When the mechanical switch is in operation and the static switch is conducting, the on-state voltage drop is so low that arcing does not occur.

Another advantage of a hybrid switch is that if the static switch becomes faulty, switching can still be carried out using a mechanical switch until the static switch is repaired. However, if an open-circuit fault occurs at the load, the full source voltage could be present across this open circuit. This dangerous situation can be avoided by including an isolating switch connected in series with the static switch to provide isolation between the load and the line (see Figure 12.7).

Figure 12.7
Hybrid switch

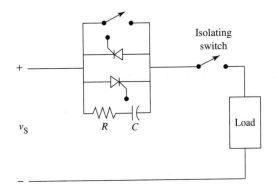

12.6 The Solid-State Relay (SSR)

Semiconductor AC switches that have no contacts or moving parts are called solid-state relays (SSRs). The SSR has become increasingly popular in applications where electromagnetic relays were previously used. SSRs are used as switching elements, for the control of motor loads, in transformers, in resistance heating, and in lighting loads.

Solid-state relays can be used to control DC and AC loads. When they are used with DC loads, a power transistor connects the load to the source. In AC applications, the most commonly used device is a triac. Figure 12.8 shows the circuit of an SSR that uses a triac with an optically isolated gate circuit. It can control a large amount of power in the main circuit from a small signal from the control circuit. Usually SSRs employ optocouplers to provide electrical isolation between the control circuit and the load circuit. Optocouplers are a combination of optoelectronic devices, normally a light-emitting diode (LED) and a phototransistor or photo-SCR assembled into one package. Another useful feature of an SSR is zero voltage switching (ZVS), which reduces electromagnetic interference (EMI). Zero voltage switching means that the switching device is turned on at the first possible instant after the source voltage crosses the zero axis. The SSR is designed for single-pole applications with normally open (NO) contact.

Figure 12.8
Solid-state relay

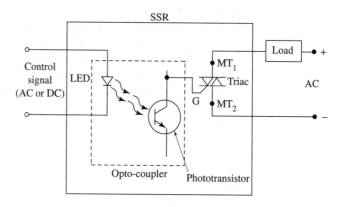

12.7 Static Tap-Changing Control

Figure 12.9(a) shows a single-phase tap changer in which the tapped voltage supply is provided by a tapped transformer. The technique involves tap changing of the transformer using static switches and continuous adjustment by phase control between taps. Ideally, the output voltage can be adjusted from zero to the maximum secondary voltage. If only SCR_3 and SCR_4 are fired, the output voltage is v_2. If only SCR_1 and SCR_2 are fired, then the output voltage is v_1.

By phase control using SCR_1 and SCR_2, the output voltage can be adjusted from zero to the full voltage v_1 of this tap. Further increases in voltage can be obtained by increasing the voltage during parts of each half-cycle using SCR_3 and SCR_4. As shown in Figure 12.9(b), SCR_1 is turned on at t_1, which is the zero voltage crossing of the positive half-cycle. SCR_3 is triggered with a phase delay at t_2. Since v_2 is more positive than v_1, SCR_3 will start conducting and this will automatically turn off SCR_1. At t_3, SCR_2 is turned on, and finally SCR_4 is fired at t_4.

Figure 12.9
Static tap changer (a) circuit
(b) current and voltage
waveforms

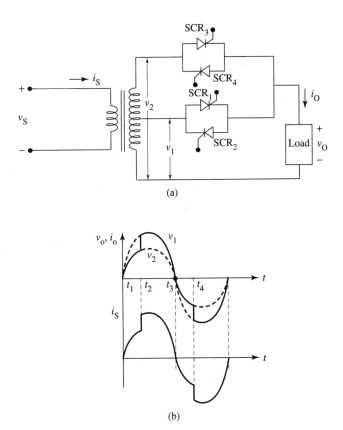

(a)

(b)

The output voltage can be adjusted continuously to its maximum value by advancing the firing of SCR_3 and SCR_4 with respect to instant t_2 and t_4, respectively. Care must be taken to ensure that SCR_2 remains off when SCR_3 is on, since such condition would short-circuit the transformer secondary winding. The same holds true for SCR_4 and SCR_1. This method of tap changing can be easily extended to transformers with multitap windings for further voltage adjustment.

12.8 The Static VAR Controller (SVC)

Consider the circuit shown in Figure 12.10(a), in which an AC switch controls power to an inductor. If the AC switch is turned on after the peak of the source voltage, the load current will flow for less than 180° and its effective value will depend on the delay angle α, that is, the reactive power drawn from the source can be continuously controlled by varying the delay angle. This circuit arrangement, which is called a thyristor-controlled inductor (TCI), can therefore be used to obtain adjustable values of inductive or lagging VARs. Unfortunately, most practical loads, being inductive in nature, also draw lagging VARs from the source. These loads require capacitive or leading VARs to compensate for the lagging power factor.

A thyristor-switched capacitor (TSC) (shown in Figure 12.10(b)) uses capacitors to generate leading VARs. However, if the firing angle is arbitrarily varied to control the VARs, the current through the AC switch flows in large pulsations as the capacitor voltage equalizes with the source voltage. These pulses can easily damage the switching device. Therefore, firing must be synchronized to occur when the instantaneous source voltage is equal to the capacitor voltage. VAR control is achieved by splitting the capacitors into banks and controlling each bank with a separate AC switch. Capacitor banks are turned on as required to provide necessary VARs. This circuit has the disadvantage of requiring an AC switch and its associated firing circuit for each bank. Also, since the capacitor bank can be switched only in steps, VAR compensation is achieved in discrete steps rather than by continuous control. Therefore, a TSC arrangement is suitable only if the load requires a fixed VAR adjustment.

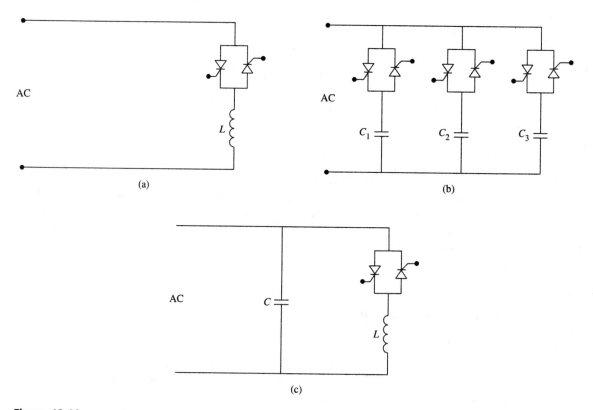

(a)

(b)

(c)

Figure 12.10
Types of VAR controllers (a) thyristor-controlled inductor (b) thyristor-switched capacitor (c) thyristor-controlled inductor with unswitched capacitor

Circuit 12.10(c) combines circuits (a) and (b). An unswitched capacitor provides the maximum leading VARs that will be required, while a phase control AC switch in series with the inductor subtracts a continuously controllable number of lagging VARs. This is the basis of the static VAR controller.

Static VAR controllers are attracting attention for use in installations with rapidly varying reactive power requirements, such as arc furnaces, where they provide quick and precise control of reactive power. The leading VARs necessary for VAR compensation are supplied by connecting a TSC across the AC lines. Figure 12.11 shows a scheme in which the capacitor banks are connected or disconnected, using SCRs to correct the load power factor. For loads with varying VAR requirements, a TCI is placed in parallel with the TCS. The capacitor banks are switched in and out to provide a fixed amount of leading VARs. Using phase control, the inductor absorbs a variable amount of lagging VARs, depending on the delay angle of the SCRs. If the lagging VARs drawn by the TCI is equal to the leading VARs supplied by the capacitor, the net reactive power will be zero and the load power actor will be unity. From this point, if the load power factor becomes lagging, the lagging VARs of the TCI can be decreased by adjusting the delay angle, thereby increasing the net leading VARs. If more leading VARs are required, another capacitor bank can be switched. In this way, the TSC provides leading VARs in discrete steps, while the TCI provides precise continuous control between steps. The same technique can be applied to three-phase circuits.

Figure 12.11
Static VAR controller

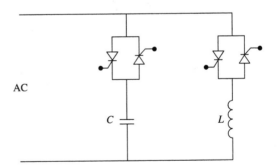

AC

12.9 Problems

12.1 What is a static switch? List some devices that the static switch has replaced.

12.2 What are the advantages and disadvantages of using a static switch versus an electromechanical switch?

12.3 What is an SSR?

12.4 What is meant by zero voltage switching?

12.5 What is opto-isolation?

12.6 Describe the principle of operation of a static tap changer.

12.7 Describe the principle of operation of a static VAR controller.

12.8 Define TCI.

12.9 Define TSC.

12.10 Equations

$$V_{o(RMS)} = \frac{V_m}{\sqrt{2}} \left[1 - \frac{\alpha}{\pi} + \frac{\sin 2\alpha}{2\pi} \right]^{1/2}$$ 12.1

$$I_{RMS} = I_m / \sqrt{2}$$ 12.2

$$I_{SCR(avg.)} = \frac{I_m}{\pi}$$ 12.3

$$I_{SCR(avg.)} = \sqrt{2} \, I_{RMS} / \pi = 0.45 \, I_{RMS}$$ 12.4

$$I_{SCR(RMS)} = \frac{I_m}{2}$$ 12.5

$$V_{FB} \geq V_{S(m)} = \sqrt{2} \, V_s$$ 12.6

$$P_o = \frac{V^2_{o(RMS)}}{R}$$ 12.7

Index